T0325454

SMILODON

SMILODON

The Iconic Sabertooth

EDITED BY LARS WERDELIN, H. GREGORY MCDONALD,
AND CHRISTOPHER A. SHAW

JOHNS HOPKINS UNIVERSITY PRESS　BALTIMORE

© 2018 Johns Hopkins University Press
All rights reserved. Published 2018
Printed in the United States of America on acid-free paper
9 8 7 6 5 4 3 2 1

Johns Hopkins University Press
2715 North Charles Street
Baltimore, Maryland 21218-4363
www.press.jhu.edu

Library of Congress Cataloging-in-Publication Data

Names: Werdelin, Lars, editor. | McDonald, H. Gregory (Hugh
    Gregory), 1951- editor. | Shaw, Christopher A., editor.
Title: Smilodon : the iconic sabertooth / edited by Lars Werdelin,
    H. Gregory McDonald, and Christopher A. Shaw.
Description: Baltimore : Johns Hopkins University Press, 2018. |
    Includes bibliographical references and index.
Identifiers: LCCN 2017039401| ISBN 9781421425566 (hardcover :
    alk. paper) | ISBN 9781421425573 (electronic) | ISBN 1421425564
    (hardcover : alk. paper) | ISBN 1421425572 (electronic)
Subjects: LCSH: Smilodon. | Saber-toothed tigers.
Classification: LCC QE882.C15 S65 2018 | DDC 569/.75—dc23
LC record available at https://lccn.loc.gov/2017039401

A catalog record for this book is available from the British Library.

Special discounts are available for bulk purchases of this book.
For more information, please contact Special Sales at 410-516-6936
or specialsales@press.jhu.edu.

Johns Hopkins University Press uses environmentally friendly book
materials, including recycled text paper that is composed of at least
30 percent post-consumer waste, whenever possible.

Contents

Contributors

John P. Babiarz
Babiarz Institute of Paleontological Studies, Inc.
Mesa, Arizona, United States

Wendy J. Binder
Loyola Marymount University
Los Angeles, California, United States

Charles S. Churcher
Royal Ontario Museum
Toronto, Ontario, Canada and
University of Toronto
Toronto, Ontario, Canada

Larisa R. G. DeSantis
Vanderbilt University
Nashville, Tennessee, United States

Robert S. Feranec
New York State Museum
Albany, New York, United States

Therese Flink
Huddinge, Sweden

James L. Knight
South Carolina State Museum
Columbia, South Carolina, United States

Margaret E. Lewis
Stockton University
Galloway, New Jersey, United States

Larry D. Martin (Deceased)
Museum of Natural History, University of Kansas
Lawrence, Kansas, United States

H. Gregory McDonald
Bureau of Land Management
Salt Lake City, Utah, United States

Julie A. Meachen
Des Moines University
Des Moines, Iowa, United States

William C. H. Parr
University of New South Wales
Randwick, New South Wales, Australia

Ashley R. Reynolds
University of Toronto
Toronto, Ontario, Canada

Kevin L. Seymour
Royal Ontario Museum
Toronto, Ontario, Canada

Christopher A. Shaw
Idaho Museum of Natural History, Idaho State University
Pocatello, Idaho, United States and
Natural History Museum of Los Angeles County
Los Angeles, California, United States

C. S. Ware
Natural History Museum of Los Angeles County
Los Angeles, California, United States and
Denver Museum of Nature and Science
Denver, Colorado, United States

Lars Werdelin
Swedish Museum of Natural History
Stockholm, Sweden

H. Todd Wheeler
Natural History Museum of Los Angeles County
Los Angeles, California, United States

Stephen Wroe
University of New England
New South Wales, Australia

M. Aleksander Wysocki
University at Buffalo
Buffalo, New York, United States

Preface

As people slowly gathered on the tarmac, you could sense the excitement building on May 14, 2008, a chilly early spring morning in the northern Rocky Mountains. Paleobiologists from Europe, South America, and across the United States had been brought together for the first International Sabertooth Workshop, a three-day meeting supported by the National Science Foundation, cohosted by the Idaho Museum of Natural History and the Department of Biological Sciences at Idaho State University, with Bill Akersten and Lars Werdelin as coordinators of the workshop. The planned activities included the testing of a machine that might simulate the bite of the iconic prehistoric predator, *Smilodon fatalis*. University faculty, students, and members of the public were also invited to watch science and engineering in action. Under tarps in the parking lot adjacent to the university science complex lay one elk carcass and one mule deer carcass, recent casualties of highway automobile collisions, provided by the Idaho Department of Fish and Game, to test the biting mechanism (termed 'Robocat' by its inventor, Todd Wheeler). The mechanism, modeled from specimens of *Smilodon* from Rancho La Brea, was mounted on a Caterpillar tractor and carefully positioned to function at various angles and from different directions. For each bite, scientists took careful aim with the steel mechanical jaws at one of the two alternative targets, the neck or the abdomen of the expired elk. After a few hours and numerous bites of information, there was finally a consensus: either target was plausible. Most sabertooth cat specialists (for whatever reason) still favor one site over the other, but as so aptly stated by one colleague, "We know they bite like that, only faster." By the time the carcass biting was concluded, most of the gathered crowd had dispersed, looking for their own bite to eat.

The International Sabertooth Workshop (May 12–14 2008) was a think tank to discuss a number of issues related to sabertooth carnivores: (1) free access to databases and/or construction of a separate dedicated database; (2) areas of further research potential for anatomical and physiological analyses, hunting and other proposed behaviors, past environments, and extinction; (3) educational outreach with the creation of teaching kits and traveling exhibits; and (4) the production of scientific and lay publications to help disseminate up-to-date knowledge about these fascinating creatures. Folded into the workshop was a symposium featuring the latest research on sabertooth carnivores and an interactive exhibition on sabertooth cat paintings and sculpture by renowned artist Mark Hallett, along with the bite test previously mentioned.

This book grew from a seed planted during this workshop, and many of its authors were attendees. However, it was at the 71st annual meeting of the Society of Vertebrate Paleontology held in Las Vegas in 2011 that the book began to take shape. The original intention was to summarize and update some of the classic research in addition to presenting new studies of the iconic sabertooth cat, *Smilodon*. As such, this volume is diverse in its contributions and includes summaries of the history of discoveries, phylogenetics, killing biomechanics and engineering, skull and postcranial morphology, and paleopathology. New studies include two faunas containing *Smilodon* in South America (Peru and Chile) and one in South Carolina, as well as a fascinating report on ontogeny and growth in *Smilodon* and yet another on the dietary ecology of this animal. At the time of the first workshop, it was proposed that others would follow, but none have yet been scheduled. Perhaps this book will provide the motivation to organize another, similar workshop to discuss the foundation laid by the first and to scrutinize what progress has been made since that cool spring morning in Pocatello, Idaho.

The Swedish Museum of Natural History is gratefully acknowledged for providing funds to defray part of the cost to produce this volume.

SMILODON

1

Smilodon: A Short History of Becoming the Iconic Sabertooth

H. GREGORY MCDONALD

The enlargement of the canine has occurred independently in a number of mammalian lineages, including the carnivores, with the felids, nimravids, a creodont (*Machaeroides simpsoni*), and even a member of the Sparassodonta in South America (*Thylacosmilus atrox*); there are even some herbivores, such as the uintatheres as well as fossil (*Blastomeryx primus*) and living cervids like the water deer (*Hydropotes inermis*). Although the function of the enlarged canine was certainly different between the carnivores and herbivores, in both groups the role of the enlarged canine in each fossil taxon's ecology has resulted in much discussion and disagreement. Part of this discussion is because the degree of enlargement of the carnivoran canine varies considerably among taxa, resulting in various descriptors such as 'dirk-tooth,' 'scimitar-tooth,' and the best-known term, 'saber-tooth.'

Since all sabertooth carnivorans are extinct and lack any modern functional equivalents, the enlarged canine so distinctive of the 'sabertooth cats'—a term also often used broadly to include the nimravids— has attracted much attention and speculation in terms of its function and its relationship to the species' ecology. This functional niche has been filled by various groups of placental carnivores, starting with the earliest appearance of the creodont *Machaeroides simpsoni* in the Eocene (Dawson et al., 1986). Many of the questions center on both its function and adaptive value. The well-known *Smilodon* was among the earliest of the sabertooth cats to be described, but it was not the first. That honor goes to *Megantereon* (Croizet and Jobert, 1828), followed by *Machairodus* (Kaup, 1833). *Smilodon* (Lund, 1842) was the third genus of sabertooth, still considered valid today, to be described.

Perhaps the best known of the sabertooths, the genus *Smilodon* was first described in 1842 by P. W. Lund (1801-80) as *Smilodon populator*. However, the material on which the genus was based was not the first remains of this animal to be discovered. Lund had previously referred specimens now known to be the remains of *Smilodon* to a new species of *Hyaena*, *H. neogaea* (Lund, 1841a). Although the name *Smilodon neogaea* has been used by some workers, Lund provided neither a description nor definition of the taxon, nor did he list the specimens on which the name was based. As no illustrations of these specimens were provided, the name is considered a *nomen nudum* (Berta, 1985).

ABBREVIATIONS

AMNH, American Museum of Natural History New York; **ANSP,** Academy of Natural Sciences of Drexel University, Philadelphia; **MACN,** Museo Argentino de Ciencias Naturales Bernardino Rivadavia, Buenos Aires; **UCMP,** Museum of Paleontology, University of California, Berkeley; **WFI,** Wagner Free Institute

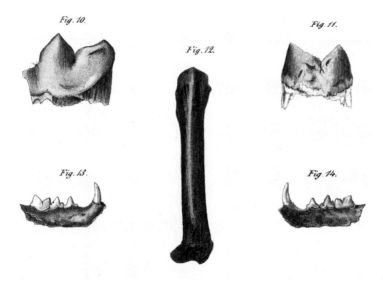

Figure 1.1. Part of Plate 26 from Lund (1841b) of dentition originally described as *Felis protopanther* (10 = ZMUC 508; 11 = ZMUC 507), the earliest illustration of *Smilodon* remains.

of Science of Philadelphia, Philadelphia; **ZMUC,** Zoological Museum, University of Copenhagen, Copenhagen.

Later, Lund (1841b) published a faunal list and used the name *Felis protopanther*, which he apparently wanted to apply to a fossil jaguar. This time he illustrated the two teeth he referred to as *F. protopanther* (Lund 1841b, Pl. 26, fig. 10-11) (Fig. 1.1), and these specimens are in the Lund collection of the Zoological Museum at the University of Copenhagen (ZMUC). ZMUC 508 (Pl. 26, fig. 10) is a right upper fourth premolar, and ZMUC 507 (Pl. 26, fig. 11) is a right lower first molar. The two illustrated teeth are clearly from *Smilodon* (Seymour, pers. comm., 2011). As with many of Lund's types, a specific specimen on which names were based was not identified and the illustrations are the only record for the type of *F. protopanther*. Because there are no published catalog numbers or descriptions that permit identification of any other specimens to which the name may be applied, these illustrations have priority for identity of the name *F. protopanther*. Berta and Marshall (1978) and Berta (1985) considered *Felis protopanther* a junior synonym of *Smilodon populator*. However, because the species name *protopanther* was published and illustrated one year before the name *Smilodon populator* and the specimens on which the name is based are still preserved in the Zoological Museum at

the University of Copenhagen, one could argue that *Smilodon protopanther* should be the earliest available name, although this species name should probably be formally suppressed in order to maintain taxonomic stability.

The type material of *Smilodon populator* consists of the right upper first (ZMUC 1/1845:443) and third incisor (ZMUC 1/1845:439), a fragmentary right upper canine (ZMUC 1/1845:1113), left second (ZMUC 1/1845:170), fourth metacarpal (ZMUC 1/1845:171), and right fifth metacarpal (ZMUC 1/1845:331) (Fig. 1.2). The assumption is that all of these bones are from the same individual. While the descriptive terms 'dirktooth,' 'scimitar-tooth,' and 'saber-tooth' are all are based on canine shape, in Lund's original description of *Smilodon* he specifically states that the term 'Smilo- (saber) and odon- (tooth)' is based on the form of the incisors. It was not until 1846, four years after the original description of *Smilodon* by Lund (1842), that a complete canine of the genus clearly illustrated its hypertrophied nature (Fig. 1.3), and since then the term 'saber' has been used in reference to the canine rather than the incisors, as originally described by Lund. Along with the canine, the first illustration of the lower jaw was also provided (Fig. 1.4).

The name *Smilodon* is well known today, but it was not readily accepted in the scientific literature. Muñiz (1845) proposed *Munifelis bonaerensis* for a nearly complete skeleton (MACN 46) of *Smilodon*

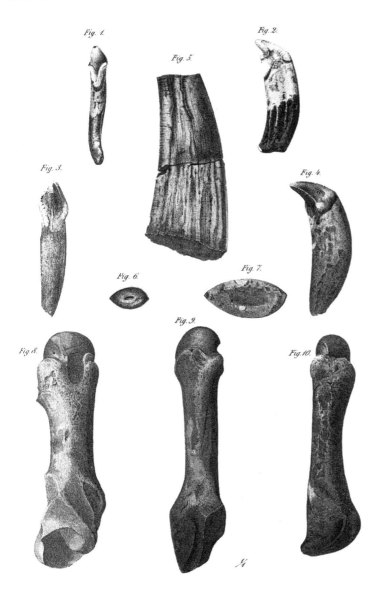

Figure 1.2. Plate 37 from Lund (1842) showing the incisors (1, 2, 4) on which Lund based the genus name *Smilodon*, a partial upper canine (3, 5, 6) and metacarpal (8-10) (ZMUC 1/1845:443) and third incisor (ZMUC 1/1845:439), a fragmentary right upper canine (ZMUC 1/1845:1113), left second (ZMUC 1/1845:170), and fourth metacarpal (ZMUC 1/1845:171) and right fifth metacarpal (ZMUC 1/1845:331).

populator from Argentina. Desmarest (1853) listed it as *Felis smilodon*, although by 1860 he did accept the genus, but with the species *S. blainvillii,* and this appears to be the first acceptance of *Smilodon* as a valid genus. Herluf Winge (1895), in the volume on the carnivores collected by Lund, referred to the material of *Smilodon* from Lapa do Bahu, Escrivania Nr. 1 and 5, and Lapa Vermela only in a cursory fashion and placed the specimens in the Old World genus *Machaerodus* as *M. neogaeus.* Ameghino (1888) considered *Smilodon* a valid South American genus although he described a new species, *S. ensenadensis,*

which, in 1889, he referred to *Machaerodus*, followed by the description of *Smilodon crucians* in 1904; he finally placed *M. bonaerensis* in *Smilodon* in 1907 and used *Smilodon* in subsequent publications (Ameghino 1888, 1889, 1904, 1907).

While clearly recognized as widespread in South and North America today (Berta, 1985; Kurtén and Werdelin, 1990), the recognition of *Smilodon* in North America took time. The earliest report of what is now considered to be remains of *Smilodon* was by Joseph Leidy (1823-91) in 1868 in a brief paper on some fossils "reported to have been obtained from

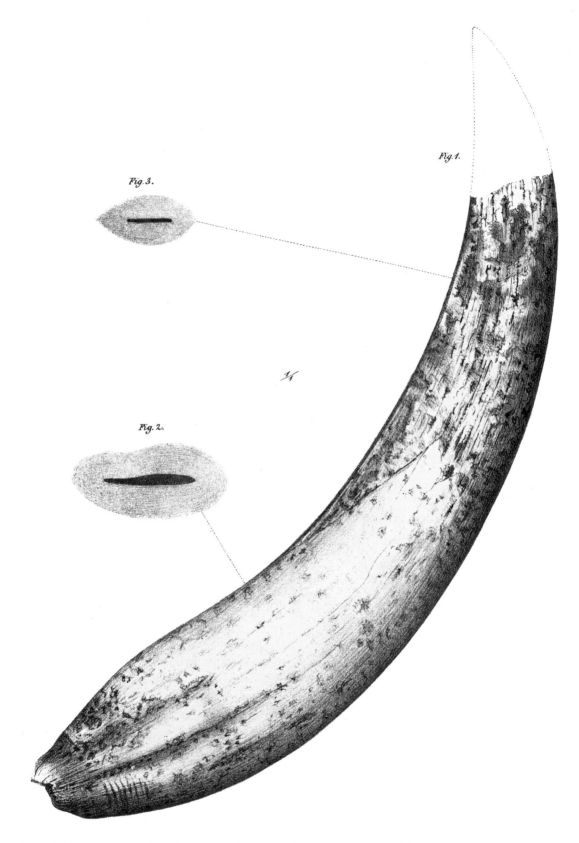

Figure 1.3. Plate 47 from Lund (1846) of the first relatively complete upper canine of *Smilodon populator*.

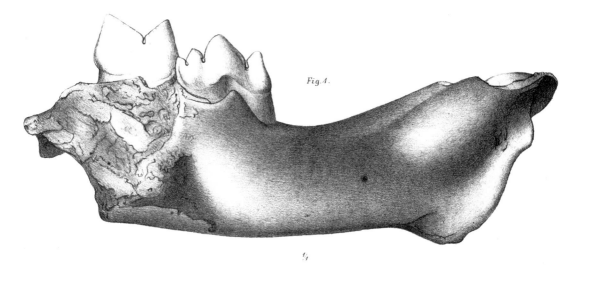

Fig. 1.

Fig. 2.

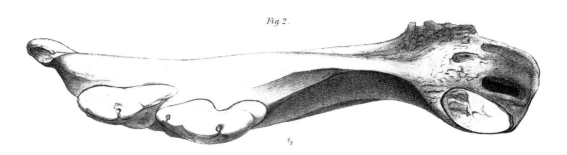

Figure 1.4. Plate 48 from Lund (1846) of partial right mandible of *Smilodon populator*.

blue clay and sand, beneath a bed of bitumen, in Harden Co. (Hardin County of today), Texas, and were donated to the New York Lyceum of Natural History by Mr. Robertson." The specimen, an upper fourth premolar in a maxillary fragment, was originally described as *Felis (Trucifelis) fatalis*. Later Leidy (1869) redescribed the specimen as *Trucifelis fatalis* and illustrated the specimen (Fig. 1.5). *Smilodon fatalis* has been applied to the Late Pleistocene species in North America by Slaughter (1963) and subsequent workers (Kurtén and Werdelin, 1990); however, Berta (1985) argued that all Late Pleistocene *Smilodon* in North and South America should be referred to a single species, *S. populator*. Babiarz et al. (chapter 5) propose a different application of the name *S. fatalis* for some *Smilodon* in North America. The type specimen of *Smilodon fatalis* is now in the American Museum of Natural History (AMNH 10395), making it the only known surviving fossil specimen not destroyed in a fire of the New York Lyceum (the predecessor of the New York Academy of Science) in May 1866. Its survival was fortuitous as the specimen was apparently on loan to Leidy at the time of the fire.

Leidy (1889a), at the February 19, 1889 meeting of the Academy of Natural Sciences of Philadelphia,

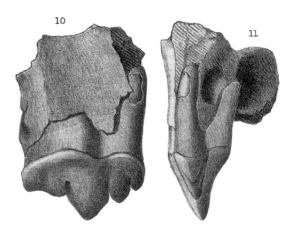

10

11

Figure 1.5. Figure from Leidy (1869) of type specimen of *Smilodon fatalis* from Hardin County, Texas (AMNH 10395). See also chapter 5.

reported a skull at the Wagner Free Institute of Science of Philadelphia (WFI 4072). Joseph Wilcox had collected the skull from a rock crevice exposed in a limestone quarry near Ocala, Marion County, Florida. Leidy considered it a distinct species, *floridanus*, but could not decide whether it should be placed in *Drepanodon* or *Smilodon*. The first genus is now considered a junior synonym of *Ursus*, although it had also been applied to another sabertooth carnivore, *Hoplophoneus* (McKenna and Bell, 1997). Later, in the *Transactions of the Wagner Free Institute of Science of Philadelphia*, Leidy (1889b) described the specimen in detail, provided an illustration (Fig. 1.6), and assigned *floridanus* to *Machairodus*. Leidy provided a list of measurements of the skull of *M. floridanus* to those taken from casts of specimens from Brazil referred to as *Machairodus neogaeus* by H. M. D. de Blainville (1839–64) and with the specimen from Argentina originally described as *Munifelis bonaerensis* but later redescribed by Burmeister (1866) as *Machairodus bonaerensis*. Leidy (1889b) eventually assigned his new species to *Machairodus*, but in his original paper he had at least noted that *Smilodon* is another known genus of sabertooth. Like Blainville, Leidy regarded the South American forms as one species and thought it probable that the Florida skull might also belong to that species. So it is not surprising that Leidy referred his new species to *Machairodus* rather than *Smilodon* because this was the genus being utilized for the South American

saber cat, although Leidy did maintain that the North American form is a separate species. It should also be noted that in both papers (Leidy, 1889a, 1889b) he also used the term 'Sabre-tooth tiger' and seems to have established the common name that is often applied to *Smilodon*. Unfortunately, this common name has caught on with the public and continues to confuse the general public about the relationship of *Smilodon* to modern felids and has no doubt biased how artists have depicted the animal.

An older species of *Smilodon* from Port Kennedy Cave was described by E. D. Cope (1840–97) as *Smilodon gracilis* (ANSP 47.5) based on a fragment of an upper canine (Cope, 1880). This appears to be the earliest actual assignment of a North American sabertooth cat to *Smilodon*. As Cope noted (p. 857), "Having established the existence of the genus *Smilodon* as a contemporary of the sloths during the Pliocene period of North America, it becomes probable that the species of the caves is also to be referred to it." Cope seems to have considered *Smilodon* a geologically older taxon and not the same as the sabertooth referred to as *Machaerodus* in the Late Pleistocene. However, in a subsequent paper Cope (1899) changed his mind regarding *gracilis* as a species of *Smilodon* and placed it in *Machaerodus*. It should be noted that the correct spelling is *Machairodus* (Kaup, 1833) but that many early workers such as Cope used the alternative spelling *Machaerodus*. Merriam (1905) referred to this specimen as *Machaerodus gracilis* and Brown (1908) referred the species to his genus *Smilodontopsis* from the Conard Fissure, Arkansas along with two new species, *Smilodontopsis troglodytes* (AMNH 11786) and *Smilodontopsis conardi* (AMNH 11790), which he considered to be similar to *Smilodon*.

The first clear recognition of *Smilodon* as a Late Pleistocene North American taxon was by Bovard (1907) with the description of *Smilodon californicus* (UCMP 10210) from the asphalt deposits of Rancho La Brea, California. Matthew (1918) described *Smilodon nebrascensis* (AMNH 17351) from the Hay Springs Quarry of Sheridan County, Nebraska. Lull (1921) identified a braincase of *Smilodon* from sand pits near Dallas, Texas as *S.* cf. *fatalis*, because the specimen could not be directly compared to the type. Merriam and Stock (1932), in their monograph on the Felidae

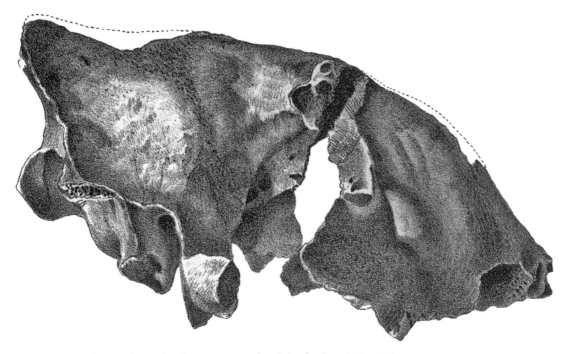

Figure 1.6. Figure from Leidy (1889b) of type specimen of *Smilodon floridanus* (WFI 4072).

of Rancho La Brea, firmly documented the single largest sample of *Smilodon* in North America. Their taxonomy recognized three subgenera: *Trucifelis*, with the species *fatalis*, *floridanus*, *californicus,* and *nebrascensis*; *Smilodontopsis* with *troglodytes* from the Conard Fissure and *gracilis* from Port Kennedy Cavern and *Dinobastis*, which is now considered a junior synonym of *Homotherium* (Slaughter, 1963).

Slaughter (1960) described *Smilodon trinitensis* although he later (1963) synonymized it with *Smilodon fatalis* along with *Smilodontopsis conardi*, *Smilodontopsis troglodytes*, and *Smilodon nebrascensis*. Kurtén (1963), following the taxonomy of Slaughter (1963), proposed an evolutionary relationship between *S. gracilis* and *S. fatalis,* with the former giving rise to the later. He maintained *S. floridanus* and *S. californicus* as separate species, with the later species considered to be a more progressive form. Webb (1974) reviewed *Smilodon floridanus* and *Smilodon californicus* and considered them to be the same species, with the former having priority. Kurtén and Anderson (1980) extended the lineage with *S. gracilis* as intermediate between the older *Megantereon hesperus* and the later *S. fatalis*. Martin

(1980), in contrast, placed *gracilis* in *Megantereon*. Berta (1985), going back to Leidy and other early workers, considered all Late Pleistocene *Smilodon* in North and South America to be a single species and considered all of the proposed species to be junior synonyms of *S. populator* but still derived from the earlier *S. gracilis*. Kurtén and Werdelin (1990) considered the North and South American Late Pleistocene forms to be distinct species, with the South American species east of the Andes as *S. populator* and the North American species as *S. fatalis* also for specimens from tar pits in Peru and Ecuador, thus extending its range along the western coast of South America. The presence of two different species in different parts of South America suggests two separate dispersal events of *Smilodon* into South America, an earlier one in the Middle Pleistocene (Ensenadan SALMA) which gave rise to *Smilodon populator* and a later event in the Late Pleistocene (Lujanian SALMA) in which *Smilodon fatalis* extended its range into northwestern South America. Support for two separate dispersal events is given by the presence of *Smilodon gracilis* from the early to middle Pleistocene El Breal de Orocual asphalt deposits in Monagas state, northeastern Venezuela (Rincón et al., 2011).

Prior to the discovery of fossil vertebrates at Rancho La Brea and the extensive excavations that followed, with the rare exception of the type of *Munifelis bonaerensis* based on a nearly complete skeleton (MACN 46), most of the known remains of *Smilodon* consisted of fragmentary specimens. In 1907, the staff at the Museo Argentino de Ciencias Naturales "Bernardino Rivadavia" mounted a cast of the *Smilodon* skeleton found by Muñiz, but the original skeleton was never exhibited; the skull and the mandible appeared in a separate display. Ameghino (1907) included a photo in an article published about this mounted skeleton. Nearly a hundred years after its original description by Muñiz, Méndez-Alzola (1941) described the skeleton in detail. The original cast of the skeleton was later removed from exhibition and replaced by a modern cast. A photo of the old cast before it was dismounted is provided in Figure 1.7. It is not exactly in the original pose, as it was slightly modified in the 1990s (Alejandro Kramarz, pers. comm., February 1, 2016). A plaster cast of the skeleton of *Smilodon populator* (MLP 7-47) is on exhibit at the Museo de la Plata (Sala VIII), established in 1888. The cast was probably sent to the museum by Florentino Ameghino, who at the time was director of the Museo Argentino de Ciencias Naturales (1902-11). A label on the Museo de la Plata cast says (translation): "The nearly complete skeleton was completed with casts from the MLP and the American Museum of Natural History." The entrance to the museum is flanked on either side by reclining statues of *Smilodon* (Fig. 1.8) sculpted by Victor de Pol (1865-1925). In front of the main entrance of the museum and behind the Francisco Pascasio Moreno bust is a mural of *Smilodon* painted by the French landscape architect Émile Bonnet Coutaret (1863-1949) in 1922. (Marcelo A. Reguero, pers. comm., May 2016).

In 1875, William Denton, a geologist/paleontologist, recognized the presence of animal remains from the asphalt deposits at Rancho La Brea, California, and was presented with a canine of *Smilodon* by the owner, Henry Hancock. Unfortunately, the current location of this canine is unknown. Denton (1877) reported on the site and described the canine in a

Figure 1.7. Mounted cast of skeleton of *Smilodon populator* (type of *Munifelis bonaerensis*) in the Museo Argentino de Ciencias Naturales "Bernardino Rivadavia." Photo courtesy of Alejandro Kramarz.

Figure 1.8. Reconstruction of *Smilodon populator* at entrance to the Museo de Ciencias Naturales de La Plata. Photo courtesy of Sergio Vizcaíno.

letter published by the Boston Society of Natural History (p. 1859: "Major Hancock presented me with what I found to be a canine tooth of *Machairodus*, a great saber-toothed feline. . . . The tooth is nine and a half inches in length, measured along the curve, and the breadth of the crown at the base is an inch and three-quarters, being larger than any tooth of the European *Machairodus*, whose measurement I have been able to find. The crown of the tooth is broken, and its entire length could not have been less, I think, than eleven inches. . . . The Californian tooth is closely serrated on both the concave and convex sides."

There was no follow-up at Rancho La Brea to this initial discovery until 1901 when W. W. Orcutt visited the site; by 1905 Orcutt had accumulated a modest collection of the fauna, including additional remains of *Smilodon*. In 1906 the Museum of Paleontology at the University of California, Berkeley, carried out a preliminary exploration of the site and J. C. Merriam subsequently returned in 1912 for a larger excavation, which lasted into 1913. J. Z. Gilbert started excavating at Rancho La Brea in 1907 under the auspices of the Los Angeles High School. The Southern California Academy of Science initiated work in 1909, and their efforts resulted in the creation of the Museum of History, Science, and Art (now the Natural History Museum of Los Angeles County), which for many years used a representation of *Smilodon* as their logo. The Southern California Academy collection was transferred to the museum in 1911 and was incorporated into the displays when the museum opened in 1913. Its involvement at Rancho La Brea is commemorated by the use of the skull of *Smilodon* as part of the academy's logo.

In 1924 G. Allan Hancock, the son of Major Hancock, who had presented the *Smilodon* canine to Denton, donated 23 acres containing the tar pits to Los Angeles County, and Hancock Park was established. As part of the park development a 'Pleistocene Zoo' composed of life-size concrete statues of the fauna from Rancho La Brea was built for the park, sculpted by J. L. Roop (1869-1932); these statues included a pair of fighting *Smilodon* (Fig. 1.9) and a *Smilodon* standing on the back of a bison trapped in the tar (McNassor, 2011). Further popularization of *Smilodon* occurred when, on September 25, 1973, *Smilodon californicus* was officially made California's state fossil. However, the legislation did not provide for subsequent taxonomic changes as when in the following year Webb (1974) made *Smilodon californicus* a junior synonym of *Smilodon floridanus*, resulting in the California state fossil being named after Florida.

After the University of California, Berkeley, excavation (1912-13), the newly created natural history museum in Los Angeles was given sole right in 1913 to excavate for two years. In 1924, Major Hancock deeded the land to Los Angeles County. Additional museum excavations took place in 1929, with a hiatus until 1969 when excavation resumed at Pit 91 and has continued to the present, complemented by other excavations resulting from construction projects in the vicinity of Hancock Park. Consequently, additional remains of *Smilodon* continued to be found, and in 1932 John C. Merriam

Figure 1.9. Reconstruction of two *Smilodon fatalis* fighting, with the sculptor J. L. Roop. Made for Hancock Park, location of the La Brea tar pits. Photo courtesy of the La Brea collections, Los Angeles County Museum of Natural History.

and Chester Stock published a monograph on the fossil cats of Rancho La Brea that included the first comprehensive description of the skeleton of *Smilodon*. Marcus (1960) reported 987 individuals of *Smilodon* from just the 7 major pits at Rancho La Brea, and Miller (1968) utilized over 2,100 skulls from La Brea in his study of the age distribution of *Smilodon* from Rancho La Brea. Despite ongoing discoveries at the site, only a single semi-articulated skeleton representing 71% of the animal has been recovered, lacking only most of the bones of the hind feet (Cox and Jefferson, 1988). The specimen was recovered from a thin tabular asphaltic deposit discovered as part of salvage operations during the construction of the Page Museum in Hancock Park.

The excavations at Rancho La Brea have recovered literally thousands of *Smilodon* bones. While Rancho La Brea is the premier locality for *Smilodon fatalis* in North America, the species is known from more than forty Rancholabrean sites distributed across Alberta, Arkansas, California, Florida, Idaho, Louisiana, Mexico, Nebraska, New Mexico, Oregon, Tennessee, Texas, and Utah (Kurtén and Anderson, 1980). There is only a single record in Central America, the Hormiguero quarry in El Salvador (Stirton and Gealey, 1949) but its range extends southward into Venezuela, Ecuador, and Peru. Most records consist of isolated bones and teeth and only rarely of a partial skeleton. The large sample from Rancho La Brea quickly transformed *Smilodon* from an obscure taxon represented by bits and pieces to one for which virtually every bone is known, including the vestigial clavicles (Hartstone-Rose et al., 2012).

Eugene J. Fischer, a preparator and taxidermist at the Los Angeles Museum of History, Science, and Art, mounted the first full skeleton of a North American *Smilodon*. Two composite skeletons were mounted in 1910 and 1911, utilizing specimens from Rancho La Brea, one for the museum, which opened to the public in 1913, and a second for Los Angeles High School (Scott, 1992). A third skeleton of *Smilodon* was mounted by Fischer in 1920. Although *Smilodon* is the second most common carnivore from Rancho La Brea, like virtually all of the rest of the fauna it is represented by disarticulated remains, and associated skeletons or even parts of skeletons are rare (Cox and Jefferson, 1988). The volume of

material recovered from Rancho La Brea allowed the Museum of Paleontology at University of California, Berkeley, and the Natural History Museum of Los Angeles County to send complete (although composite) skeletons of *Smilodon* to multiple museums around the world (e.g., Barthel, 1966). Today at least twenty-five skeletons of *Smilodon fatalis* from Rancho La Brea are on display or have been on display at museums around the world (C. A. Shaw, pers. comm., 2016). The many composite original bones of skeletons exchanged with museums initially made a major contribution to museum visitors' familiarity with *Smilodon*; in recent years its continued popularity for museum exhibits has been greatly augmented by the active production of casts of skeletons based on specimens from Rancho La Brea. This in turn has certainly contributed to making it the best-known Pleistocene carnivore and firmly establishing it in the popular mind and culture.

Remains of *Smilodon* continue to be discovered (Scott and Springer, 2016) and some of these discoveries in turn continue to spark interest in the animal through its incorporation into popular culture. In May 1971, during an excavation in downtown Nashville, Tennessee for what is today the 28-story First American Center, construction workers drilled through 20 feet of solid rock before coming to a soft muddy area. Further digging revealed a cave containing the canine and other parts of the skeleton of an individual of *Smilodon* (Guilday, 1977). The cave, located beneath the building, has been preserved under concrete for historic and educational purposes. Twenty-six years after this discovery, The Predators, Nashville's hockey team, unveiled their logo, a 'saber-toothed tiger,' at the First American Center on September 25, 1997.

As we hope this volume shows, we still have much to learn about the paleoecology, natural history, functional anatomy, behavior, and evolutionary history of *Smilodon*, and there is certainly more information to extract from the fossil record (McCalla et al., 2003; Carbone et al., 2009). As can be seen from these contributions, new technologies such as stable isotope analysis (Prevosti and Martin, 2013; Bocherens et al., 2016) or the recovery of ancient DNA (Barnett et al., 2005) provide new approaches to elucidating its natural history. We hope this volume will not only serve as useful reference for future research but may also prove to be equally useful to artists (e.g., Antón et al., 1998) and animators who produce images of these animals not only for documentaries but for animated movies, and continue to make *Smilodon* a part of the popular culture. Indeed, *Smilodon* has even appeared on the stamps of multiple countries (http://www.owasu.org/primer_files/sabretooth.pdf). We also hope that this volume will be of interest to members of the public who want to know more about *Smilodon* and that it will enhance their understanding of this iconic sabertooth cat.

ACKNOWLEDGMENTS

Kasper Hansen kindly provided the updated catalog numbers and information on the type material of *Smilodon popular* in the Zoological Museum at the University of Copenhagen. Alejandro Kramarz, Museo Argentino de Ciencias Naturales "Bernardino Rivadavia" graciously provided information on the history of the mounted *Smilodon* collected by Muñiz. The author thanks Marcelo A. Reguero, Museo de La Plata, for his efforts in providing information on the *Smilodon* skeleton and murals at his museum, and Sergio Vizcaíno for providing the photograph of the *Smilodon* statue at the entrance to the museum. The efforts of Richard Hulser and Kim Walters in support of this chapter are greatly appreciated.

REFERENCES

Ameghino, F. 1888. Rapidas diagnosis de algunos mamíferos fósiles nuevos de la Republica Argentina. F. Ameghino Obras Completas 5:471-480.

Ameghino, F. 1889. Contribución al conocimiento de los mamíferos fósiles de la Republica Argentina, obra escrita bajo los auspicios de la Academia nacional de Ciencias de la Republica Argentina para presentarla a la Exposición Universal de Paris de 1889. Actas de la Academia Nacional de Ciencias, Córdoba 6:1-1027.

Ameghino, F. 1904. Nuevas especies de mamíferos cretáceos y terciarios de la República Argentina. Anales de la Sociedad cientifica Argentina 56-58:1-142.

Ameghino, F. 1907. Sobre dos esqueletos de mamiferos fósiles armados recientemente en el Museo Nacional. Imprenta de Juan A. Alsina, Buenos Aires.

Antón, M., R. García-Perea, and A. Turner. 1998. Reconstructed facial appearance of the sabretoothed felid *Smilodon*. Zoological Journal of the Linnean Society 124:369-386.

Barnett, R., I. Barnes, M. J. Phillips, L. D. Martin, C. R. Harington, J. A. Leonard, and A. Cooper. 2005. Evolution of the extinct sabretooths and the American cheetah-like cat. Current Biology 15:589-590.

Barthel, K. W. 1966. Mounting a skeleton of *Smilodon californicus* Bovard. Curator 9:119-124.

Berta, A. 1985. The status of *Smilodon* in North and South America. Natural History Museum of Los Angeles County, Contributions in Science 370:1-15.

Berta, A. and L.G Marshall. 1978. The South American carnivora; pp. 1-48 in F. Westphal (ed.), Fossilium Catalogus I: Animalia. W. Junk, The Hague, 125.

Blainville, H. M. D. de. 1839-1864. Ostéographie. Ou description iconographique comparée du squelette et du système dentaire des mammifères récents et fossiles pour servir de base à la zoologie et à la géologie. Volume 2. J.-B. Baillière, Paris, France, 196 pp.

Bocherens, H., M. Cotte, R. Bonini, D. Scian, P. Straccia, L. Soibelzon, and F. J. Prevosti. 2016. Paleobiology of sabretooth cat *Smilodon populator* in the Pampean region (Buenos Aires Province, Argentina) around the last glacial maximum: insights from carbon and nitrogen stable isotopes in bone collagen. Palaeogeography, Palaeoclimatology, Palaeoecology 449:463-474.

Bovard, J. F. 1907. Notes on Quaternary Felidae from California. University of California Publications, Bulletin of the Department of Geology 5:155-170.

Brown, B. 1908. The Conard fissure, a Pleistocene bone deposit in northern Arkansas: with descriptions of two new genera and twenty new species of mammals. Memoirs of the American Museum of Natural History 9:158-208.

Burmeister, G. 1866. Lista de los mamíferos fósiles del terreno diluviano. Anales del Museo Público de Buenos Aires 1:121-232.

Carbone, C., T. Maddox, P. J. Funston, M. G. L. Mills, G. F. Grether, and B. Van Valkenburgh. 2009. Parallels between playbacks and Pleistocene tar seeps suggest sociality in an extinct sabretooth cat, *Smilodon*. Biology Letters (Paleontology) 5:81-85.

Cope, E. D. 1880. On the extinct cats of North America. American Naturalist 14:833-858.

Cope, E. D. 1899. Vertebrate remains from Port Kennedy bone deposit. Journal of the Academy of Natural Sciences, Philadelphia 11:194-267.

Cox, S. M., and G. T. Jefferson. 1988. The first individual skeleton of *Smilodon* from Rancho La Brea. Current Research in the Pleistocene 5:66-67.

Dawson, M. R., R. K. Stucky, L. Krishtalka, and C. C. Black. 1986. *Machaeroides simpsoni*, new species, oldest known sabertooth creodont (Mammalia), of the Lost Cabin Eocene. Special paper no. 3, Rocky Mountain Geology 24:177-182.

Denton, W. 1877. On the asphalt bed near Los Angeles, California. Proceedings of the Boston Society of Natural History 18:185-186.

Desmarest, E. 1853. Carnassiers; pp. 222-224 in J. C. Chenu (ed.), Encyclopédie d'Histoire Naturelle, Carnassiers, 2:ème partie. Marescq & Compagnie, Paris.

Guilday, J. E. 1977. Sabertooth cat, *Smilodon floridanus* (Leidy), and associated fauna from a Tennessee cave (40DV40)—the First American Bank Site. Journal of the Tennessee Academy of Science 52:84-94.

Hartstone-Rose, A., R. C. Long, A. B. Farrell, and C. A. Shaw. 2012. The clavicles of *Smilodon fatalis* and *Panthera atrox* (Mammalia, Felidae) from Rancho La Brea, Los Angeles, California. Journal of Morphology 273:981-991.

Kurtén, B. 1963. Notes on some Pleistocene mammal migrations from the Palaearctic to the Nearctic. Eiszeitalter und Gegenwart 14:96-103.

Kurtén, B., and E. Anderson. 1980. Pleistocene mammals of North America. Columbia University Press, New York.

Kurtén, B., and L. Werdelin. 1990. Relationships between North and South American *Smilodon*. Journal of Vertebrate Paleontology 10:158-169.

Leidy, J. 1868. Notice of some vertebrate remains from Harden Co., Texas. Proceedings of the Academy of Natural Sciences of Philadelphia 20:174-176.

Leidy, J. 1869. The extinct mammalian fauna of Dakota and Nebraska, including an account of some allied forms from other localities, together with a synopsis of the mammalian remains of North America. Journal of the Academy of Natural Sciences of Philadelphia, 2nd ser. 7:8-472.

Leidy, J. 1889a. The saber-tooth tiger of Florida. Proceedings of the Academy of Natural Sciences of Philadelphia 41:29-31.

Leidy, J. 1889b. Description of mammalian remains from a rock crevice in Florida. Transactions of the Wagner Free Institute of Science of Philadelphia 2:13-17.

Lull, R. S. 1921. Fauna of the Dallas sand pits. American Journal of Science, 5th series 2:159-176.

Lund, P. W. 1841a. Blik paa Brasiliens Dyreverden för Sidste Jordomvæltning. Anden Afhandling: Pattedyrene. Det Kongelige Danske videnskabernes naturvidenskabelige og mathematiske afhandlinger 8:61-144. [Danish]

Lund, P. W. 1841b. Tillaeg til de to sidste afhandlinger over Brasiliens dyreverden för sidste jordomvæltning. Det Kongelige Danske videnskabernes naturvidenskabelige og mathematiske afhandlinger 8:273-296. [Danish]

Lund, P. W. 1842. Blik paa Brasiliens Dyreverden för sidste jordomvæltning. Fjerde afhandling: Fortsættelse af pattedyrene. Det Kongelige Danske videnskabernes selskabs naturvidenskabelige og mathematiske afhandlinger 9:137-208. [Danish]

Lund, P.W. 1846. Meddelelse af det Udbytte de i 1844 undersögte knoglehuler have afgivet til kundskaben om Brasiliens Dyreverden för sidste jordomvæltning. Lagoa Santa, d. 22de November 1844. Det Kongelige Danske videnskabernes selskabs naturvidenskabelige og mathematiske afhandlinger 12:1-94. [Danish]

Marcus, L. F. 1960. A census of the abundant large Pleistocene mammals from Rancho La Brea. Natural History Museum of Los Angeles County, Contributions in Science 38:1-11.

Martin, L. D. 1980. Functional morphology and the evolution of cats. Transactions of the Nebraska Academy of Sciences 7:141-154.

Matthew, W. D. 1918. Contributions to the Snake Creek Fauna with notes upon the Pleistocene of western Nebraska. Bulletin of the American Museum of Natural History 38:183-229.

McCalla, S., V. Naples, and L. D. Martin. 2003. Assessing behavior in extinct animals: was *Smilodon* social? Brain Behavior and Evolution 61:159-164.

McKenna, M. C., and S. K. Bell. 1997. Classification of Mammals: Above the Species Level. Columbia University Press, New York, 631 pp.

McNassor, C. 2011. Los Angeles's La Brea Tar Pits and Hancock Park. Images of America. Arcadia Publishing, Charleston, South Carolina, 129 pp.

Méndez-Alzola, R. 1941. El *Smilodon bonaërensis* (Muniz): estudio osteológico y ostrométrico del gran tigre fósil de la Pampa comparado con otros félidos actuales y fósiles. Anales del Museo Argentino de Ciencias Naturales 40:135-252.

Merriam, J. C. 1905. A new saber-tooth from California. University of California Publications, Bulletin of the Department of Geology 4:171-175.

Merriam, J. C., and C. Stock. 1932. The Felidae of Rancho La Brea. Carnegie Institution of Washington Publication No. 422:1-232.

Miller, G. J. 1968. On the age distribution of *Smilodon californicus* Bovard from Rancho La Brea. Natural History Museum of Los Angeles County, Contributions in Science 131:1-17.

Muñiz, F. J. 1845. Descripción del *Munifelis bonaerensis*. Gaceta Mercantil 6603:1-2.

Prevosti, F. J., and Martin, F. M. 2013. Paleoecology of the mammalian predator guild of Southern Patagonia during the latest Pleistocene: ecomorphology, stable isotopes, and taphonomy. Quaternary International 305:74-84.

Rincón, A. D., F. J. Prevosti, and G. E. Parra. 2011. New saber-toothed cat records (Felidae, Machairodontinae) for the Pleistocene of Venezuela, and the great American biotic interchange. Journal of Vertebrate Paleontology 31:468-478.

Scott, E. 1992. Mount Up! The skeletons of Rancho La Brea on display. Terra 31:11-13.

Scott, E., and K. B. Springer. 2016. First records of *Canis dirus* and *Smilodon fatalis* from the Late Pleistocene Tule Springs local fauna, upper Las Vegas Wash, Nevada. PeerJ 4:e2151; DOI 10.7717/peerj.2151.

Slaughter, B. H. 1960. A new species of *Smilodon* from a Late Pleistocene alluvial terrace deposit of the Trinity River. Journal of Paleontology 34:486-492.

Slaughter, B. H. 1963. Some Observations Concerning the Genus *Smilodon*, With Special Reference to *Smilodon fatalis*. The Texas Journal of Science 15:68-81.

Stirton, R. A., and W. K. Gealey. 1949. Reconnaissance geology and vertebrate paleontology of El Salvador, Central America. Geological Society of America Bulletin 60:1731-1754.

Webb, S. D. 1974. The Status of *Smilodon* in the Florida Pleistocene; pp.149-153 in S. D. Webb (ed.), Pleistocene Mammals of Florida. University Presses of Florida, Gainesville, Florida.

Winge, H. 1895. Jordfundne og nulevende rovdyr (Carnivora) fra Lagoa Santa, Minas Geraes, Brasilien. Med udsigt over rovdyrenes indbyrdes slægtskab. E Museo Lundi 2(2):1-103. [Danish]

2 The Phylogenetic Context of *Smilodon*

LARS WERDELIN AND THERESE FLINK

Introduction

Although *Smilodon* is an icon among fossil organisms, invoked for its power and beauty as well as for its recent extinction (and for its sharp, nasty teeth), it did not spring fully formed straight from the head of some feline Zeus. There were other animals with similar but less extreme adaptations to which it was related, more or less closely. Which of these animals were closely related to *Smilodon* and the exact nature of their relationship is subject to some controversy. On the one hand are traditional presentations (e.g., Werdelin et al., 2010; Antón, 2013) that subdivide the sabertooth felids (subfamily Machairodontinae within the family Felidae) into two major groups, the tribes Homotherini (commonly called scimitar-tooth cats) and Smilodontini (called dirk-tooth cats). The two common names are misleading and are not used further here. As the name implies, *Smilodon* is a member of the Smilodontini. On the other hand, there are quantitatively based phylogenetic analyses of sabertooths that differ greatly among themselves and with the traditional view of relationships of felids with a sabertooth morphology (e.g., Christiansen, 2013; Wallace and Hulbert, 2013).

Regardless of the precise phylogeny espoused, the published literature gives differing views of the range of taxa that should be included within the Smilodontini, from a very restrictive set of genera and species living between the Early Pliocene and Late Pleistocene, to a very inclusive set that stretches back to the latest Middle Miocene (Berta, 1987; Antón and Galobart, 1999; Antón et al., 2004). This chapter focuses on the smaller set for two reasons: the exact membership of the more inclusive set is uncertain, and including all these taxa would make this chapter too long. We thus do not consider taxa such as *Paramachaerodus* or *Promegantereon* in the following presentation of the Smilodontini. The latter is, however, included in the phylogenetic analysis and we return to the issue of inclusion in the discussion section.

ABBREVIATIONS

AMNH, American Museum of Natural History, New York; **ANSP,** Academy of Natural Sciences of Drexel University, Philadelphia; **IGF,** Institute of Geology, University of Florence, Italy; **KNM,** National Museums of Kenya, Nairobi; **NHMUK,** Natural History Museum, London; **TM,** Ditsong National Museum of Natural History, South Africa (formerly Transvaal Museum); **UF,** Florida State Museum, University of Florida, Gainesville; **USNM,** United States National Museum of Natural History, Smithsonian Institution, Washington, DC; **VM,** Venta Micena Collection, Museo de Prehistoria y Paleontología, Orce, Spain; **ZMUC,** Zoological Museum, Copenhagen, Denmark

The Smilodontini: A Taxonomic Review

Family FELIDAE Fischer, 1817
Subfamily MACHAERODONTINAE Gill, 1872
Tribe SMILODONTINI Kretzoi, 1929

We here accept the existence of a monophyletic Tribe Smilodontini, although this taxon in its original conception (Kretzoi, 1929) (as subfamily Smilodontinae) was more akin to the view of Christiansen (2013) and not monophyletic in the sense presented here and in other traditional reviews.

Genus *RHIZOSMILODON* Wallace and Hulbert, 2013
RHIZOSMILODON FITEAE Wallace and Hulbert, 2013

Megantereon hesperus (partim) (Berta and Galiano, 1983)
Machairodontinae gen. et sp. indet. (Webb et al., 2008)

Holotype—UF 124634, partial right ramus including mandibular symphysis, c1, m1, and alveoli for p3 and p4.

Type locality—See Wallace and Hulbert, 2013.

Diagnosis—(Wallace and Hulbert, 2013) Nearly complete verticalization of mandibular symphysis; presence of a weak but clear mandibular flange; suggestion of crenulations on at least lower canines; lower canine remains large but is moderately compressed laterally; nonprocumbent incisor arcade; incisors small; P3 and P4 not aligned (implied by offset of p4 and m1); lack of p2; p3 elongate, but smaller than p4 (2/3 length); posterior lean of p3 and p4 toward m1; lack of anterior accessory cusp on p3; presence of a small posterior accessory cusp on p3; anterior accessory cusp on p4 small; p3 and p4 offset (not in straight line); p4 long and bladelike, yet with distal widening around posterior accessory cusp; p4 and m1 also offset; m1 talonid variably present as small accessory cuspid to slight raised bump at the base of the tooth, never fully formed metaconid; m1 robust as in *Smilodon* with widest point at or anterior to the carnassial notch; and m1 paraconid remains shorter than protoconid.

The material assigned to this taxon by Wallace and Hulbert (2013) has had a somewhat checkered history. The first specimen in the *Rhizosmilodon fiteae*

hypodigm to be described was UF 22890, assigned to *Megantereon hesperus* by Berta and Galiano (1983). If correct, this would have been the earliest record of *Megantereon,* suggesting that this genus was of North American origin. This assignment drew criticism from several authors (e.g., Turner, 1987), because the specimen, a partial left ramus with p3-m1 (broken), lacked a deep mandibular flange, an important character of *Megantereon*. Turner (1987) suggested affinity with *Dinofelis*.

Some years later, Hulbert et al. (2001) mentioned and illustrated a new specimen belonging to the same taxon as UF 22890. This was UF 124634, now the holotype of *R. fiteae*. Later, Webb et al. (2008) again mentioned this specimen, noting that other specimens had been recovered in the meantime and that the affinities of these taxa were under study. Hodnett (2010) again illustrated UF 22890 and UF 124634, suggesting potential affinities with *Paramachairodus* sp. based on USNM 244453, a specimen from Arizona. Thus, all these authorities (excepting Turner, 1987) agreed that the material belonged in Smilodontini but disagreed on its precise affinities within the tribe. Finally, Wallace and Hulbert (2013) formalized this, placing their new taxon *Rhizosmilodon* as sister taxon to *Megantereon + Smilodon*.

Genus *MEGANTEREON* Croizet and Jobert, 1828

Type species—*Ursus cultridens* (partim) Cuvier, 1824

The number of species of *Megantereon* is a matter of considerable debate. There are at least twelve described species, but there is no consensus regarding which of these are actually valid. Current opinion ranges from two (e.g., Palmqvist et al., 2007) to six or seven (Lewis and Werdelin, 2010). Herein we do not address this issue directly but use the scheme of Lewis and Werdelin (2010) with the remaining species in the synonymies of the taxa accepted there.

MEGANTEREON CULTRIDENS (Cuvier, 1824)

Megantereon megantereon (Croizet and Jobert, 1828)
Felis megantereon (Bravard, 1828)
Megantereon macroscelis (Pomel, 1853)

Lectotype—IGF 816, upper right canine.

Type locality—Tasso Faunal Unit of the Upper Valdarno, Italy (Ficcarelli, 1979).

Diagnosis—(Turner, 1987) Medium-sized, sexually dimorphic cat; short, high skull, triangular in profile with convex dorsal outline; extended glenoid apophyses; limbs and feet short and powerful, the forelimbs more so than the hind limbs; tail reduced in length; upper canines elongated, compressed, and curved; crenulations absent on all teeth; long postcanine diastema; upper and lower second premolars lost; upper and lower third premolar present and functional but very variable in size; P4 protocone developed but variable in size; M1 present but reduced; p3-m1, backward raking; main cusp of p4 high crowned; m1, lacking talonid or metaconid, paraconid shorter than protoconid and forming a relatively acute angle with it; mandible with short coronoid process, strong and vertical symphysis with enlarged flange extending well below the ventral margin of the horizontal ramus; single large mental foramen.

This is the European Pliocene and Early Pleistocene *Megantereon*. Sardella (1998) distinguished between an early, less derived form, exemplified by the material from the Montagne de Perrier described by Croizet and Jobert (1828) and a later, larger, and more derived form exemplified by material from St.-Vallier, Senèze, and the Upper Valdarno (Viret, 1954; Ficcarelli, 1979; Christiansen and Adolfssen, 2007). These may represent distinct taxa at some level or merely a modest anagenetic change. Turner (1987) considered this the only species of *Megantereon*, while Palmqvist et al. (2007) considered it to be one of two species, the other being *M. whitei*.

Megantereon falconeri Pomel, 1853

Drepanodon sivalensis (partim) (Falconer and Cautley in Murchison, 1868)
Megantereon nihowanensis (Teilhard de Chardin and Piveteau, 1930)
Megantereon inexpectatus (Teilhard de Chardin, 1939)
Megantereon lantianensis (Hu and Qi, 1978)

Lectotype—NHMUK M16557 partial mandibular ramus with p4-m1.
Type locality—Siwalik Hills, India.
Diagnosis—(translated from Pomel [1853]) A little larger than *macroscelis* [=*M. cultridens*]; the canine is modest; the first lower molar lacks distal accessory

cusps and is small; the mandibular flange is very large.

Megantereon falconeri is here considered the sole Asian species of *Megantereon*. It was named by Pomel (1853) based on material in the collections of the Geological Society of London (now in the NHMUK). The subsequent taxonomic history of Siwalik *Megantereon* is confusing, as is so much related to Siwalik Carnivora. However, Matthew (1929) selected NHMUK M16557 as the lectotype of *M. falconeri*, and this specimen undoubtedly belongs in the genus *Megantereon*. Further analysis of the validity of Siwalik *Megantereon* was presented by Petter and Howell (1982) and we refer to that paper for detailed information.

Megantereon falconeri is poorly characterized, and all that can really be said at this point is that it seems to have been larger than typical *M. cultridens*, leading most authorities to synonymize the two. However, we here prefer to recognize *M. falconeri* as the senior synonym of Asian *Megantereon* pending detailed study. The three species of the genus described from China seem to conform to *M. falconeri*, but should further analysis show them to be distinct, *M. nihowanensis* (Teilhard and Piveteau, 1930) has priority.

MEGANTEREON HESPERUS (Gazin, 1933)

Machairodus ? hesperus (Gazin, 1933)
Machairodus sp. (Hibbard, 1937)
Megantereon hesperus (Schultz and Martin, 1970)

Holotype—USNM 12614, partial right ramus with m1.
Type locality—Hagerman Local Fauna, Glenns Ferry Formation, Twin Falls County, Idaho (Bjork, 1970).
Diagnosis—(Berta and Galiano, 1983) *Megantereon hesperus* can be distinguished from all other species of *Megantereon* in possessing primitively a larger lower canine and P3, and by the development of a prominent groove on the medial surface of the mandible and a stronger dorsoventral ridge marking the anterolateral rim of the mandibular flange.

A rare species, comprehensively discussed by Berta and Galiano (1983), although one specimen has since been reassigned to *Rhizosmilodon* (see above). It is known from only a handful of localities in North America, but ranges in space and time from Idaho (Blancan) in the northwest to Florida (Hemphillian) in the southeast.

MEGANTEREON WHITEI (Broom, 1937)

Megantereon gracile (Broom, 1948)
Megantereon eurynodon (Ewer, 1955)

Holotype—TM 856, partial left ramus with p3-m1.
Type locality—Schurveberg, Gauteng, South Africa.

Diagnosis—(modified from Werdelin and Lewis, 2013) A *Megantereon* with entire cheek dentition reduced relative to skull size. Anterior cheek dentition especially reduced. Mandibular symphysis more vertically oriented than in other *Megantereon*.

This African *Megantereon* has now definitively been shown to be a species distinct from *M. cultridens* (Palmqvist et al., 2007). It is known from both southern and eastern Africa but *Megantereon* has not yet been found in North Africa.

MEGANTEREON ADROVERI Pons Moya, 1987

Holotype—VM 2264, left p4; fragmentary left hemimandible with p4 and m1; fragmentary hemimandible with teeth; fragmentary hemimandible with m1; fragmentary hemimandible with c (syntypes).
Type locality—Venta Micena, Spain.
Diagnosis—Dentition, especially p3-p4, reduced and diastema longer relative to *M. cultridens*. Differs from *M. whitei* in dentition not being reduced relative to overall skull size.

There is no doubt that in the middle and late parts of the Early Pleistocene there was a species of *Megantereon* in Europe that differed from typical *M. cultridens* (Martínez-Navarro and Palmqvist, 1995, 1996). Pons Moya (1987) assigned material of this taxon from Venta Micena to a new subspecies, *M. c. adroveri*, whereas Martínez-Navarro and Palmqvist (1995, 1996) saw similarities to African *M. whitei* and synonymized the two. Other authors have disagreed with the latter assessment (Hemmer, 2001; Werdelin and Lewis, 2002; Lewis and Werdelin, 2010). The arguments need not be repeated here and we

consider *M. adroveri* a European form that evolved a reduced dentition in parallel with *M. whitei*.

MEGANTEREON EKIDOIT Werdelin and Lewis, 2000

Holotype—KNM-ST 23812, complete right hemimandible.
Type locality—South Turkwell, Nachukui Formation, West Turkana, Kenya.
Diagnosis—(Werdelin and Lewis, 2000) A *Megantereon* with a slender mandibular ramus, large salivary gland pit on the anteromedial face of the ramus, small masseteric and mental foramina, and well-developed, hook-shaped coronoid process.

This is the oldest African *Megantereon* and coeval with the oldest European representatives of the genus (older putative records of *Megantereon* have been discounted; cf. Lewis and Werdelin, 2010). Some authors have synonymized this species with *M. whitei* (e.g., Palmqvist et al., 2007), but in doing so they have consistently failed to account for the characters on which the species is actually diagnosed (see Lewis and Werdelin, 2010 for discussion).

Genus SMILODON Lund, 1842

Smilodontopsis (Brown, 1908)
For additional synonyms, see Berta (1985)

Type species—*Smilodon populator* Lund, 1842

SMILODON POPULATOR Lund, 1842

Hyaena neogaea (Lund, 1841a) (nomen nudum)
Felis protopanther (Lund, 1841b) (nomen dubium)
For additional synonymy, see Berta (1985)

Syntypes—Right upper first (ZMUC 1/1845:443) and third incisor (ZMUC 1/1845:439), a fragmentary right upper canine (ZMUC 1/1845:1113), left second (ZMUC 1/1845:170), and fourth metacarpal (ZMUC 1/1845:171) and right fifth metacarpal (ZMUC 1/1845:331).
Type locality—Lagoa Santa caves, Minas Gerais, Brazil.
Diagnosis—(Berta, 1985; Kurtén and Werdelin, 1990) *Smilodon* distinguished from the North American *S. fatalis* by generally larger size, narrower skull with cranial part elongated relative to facial part, high nasals resulting in nearly straight dorsal profile, marked angle between mastoid and occipital plane,

more graviportal limb bones, and extremely massive metapodials; from *S. gracilis* by large, robust skull with broad muzzle, upper canines strongly recurved, upper canines and cheek teeth with finely serrated anterior and posterior margins, P4 with very reduced protocone, mandible usually with a single, large mental foramen, and mandibular flange greatly reduced.

When first encountered, *S. populator* was thought by Lund (1841a) to be the remains of a giant hyena and hence he named it *Hyaena neogaea*. However, this name was not accompanied by any description distinguishing the taxon from related taxa and is therefore a nomen nudum. Soon afterward, Lund (1842) realized that he was dealing with a very large felid with saber-like canines and gave it the appropriate name *Smilodon* (knife-tooth) *populator* (plunderer). *Smilodon populator* was larger than the North American *S. fatalis* and may have been the heaviest felid of all time. Some workers, notably Berta (1985) have considered the two species synonymous, but Kurtén and Werdelin (1990) presented arguments for their distinction. This included the more raised nasals of *S. populator*, the realignment of the mastoid and occipital planes of the posterior part of the skull, the relatively narrower skull with longer neurocranium of *S. populator*, and the more robust metapodials of the latter. This view has since been accepted.

On the basis of these features and others, Kurtén and Werdelin (1990) also argued that the *Smilodon* from northwest South America (Peru and other sites west of the Andes) belong in *S. fatalis*, while *S. populator* is restricted to sites to the east and south. For additional information on the distribution of these taxa in South America see, for example, chapter 3.

SMILODON FATALIS (Leidy, 1868)

Smilodon floridanus (Leidy, 1889)
Smilodon californicus (Bovard, 1907)
Smilodontopsis (Brown, 1908)
For additional synonymy see McDonald (chapter 1) and Berta (1985)

Holotype—AMNH 10395, right maxilla fragment with P4.

Type locality—Hardin County, Texas (Leidy, 1868, 1869).

Diagnosis—On average smaller than *S. populator*, with relatively broader skull, less raised nasals, more obtuse angle between mastoid and occipital plane, more slender build; differs from *S. gracilis* in larger size, more robust skull with broad muzzle, upper canines more strongly recurved, mandibular flange greatly reduced.

This is the typical *Smilodon* as depicted in numerous popular publications. It is the felid for which the most individuals and specimens are known in the fossil record with a minimum number of individuals of ≧1,000 in the census of Marcus (1960), a number that has increased considerably since. Despite this, the status of the name *Smilodon fatalis* is uncertain. The holotype P4 may not be adequately diagnostic, as the material from South Carolina (see chapter 5) suggests. However, addressing this issue is beyond the scope of this contribution. Suffice it to say that the species currently designated *S. fatalis* is exceedingly well represented in the fossil record and described in detail in several publications (e.g., Merriam and Stock, 1932; Tejada-Flores and Shaw, 1984; Shaw and Tejada-Flores, 1985). In addition, most of the chapters in this volume concern *S. fatalis*.

SMILODON GRACILIS Cope, 1880

Uncia mercerii (Cope, 1895)

Holotype—ANSP 46, upper canine root.
Type locality—Port Kennedy Cave, Montgomery County, Pennsylvania.

Diagnosis—Differs from other *Smilodon* in somewhat smaller size, canines less recurved, mastoid process less developed, less developed P3 posterior cingulum, jugal postorbital process less developed, less reduced p3.

Smilodon gracilis was described by Cope (1880) on the basis of a canine root (and small fragment of crown). In subsequent publications the same author (Cope, 1895, 1899) added topotypic specimens and information to the original description, while increasing taxonomic confusion by assigning some material to *Uncia mercerii* and assigning *S. gracilis* to *Machairodus*. Brown (1908) identified only a single species at Port Kennedy Cave and assigned it to his new genus *Smilodontopsis*, which has subsequently been synonymized with *Smilodon*.

In addition to the Port Kennedy material, remains of *Smilodon gracilis* have been reported at various locations in Florida (Kurtén, 1965; Webb, 1974; Churcher, 1984; Berta, 1987). The most complete sample of *S. gracilis* is found at Leisey Shell Pit in the vicinity of Ruskin, Florida (Berta, 1995) and has provided the most detailed information on this species so far. Recently, *S. gracilis* has also been identified from northeastern South America (Rincón et al., 2011).

Phylogenetic Analysis of Machairodontinae

Phylogenetic hypotheses for the subfamily Machairodontinae have been of two forms. Most have been traditional, experience-based hypotheses formulated around key characters and intangible similarity. Recent examples include Werdelin et al. (2010) and Rincón et al. (2011). A few (Geraads et al., 2004; Salesa et al., 2010; Sakamoto and Ruta, 2012; Christiansen, 2013; Wallace and Hulbert, 2013) have been quantitative cladistic hypotheses based on character coding and computer analysis, but of these, only Christiansen (2013) has taken a comprehensive view of the subfamily.

Differences between the traditional view and that obtained by Christiansen (2013) are substantial. Most significant among these is that Christiansen (2013) did not retrieve a monophyletic Smilodontini, with *Smilodon* spp. instead being the sister taxon to the Homotherini and *Megantereon* the successive sister taxon to that clade. Christiansen (2013) also did not retrieve a monophyletic Metailurini or a monophyletic *Dinofelis*. These issues are important topics of debate, and the hypotheses of relationship generated by Christiansen (2013) must therefore be tested within a cladistic framework.

We have here attempted such a test. We have modified the taxon list somewhat, specifically by adding *Rhizosmilodon* and two species of Felinae, *Panthera leo* and *Lynx rufus*, as outgroups in place of *Dinictis felina*, which has been removed along with *Pseudaelurus* spp., which we consider non-monophyletic (Werdelin et al., 2010). The character list is based on that of Christiansen (2013), but some characters

have been removed because they merely specified the position of *D. felina* and were uniform in the ingroup. In addition, some characters from Christiansen (2013) were modified and/or recoded. Finally, we also added a few characters from Salesa et al. (2010) and Wallace and Hulbert (2013). The character and state list is given in the appendix, and the full matrix is available at morphobank.org, project number 2377.

For the analyses we used TNT version 1.5beta provided by the Willi Hennig Society (Goloboff et al., 2008a). We analyzed the matrix with implicit enumeration and all characters unordered. We ran two separate analyses: a traditional one using equal weights for all characters, and one using the implied weights option available in TNT with the default values set. The first analysis yielded two trees of 125 steps with consistency index (CI) 0.608 and retention index (RI) 0.767. These trees are shown in Figure 2.1. The second analysis also yielded two trees, this time of length 129 steps, CI 0.589, RI 0.748, weight 9.163. These trees are shown in Figure 2.2. Note that support for all nodes in the trees is weak, except for the node leading to Felidae.

Results

The two trees obtained in the first analysis differ only in that in tree 1 (Fig. 2.1A) *Xenosmilus* is the sister taxon to *Homotherium serum*, while in tree 2 (Fig. 2.1B) *Xenosmilus* is sister taxon to the two species of *Homotherium*. The character state transitions in the two trees are given in Table 2.1. Noteworthy aspects of the two trees are that *Megantereon* spp. and *Smilodon* spp. are sister taxa, thus forming a restricted Smilodontini. Homotherini is sister taxon to this restricted Smilodontini and *Rhizosmilodon* is sister taxon to all of them. The next group of taxa is a paraphyletic *Dinofelis*, which is congruent with the results of Christiansen (2013). Unlike in the latter analysis, however, *Metailurus* forms a monophyletic clade with *Machairodus aphanistus* and *Promegantereon ogygia*. Lower down still in the trees, *Nimravides pedionomus* falls outside the Machairodontinae, more or less in the position where *Pseudaelurus* spp. falls in Christiansen's (2013) analysis.

The two trees obtained from the second analysis differ in the position of *N. pedionomus*. In tree 1

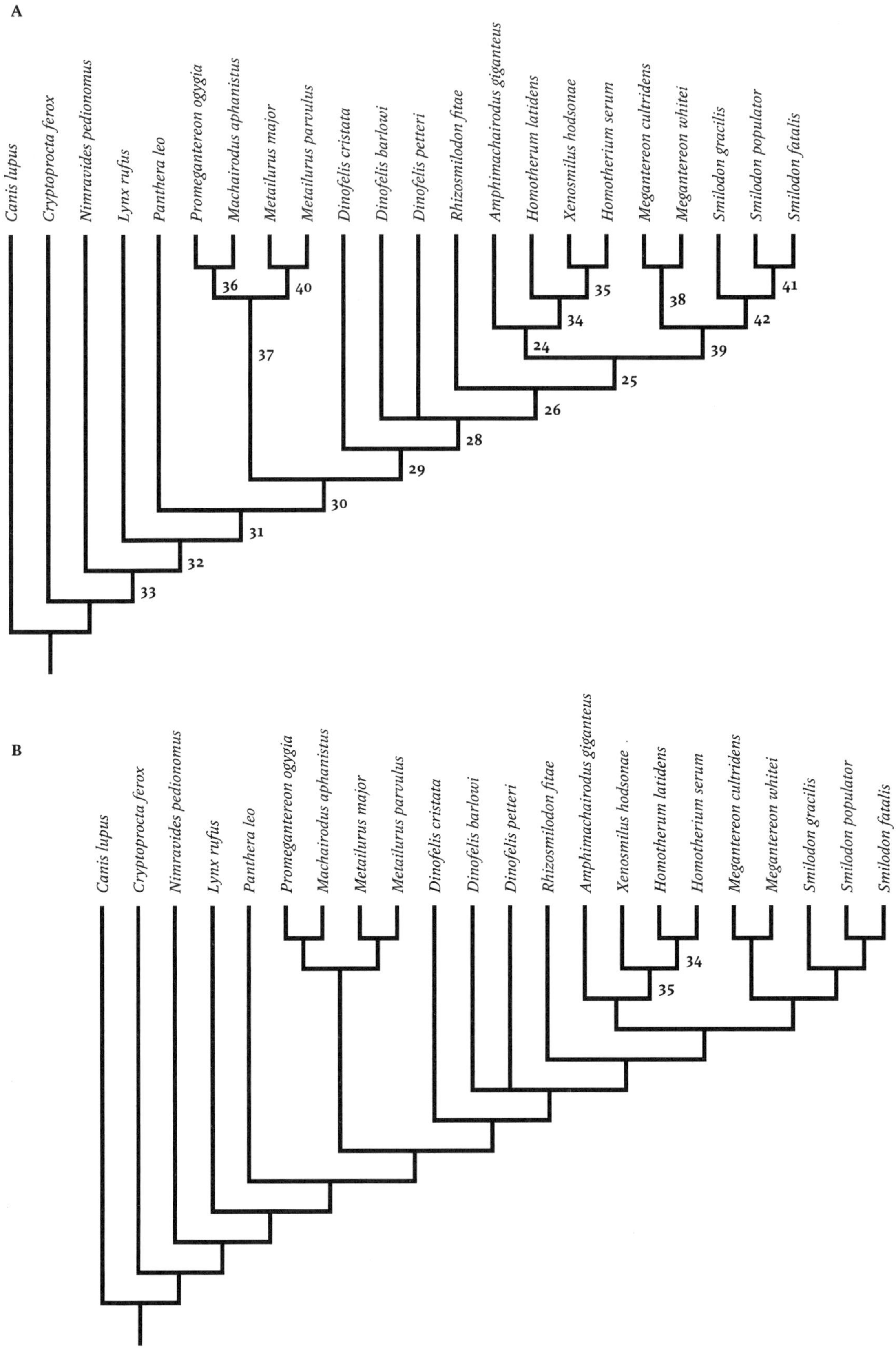

Figure 2.1. The two trees retrieved in the analysis of unweighted characters. In B only the nodes not present in A are numbered.

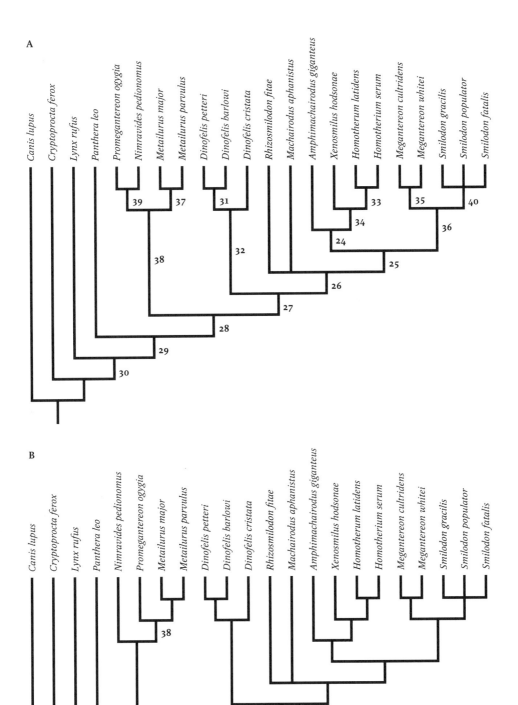

Figure 2.2. The two trees retrieved in the analysis using implied weights. In B only the nodes that differ from A are numbered.

Table 2.1. Synapomorphy lists for the two trees retrieved by the unweighted analysis

Node Number	Synapomorphy List Tree 1	Synapomorphy List Tree 2
24	1 [1→0]	1 [1→0]
	3 [0→2]	3 [0→2]
	14 [0→1]	
25	4 [1→2]	4 [1→2]
26	41 [0→1]	41 [0→1]
	48 [0→1]	48 [0→1]
28	13 [0→1]	13 [0→1]
29	1 [2→1]	1 [2→1]
	16 [1→2]	16 [1→2]
	21 [1→2]	21 [1→2]
	34 [0→1]	34 [0→1]
	38 [0→1]	38 [0→1]
30	4 [0→1]	4 [0→1]
	35 [0→1]	35 [0→1]
31	16 [0→1]	16 [0→1]
	31 [2→1]	31 [2→1]
	33 [0→1]	33 [0→1]
32	7 [0→1]	7 [0→1]
	26 [0→1]	26 [0→1]
	44 [0→1]	44 [0→1]
33	15 [0→1]	15 [0→1]
	17 [0→1]	17 [0→1]
	27 [0→1]	27 [0→1]
	40 [0→1]	40 [0→1]
34	13 [1→3]	22 [0→1]
	16 [2→0]	
	17 [1→2]	
	38 [1→2]	
35	14 [1→2]	13 [1→3]
		16 [2→0]
		17 [1→2]
		38 [1→2]
36	26 [1→0]	26 [1→0]
	36 [0→1]	36 [0→1]
37	2 [0→1]	2 [0→1]
	23 [1→0]	23 [1→0]
38	15 [1→3]	15 [1→3]
	16 [2→1]	16 [2→1]
	23 [1→0]	23 [1→0]
	38 [1→0]	38 [1→0]
	41 [1→2]	41 [1→2]
39	34 [1→0]	34 [1→0]
	46 [0→1]	46 [0→1]
	48 [1→2]	48 [1→2]
40	17 [1→2]	17 [1→2]
41	14 [0→1]	14 [0→1]
42	3 [0→1]	3 [0→1]
	17 [1→2]	17 [1→2]
	33 [2→3]	33 [2→3]

(Fig. 2.2A) it is sister taxon to *P. ogygia*, with those two taxa sister to *Metailurus* spp. In tree 2 (Fig. 2.2B) it is sister to a clade formed of (*P. ogygia* (*Metailurus major* (*M. parvulus*))). In this analysis, *Xenosmilus* is always sister to *Homotherium* spp. *Megantereon* and *Smilodon* again form a restricted Smilodontini, sister to Homotherini. This clade forms an unresolved trichotomy with *Rhizosmilodon* and *M. aphanistus*. The next sister group down is a monophyletic *Dinofelis*, with the *Nimravides* + *Promegantereon* + *Metailurus* clade as the basal Machairodontinae.

Discussion

This discussion focuses on a few of the more important or surprising nodes retrieved in the phylogenetic analyses. It is worth remembering that the topologies obtained are not highly supported, so that any statements made here should be viewed as tentative working hypotheses. This is especially the case with the second analysis, using implied weights, where it should be emphasized that the weighting function may inflate the number of resolved nodes, although this is not necessarily the case (see discussion in Goloboff, 1993; Goloboff et al., 2008b). It should also be noted that in some cases 'obvious' synapomorphies are not listed below or in Tables 2.1 and 2.2 because they are not unambiguously resolved due to missing data. This is the case with the homothere and smilodontine upper canine morphologies, which are central to the traditional systematics of derived machairodonts, but which have multiple resolutions along the backbone of the trees here.

First Analysis: Uniform Weights

Node 39 (Table 2.1), which unites *Megantereon* and *Smilodon* into a monophyletic, restricted Smilodontini is supported by three synapomorphies: 34(0) P3/P4 junction posterior to or level with infraorbital foramen (a reversal from node 29); 46(1) P3 and P4 not aligned (i.e., imbricated); 48(2): p4 distal lean extreme. The first two may be functionally related through the shortening of the tooth row and verticalization of the splanchnocranium, but they are not congruent on the trees. The latter two traits are unique synapomorphies on the trees. In the absence of the skull and upper dentition material, it is

Table 2.2. Synapomorphy lists for the two trees retrieved by the implied weights analysis

Node number	Synapomorphies Tree 1	Synapomorphies Tree 2
24	3 [01→2]	3 [01→2]
	34 [0→1]	24 [0→1]
		34 [0→1]
25	4 [1→2]	4 [1→2]
	32 [0→1]	32 [0→1]
	33 [1→2]	33 [1→2]
	35 [1→0]	35 [1→0]
	42 [0→1]	42 [0→1]
26	41 [0→1]	41 [0→1]
	48 [0→1]	48 [0→1]
27	21 [1→2]	21 [1→2]
	38 [0→1]	38 [0→1]
28	4 [0→1]	4 [0→1]
	35 [0→1]	35 [0→1]
29	16 [0→1]	31 [2→1]
	31 [2→1]	33 [0→1]
	33 [0→1]	
30	7 [0→1]	7 [0→1]
	15 [0→1]	15 [0→1]
	26 [0→1]	26 [0→1]
	27 [0→1]	27 [0→1]
	40 [0→1]	40 [0→1]
	44 [0→1]	44 [0→1]
31	13 [0→1]	13 [0→1]
32	34 [0→1]	34 [0→1]
33	22 [0→1]	22 [0→1]
34	17 [1→2]	17 [1→2]
	38 [1→2]	38 [1→2]
35	15 [1→3]	15 [1→3]
	23 [1→0]	23 [1→0]
	38 [1→0]	38 [1→0]
	41 [1→2]	41 [1→2]
36	2 [2→3]	2 [2→3]
	46 [0→1]	46 [0→1]
	48 [1→2]	48 [1→2]
37	17 [1→2]	17 [1→2]
38	2 [0→1]	24 [0→1]
	23 [1→0]	
39	26 [1→0]	2 [0→1]
		23 [1→0]
40	9 [1→2]	9 [1→2]
	17 [1→2]	17 [1→2]
	33 [2→3]	33 [2→3]
	39 [1→2]	39 [1→2]

unknown whether *Rhizosmilodon* shares the first two traits, but it does not share the third, with the p4 lean being slight rather than extreme (Wallace and Hulbert, 2013).

Node 25, which unites the restricted Smilodontini with Homotherini is supported by only one (unique) synapomorphy: 4(2) lower canine very small and incisiform. In *Rhizosmilodon* the lower canine is substantially larger (Wallace and Hulbert, 2013) and this excludes the genus from the Smilodontini in this analysis. On the other hand, node 26, which unites *Rhizosmilodon* with the restricted Smilodontini + Homotheriini is supported by two synapomorphies: 41(1) mandibular flange intermediate; 48(1) p4 distal lean intermediate.

Metailurus, *Promegantereon*, and *Machairodus* (node 37) are united by two synapomorphies: 2(1) flattened and moderately elongate upper canine; 23(0) m1 paraconid clearly lower than protoconid. The first trait is shared with *Nimravides pedionomus*. *Metailurus* (node 40) is supported by a single synapomorphy: 17(2) P2 absent (shared with node 42). *Promegantereon* and *Machairodus* (node 36) are united by two synapomorphies: 26(0) p2 present (reversal from node 32); 36(1) jugal postorbital process rounded triangle. The first of these is the primitive state of the outgroups, but changes to the derived (absent) state at node 32 and then shows a reversal in *Amphimachairodus giganteus* as well as node 36.

Finally, node 33, which excludes *N. pedionomus* from other machairodonts, is supported by three synapomorphies: 7(1) P4 protocone small but distinct; 26(1) p2 absent; 44(1) mandible horizontal ramus robust relative to length. The state in *N. pedionomus* is the primitive one in these cases and the difference from other machairodonts may reflect the long temporal separation between Eurasian machairodonts and the North American endemic *Nimravides*.

Second Analysis: Implied Weights

In this section we do not consider nodes already dealt with above because these will be broadly similar here (although the synapomorphy schemes may be slightly different) (see Tables 2.1 and 2.2).

Node 26, which allies the Miocene *Machairodus aphanistus* with Plio-Pleistocene machairodonts, is supported by two synapomorphies: 41(1) mandibular flange is present but small; 48(1) p4 distal lean is present but slight. The shifting of *M. aphanistus* to this position is due mainly to a relative downweighting of the dental characters 2 (upper canine shape) and 23 (height of m1 cusps).

Node 38 of Fig. 2.2A, uniting *Promegantereon* and *Metailurus* with *Nimravides*, is supported by two synapomorphies, 2(1) upper canine flattened and moderately long; 23(0) m1 paraconid lower than protoconid. The shift of *Nimravides* upward to this position is due to a relative downweighting of those characters for which *Nimravides* displays the out-group (plesiomorphic) condition, that is, 7, 26, and 44.

Finally, node 32, which unites a monophyletic *Dinofelis* is supported by a single synapomorphy: 34(1) P3/P4 junction anterior to infraorbital foramen, a result of the longer tooth row of *Dinofelis* spp. compared to taxa that bracket it in the phylogeny. According to this topology, 34(1) evolved in parallel between *Dinofelis* and Homotherini.

The analysis using implied weights may raise some methodological issues, and the results have several features of interest. This includes a monophyletic Smilodontini, although there is no evidence for a monophyletic Metailurini in any of the analyses. The analysis places *Nimravides* in the Machairodontinae rather than as sister to crown group Felidae as in the unweighted analysis, which is in line with some traditional hypotheses (Beaumont, 1978; Werdelin et al., 2010). Another feature that approaches traditional views is the position of *Machairodus*. In some analyses (e.g., Monescillo et al., 2014), this genus is placed in the Homotherini and this position is at least approximated in the weighted analysis (relative to the unweighted one).

Conclusions

The phylogenetic hypothesis for Machairodontinae put forward here should, of course, be considered very tentative and provisional. The stability of the trees is poor and many nodes are supported by only a single synapomorphy. Nevertheless, it does represent a step toward reconciling the traditional and cladistic views of machairodont relationships

(Werdelin et al., 2010; Christiansen, 2013). Examples of this include the sister-taxon relationship suggested herein between *Megantereon* and *Smilodon* and (in the implied weighting analysis) the position of *Machairodus aphanistus* closer to Homotherini.

Some issues still persist or have become acute, such as the nonretrieval of a monophyletic or even paraphyletic Metailurini. This may, of course, reflect the actual pattern of relationships, but within the confines of this analysis it is more likely to be due to a general lack of resolution along the stem lineage leading to Smilodontini and Homotherini. *Rhizosmilodon* is not retrieved within Smilodontini, which may be due either to the dearth of available material or to an actual position as the sister taxon to Smilodontini + Homotherini. The age of *Rhizosmilodon* is consistent with a position either within or outside Smilodontini as traditionally conceived. Recently the divergence between Smilodontini and Homotherini has been dated by mitogenomic, time-calibrated Bayesian analysis to 18 Ma, with a 95% credibility interval of 20 Ma to 16 Ma (Paijmans et al., 2017). This is considerably older than previous estimates of this divergence, including the implied estimate herein. Regardless of the accuracy of this molecular estimate it will require reconsideration of a number of taxa that are here and elsewhere considered to belong to the stem lineage leading up to the divergence.

Future work in this area must involve the identification and coding of additional taxa (especially *Paramachaerodus*) and characters. There are two areas not sufficiently or at all plumbed for phylogenetically informative characters: the deciduous dentition and the postcranial skeleton. The former is usually a source of useful characters but has not been sufficiently analyzed in the Machairodontinae, in part due to absence of juvenile specimens in some key taxa, such as *Megantereon*. The latter region, on the other hand, is known to include characters that differentiate between Smilodontini and Homotherini and has been analyzed in other taxa as well (Antón and Galobart, 1999; Salesa et al., 2006; Turner et al., 2012; Antón et al., 2013; Monescillo et al., 2014; Siliceo et al., 2014), although these analyses have not been placed in a quantitative cladistic framework as yet.

ACKNOWLEDGMENTS

The authors would like to thank R. Sardella for providing information on *Megantereon*. TNT was provided by the Willi Hennig Society, which is gratefully acknowledged. This work was supported through grants from the Swedish Research Council to L. Werdelin and an internship from the Department of Zoology, Stockholm University to T. Flink.

APPENDIX: LIST OF CHARACTERS

1. Upper incisor arcade shape: strongly parabolic (0); slightly parabolic (1); straight (2).
2. C morphology: conical-oval (rounded) in cross-section, absence or weakly developed posterior carina, crown not elongated (~13-20% of condylobasal skull length (CBL) (0); flattened in cross section, occasionally posterior carina and moderately elongate (~20-25% of CBL) (1); scimitar-toothed, greatly flattened with distinct posterior carina and moderately to somewhat elongate (~25-30% of CBL); (2); dirk-toothed, greatly flattened with distinct posterior carina and greatly to enormously elongate (>35% of CBL) (3).
3. Carinae of C: absence of carinal crenulations/serrations (0); carinal crenulations (1); carinal serrations (2).
4. c morphology: caniniform and large (0); small (1); very small and incisiform (2).
5. M2 Present (0); Absent (1).
6. M1: presence of large, multicuspid M1 situated in normal, anteroposterior direction (0); presence of reduced, multi-cuspid M1 situated transversely (1); presence of very small, knob-like M1 (2); absence of M1 (3).
7. P4 protocone: presence of large, usually cusped protocone (0); protocone small but still distinct (1); protocone very small or absent (2).
8. P4 protocone angulation: strongly anteriorly directed (0); lingually directed (1).
9. P4 protocone placement: rostral to paracone (0); at the paracone-parastyle junction (1); medial to paracone (2).
10. P4 parastyle anteroposterior length: parastyle absent or indistinct (0); parastyle large (1).
11. P4 anterolateral edge: rounded and medially directed after paracone cusp (0); straight or sinusoid after paracone cusp (1); distinctly medially directed, forming a wide, trough-like cingulum (2).
12. Relative size of P4: P4 short relative to PL (<25% of PL) (0); P4 long relative to PL (>25% of PL) (1).
13. Relative length P3/P4: P3 long relative to P4 (>60% of LP4) (0); P3 moderate (ca. 50-60% of LP4) (1); P3 short (ca. 35-50% of LP4) (2); P3 very short (<35% of LP4) (3).
14. P3 posterior cingulum: posterior cingulum forming a ridge (0); posterior cingulum forming a posterior accessory cusp (1); posterior cingulum indistinct (2).
15. P3 parastyle: indistinct or absent (0); distinct parastyle cusp lingually situated (1); large and with mesial cingulum cusp (2); secondarily lost (3).
16. Greatest width of P3: at main cusp (0); distally (1); equal (2).
17. P2: large with multiple cusps (0); present but very small, usually only with a paracone cusp (1); absent (2).
18. P1: present (0); present but very small (1); absent (2).
19. m3: present (0); absent (1).
20. m2: present (0); absent (1).
21. m1: large talonid basin with distinct hypoconid, entoconid, and metaconid (0); distinct talonid with or without a metaconid cusp (1); little if any talonid (2).
22. p4 relative length: long (>80% of m1 length) (0); short (<80% of m1 length) (1).
23. Relative height of the m1 major cusps: paraconid clearly lower than protoconid (0); paraconid equal to or taller than protoconid (1).
24. Relative size of m1: short relative to mandible length (<15%) (0); long relative to mandible length (>15%) (1).
25. p4 paraconid: paraconid distinctly asymmetrical, with gently posteriorly sloping anterior face, but posterior face with posteroventral slope from apex, followed by downward-sloping ventral part of cusp (i.e., cusp not bilaterally symmetrical about its central axis) (0); paraconid bilaterally symmetrical around its central axis; most with a distinct angle between the ventral and apical part (1).
26. p2: present (0); absent (1).
27. p1: present (0); absent (1).
28. Mandibular glenoid: facing anteriorly (0); facing ventrally (1).
29. Preglenoid process: small or absent (0); large and distinct (1).
30. Infraorbital fenestra: situated distinctly anterior to orbit (0); situated close to orbit (1).
31. Anterior rim of orbit: situated above or behind P4 metastyle (0); situated above P4 paracone or parastyle-paracone junction (1); situated in front of the P4 (2).
32. Snout area elevation compared with braincase: snout area low (0); snout area distinctly elevated (1).
33. Size of mastoid process: small, sometimes knob-like (0); large and elongate (1); much larger relative to the condition in extant pantherines and anteriorly sloping (2); enormously large, obscuring the auditory bulla in lateral view (3).
34. P3/P4 junction relative to infraorbital fenestra: posterior to or level with infraorbital fenestra (0); anterior to infraorbital fenestra (1).

35. Nasal extent past the maxilla-frontal suture: does (0); does not (1).
36. Size of the jugal postorbital process: small, rounded, often almost absent (0); rounded triangle (1); long, pointed (2).
37. Angle of the occipital condyles: low (~45-60) (0); high (~70-100) (1).
38. Relative width across the incisor arcade: narrow (12-15% of CBL) (0); wide (~16-19% of CBL) (1); extremely wide (~21-23% of CBL) (2).
39. Palatal region relative width: palate across the center of P3 comparatively narrow compared with CBL (~23-25%) (0); palatal region across center of P3 relatively wide (~27-33% of CBL) (1); palatal region across center of P3 extremely wide (~36-41% of CBL) (2).
40. Alisphenoid canal: present (0); absent (1).
41. Mandibular flange: absent (0); small (1); large (2).
42. Relative size of mandibular coronoid process (MAT) to horizontal ramus length: large (~20-30%) (0); short (~15-20%) (1).
43. Mandibular fossa termination: posterior to the carnassial (0); well anterior of the posterior edge of the carnassial, frequently terminating around carnassial saddle or even at the m1/p4 junction (1).
44. Posterior part of horizontal dentary ramus height relative to ramus length: slender (~15-17%) (0); robust (~18-23%) (1).
45. Shape of nasals: subrectangular, relatively narrow (0); rectangular, relatively wide (1); subrectangular, relatively wide (2).
46. Orientation of P3 and P4: aligned (0); not aligned (1).
47. p3: simple (0); elongate, bladelike, often nearly as long as p4 (1); generally reduced (2); very small and peglike or lost (3).
48. p4 distal lean: nonexistent (0); slight (1); extreme (2).

REFERENCES

Antón, M. 2013. Sabertooth. Indiana University Press. Bloomington, Indiana, 243 pp.

Antón, M., and A. Galobart. 1999. Neck function and predatory behavior in the scimitar toothed cat *Homotherium latidens* (Owen). Journal of Vertebrate Paleontology 19:771-784.

Antón, M., M. J. Salesa, and G. Siliceo. 2013. Machairodont adaptations and affinities of the Holarctic Late Miocene homotherin *Machairodus* (Mammalia, Carnivora, Felidae): the case of *Machairodus catocopis* Cope, 1887. Journal of Vertebrate Paleontology 33:1202-1213.

Antón, M., M. J. Salesa, J. Morales, and A. Turner. 2004. First known complete skulls of the scimitar-toothed cat *Machairodus aphanistus* (Felidae, Carnivora) from the Spanish Late Miocene site of Batallones-1. Journal of Vertebrate Paleontology 24:957-969.

Beaumont, G. de. 1978. Notes complementaires sur quelques félidés (Carnivores). Archives des Sciences, Genève 31:219-227.

Berta, A. 1985. The status of *Smilodon* in North and South America. Natural History Museum of Los Angeles County, Contributions in Science 370:1-15.

Berta, A. 1987. The sabercat *Smilodon gracilis* from Florida and a discussion of its relationships (Mammalia, Felidae, Smilodontini). Bulletin of the Florida State Museum, Biological Sciences 31:1-63.

Berta, A. 1995. Fossil carnivores from the Leisey Shell Pits, Hillsborough County, Florida. Bulletin of the Florida Museum of Natural History 37:463-499.

Berta, A., and H. Galiano. 1983. *Megantereon hesperus* from the Late Hemphillian of Florida with remarks on the phylogenetic relationships of machairodonts (Mammalia, Felidae, Machairodontinae). Journal of Paleontology 57:892-899.

Bjork, P. 1970. The Carnivora of the Hagerman Local Fauna (Late Pliocene) of southwestern Idaho. Transactions of the American Philosophical Society 60:3-54.

Bovard, J. F. 1907. Notes on Quaternary Felidae from California. University of California Publications, Bulletin of the Department of Geology 5:155-170.

Bravard, A. 1828. Monographie de la Montagne de Perrier près d'Issoire (Puy-de-Dome) et de deux espèces fossiles du genre *Felis* découvertes dans l'une de ses couches d'Alluvion. Thibaud-Landriot and Veysset, Clermont, 145 pp.

Broom, R. 1937. On some new Pleistocene mammals from limestone caves of the Transvaal. South African Journal of Science 33:750-768.

Broom, R. 1948. Some South African Pliocene and Pleistocene mammals. Annals of the Transvaal Museum 21:1-38.

Brown, B. 1908. The Conard Fissure, a Pleistocene bone deposit in northern Arkansas: with descriptions of two new genera and twenty new species of mammals. Memoirs of the American Museum of Natural History 9:157-208.

Christiansen, P. 2013. Phylogeny of the sabertoothed felids (Carnivora, Felidae, Machairodontinae). Cladistics 29:543-559.

Christiansen, P., and J. S. Adolfssen. 2007. Osteology and ecology of *Megantereon cultridens* SE311 (Mammalia, Felidae, Machairodontinae), a sabrecat from the Late Pliocene-Early Pleistocene of Senèze, France. Zoological Journal of the Linnean Society 151:833-884.

Churcher, C. S. 1984. The status of *Smilodontopsis* (Brown, 1908) and *Ischyrosmilus* (Merriam, 1918): a taxonomic review of two genera of sabretooth cats (Felidae, Machairodontinae). Royal Ontario Museum, Life Sciences Contributions 140:1-59.

Cope, E. D. 1880. On the extinct cats of America. American Naturalist 14:833858.

Cope, E. D. 1895. The fossil Vertebrata from the fissure at Port Kennedy. Proceedings of the Academy of Natural Sciences of Philadelphia 1895:446-450.

Cope, E. D. 1899. Vertebrate remains from Port Kennedy bone deposit. Journal of the Academy of Natural Sciences, Philadelphia 11:194-267.

Croizet, J. B., and A. C. G. Jobert. 1828. Recherches sur les ossemens fossiles du département du Puy-de-Dôme. Paris, 219 pp.

Cuvier, G. 1824. Recherches sur les ossemens fossiles: ou l'on rétablit les caractères de plusieurs animaux dont les révolutions du globe ont détruit les espèces. Volume 5, part 2, addendum p. 517. G. Dufour et E. D'Ocagne, Paris, 547 pp.

Ewer, R. F. 1955. The fossil carnivores of the Transvaal caves: Machairodontinae. Proceedings of the Zoological Society of London 125:587-615.

Ficcarelli, G. 1979. The Villafranchian machairodonts of Tuscany. Palaeontographica Italica 71:17-26.

Fischer, G. 1817. Adversaria zoologica, fasciculus primus. Mémoires de la Société des Naturalistes de Moscou 5:357-446.

Gazin, C. L. 1933. New felids from the upper Pliocene of Idaho. Journal of Mammalogy 14:251-256.

Geraads, D., T. Kaya, and V. Tuna. 2004. A skull of *Machairodus giganteus* (Felidae, Mammalia) from the Late Miocene of Turkey. Neues Jahrbuch für Geologie und Paläontologie, Monatshefte 2004:95-110.

Gill, T. 1872. Arrangement of the families of mammals with analytical tables. Smithsonian Miscellaneous Collections 11:1-98.

Goloboff, P. A. 1993. Estimating character weights during tree search. Cladistics 9:83-91.

Goloboff, P. A., J. S. Farris, and K. C. Nixon. 2008a. TNT, a free program for phylogenetic analysis. Cladistics 24:774-786.

Goloboff, P. A., J. M. Carpenter, J. S. Arias, and D. R. Miranda Esquivel. 2008b. Weighting against homoplasy improves phylogenetic analysis of morphological data sets. Cladistics 24:758-773.

Hemmer, H. 2001. Die Felidae aus dem Epivillafranchium von Untermassfeld. Monographien des Römisch-Germanischen Zentralmuseum 40:699-782.

Hibbard, C. W. 1937. An upper Pliocene fauna from Meade County, Kansas. Transactions of the Kansas Academy of Sciences 40:239-265.

Hodnett, J.-P. 2010. A machairodont felid (Mammalia, Carnivora, Felidae) from the latest Hemphillian (Late Miocene/Early Pliocene) Bidahochi Formation, northeastern Arizona. PaleoBios 29:64-79.

Hu, C. K., and T. Qi. 1978. Gongwangling Pleistocene mammalian fauna of Lantian, Shaanxi. Palaeontologica Sinica, New Series C 21:1-64.

Hulbert, R. C., N. Tessman, C. E. Ray, J. A. Baskin, and A. Berta. 2001. Mammalia 3: Carnivorans; pp. 188-225 in R. C. Hulbert (ed.), The Fossil Vertebrates of Florida. University Press of Florida, Gainesville, Florida.

Kretzoi, N. 1929. Materialien zur phylogenetischen Klassifikation der Aeluroïdeen. X Congrès International de Zoologie 2:1293-1355.

Kurtén, B. 1965. The Pleistocene Felidae of Florida. Bulletin of the Florida State Museum, Biological Sciences 9:215-273.

Kurtén, B., and L. Werdelin. 1990. Relationships between North and South American *Smilodon*. Journal of Vertebrate Paleontology 10:158-169.

Leidy, J. 1868. Notice of some vertebrate remains from Harden Co., Texas. Proceedings of the Academy of Natural Sciences of Philadelphia 20:174-176.

Leidy, J. 1869. The extinct mammalian fauna of Dakota and Nebraska, including an account of some allied forms from other localities, together with a synopsis of the mammalian remains of North America. Journal of the Academy of Natural Sciences of Philadelphia, 2nd ser. 7:8-472.

Leidy, J. 1889. The sabre-tooth tiger of Florida. Proceedings of the Academy of Natural Sciences of Philadelphia 41:29-31.

Lewis, M. E., and L. Werdelin. 2010. Carnivoran dispersal out of Africa during the Early Pleistocene: relevance for Hominins?; pp. 13-26 in J. G. Fleagle, J. J. Shea, F. E. Grine, and R. E. Leakey (eds.), Out of Africa 1: The First Hominin Colonization of Eurasia. Springer, New York.

Lund, P. W. 1841a. Blik paa Brasiliens Dyreverden för sidste jordomvæltning. Anden Afhandling: Pattedyrene. Det Kongelige Danske videnskabernes selskabs naturvidenskabelige og mathematiske afhandlinger 8:61-144.

Lund, P. W. 1841b. Tillæg til de to sidste afhandlinger over Brasiliens dyreverden för sidste jordomvæltning. Det Kongelige Danske videnskabernes naturvidenskabelige og mathematiske afhandlinger 8:273-296.

Lund, P. W. 1842. Blik paa Brasiliens Dyreverlden för sidste jordomvæltning. Fjerde Afhandling: Fortsættelse af Pattedyrene. Det Kongelige Danske videnskabernes selskabs naturvidenskabelige og mathematiske afhandlinger 9:137-208.

Marcus, L. F. 1960. A census of the abundant large Pleistocene mammals from Rancho La Brea. Los Angeles County Museum of Natural History, Contributions in Science 38:1-11.

Martínez-Navarro, B., and P. Palmqvist. 1995. Presence of the African machairodont *Megantereon whitei* (Broom, 1937) (Felidae, Carnivora, Mammalia) in the lower Pleistocene site of Venta Micena (Orce, Granada, Spain), with some considerations on the origin, evolution and dispersal of the genus. Journal of Archaeological Science 22:569-582.

Martínez-Navarro, B., and P. Palmqvist. 1996. Presence of the African saber-toothed felid *Megantereon whitei* (Broom, 1937) (Mammalia, Carnivora, Machairodontinae) in Apollonia-1 (Mygdonia Basin, Macedonia, Greece). Journal of Archaeological Science 23:869-872.

Matthew, W. D. 1929. Critical observations upon Siwalik mammals. Bulletin of the American Museum of Natural History 56:437-560.

Merriam, J. C., and C. Stock. 1932. The Felidae of Rancho La Brea. Carnegie Institution of Washington Publication No. 422:1-231.

Monescillo, M. F. G., M. J. Salesa, M. Antón, G. Siliceo, and J. Morales. 2014. *Machairodus aphanistus* (Felidae, Machairodontinae, Homotherini) from the Late Miocene (Vallesian, MN 10) site of Batallones-3 (Torrejón de Velasco, Madrid, Spain). Journal of Vertebrate Paleontology 34:699-709.

Murchison, C., ed. 1868. Palaeontological Memoirs and Notes of the Late Hugh Falconer, A.M., M.D., Volume 1. Fauna Antiqua Sivalensis. Robert Hardwicke, London, 590 pp.

Paijmans, J. L. A., R. Barnett, M. T. P. Gilbert, M. L. Zepeda-Mendoza, J. W. F. Reumer, J. de Vos, G. Zazula, D. Nagel, G. F. Baryshnikov, J. A. Leonard, N. Rohland, M. V. Westbury, A. Barlow, and M. Hofreiter. 2017. Evolutionary history of saber-toothed cats based on ancient mitogenomics. Current Biology. doi: 10.1016/j. cub.2017.09.033

Palmqvist, P., V. Torregrosa, J. A. Pérez-Claros, B. Martínez-Navarro, and A. Turner. 2007. A re-evaluation of the diversity of *Megantereon* (Mammalia, Carnivora, Machairodontinae) and the problem of species identification in extinct carnivores. Journal of Vertebrate Paleontology 27:160-175.

Petter, G., and F. C. Howell. 1982. Un félidé machairodonte des formations plio-pleistocènes des Siwaliks: *Megante-*

reon falconeri Pomel/ = *M. sivalensis* (F. et C.)/ (Mammalia, Carnivora, Felidae). Comptes Rendus de l'Académie des Sciences 295:281-284.

Pomel, M. 1853. Catalogue méthodique et descriptif des vertébrés fossiles découverts dans le bassin hydrographique supérieur de la Loire et surtout dans la vallée de son affluent principal, l'Allier. J.-B- Baillière, Paris, 193 pp.

Pons Moya, J. 1987. Los carnivoros (Mammalia) de Venta Micena (Granada, Espana). Paleontologia i Evolució, Memoria Especial 1:109-128.

Rincón, A. D., F. J. Prevosti, and G. E. Parra. 2011. New saber-toothed cat records (Felidae, Machairodontinae) for the Pleistocene of Venezuela, and the great American biotic interchange. Journal of Vertebrate Paleontology 31:468-478.

Sakamoto, M., and M. Ruta. 2012. Convergence and divergence in the evolution of cat skulls: temporal and spatial patterns of morphological diversity. PLoS One 7:e39752.

Salesa, M. J., M. Anton, A. Turner, and J. Morales. 2006. Inferred behaviour and ecology of the primitive sabre-toothed cat *Paramachairodus ogygia* (Felidae, Machairodontinae) from the Late Miocene of Spain. Journal of Zoology 268:243-254.

Salesa, M. J., M. Antón, A. Turner, L. Alcalá, P. Montoya, and J. Morales. 2010. Systematic revision of the Late Miocene sabre-toothed felid *Paramachaerodus* in Spain. Palaeontology 53:1369-1391.

Sardella, R. 1998. The Plio-Pleistocene Old World dirk-toothed cat *Megantereon* ex. gr. *cultridens* (Mammalia, Felidae, Machairodontinae), with comments on taxonomy, origin and evolution. Neues Jahrbuch für Geologie und Paläontologie, Abhandlungen 207:1-36.

Schultz, C. B., and L. D. Martin. 1970. Machairodont cats from the Early Pleistocene Broadwater and Lisco Local Faunas. Bulletin of the University of Nebraska State Museum 9:33-38.

Shaw, C. A., and A. E. Tejada-Flores. 1985. Biomechanical implications of the variation in *Smilodon* ectocuneiforms from Rancho La Brea. Natural History Museum of Los Angeles County, Contributions in Science 359:1-8.

Siliceo, G., M. J. Salesa, M. Antón, M. F. G. Monescillo, and J. Morales. 2014. *Promegantereon ogygia* (Felidae, Machairodontinae, Smilodontini) from the Vallesian (Late Miocene, MN 10) of Spain: morphological and functional differences in two noncontemporary populations. Journal of Vertebrate Paleontology 34:407-418.

Teilhard de Chardin, P. 1939. On two skulls of *Machairodus* from the lower Pleistocene beds of Choukoutien. Bulletin of the Geological Society of China 19:235-256.

Teilhard de Chardin, P., and J. Piveteau. 1930. Les mammifères fossiles de Nihowan (Chine). Annales de Paléontologie 19:1-134.

Tejada-Flores, A. E., and C. A. Shaw. 1984. Tooth replacement and skull growth in *Smilodon* from Rancho La Brea. Journal of Vertebrate Paleontology 4:114-121.

Turner, A. 1987. M*egantereon cultridens* (Cuvier) (Mammalia, Felidae, Machairodontinae) from Plio-Pleistocene deposits in Africa and Eurasia, with comments on dispersal and the possibility of a New World origin. Journal of Paleontology 61:1256-1268.

Turner, A., M. Antón, M. J. Salesa, and J. Morales. 2012. Changing ideas about the evolution and functional morphology of Machairodontine felids. Estudios Geológicos 67:255-276.

Viret, J. 1954. Le lœss a bancs durçis de Saint-Vallier (Drôme) et sa faune de mammifères villafranchiens. Nouvelles Archives du Muséum d'Histoire Naturelle de Lyon 4:1-200.

Wallace, S. C., and R. C. Hulbert Jr. 2013. A new machairodont from the Palmetto Fauna (Early Pliocene) of Florida, with comments on the origin of the Smilodontini (Mammalia, Carnivora, Felidae). PLoS One 8:e56173.

Webb, S. D. 1974. The status of *Smilodon* in the Florida Pleistocene; pp. 149-157 in S. D. Webb (ed.), Pleistocene Mammals of Florida. University Presses of Florida, Gainesville, Florida.

Webb, S. D., R. C. J. Hulbert, G. S. Morgan, and H. F. Evans. 2008. Terrestrial mammals from the Palmetto Fauna (Early Pliocene, latest Hemphillian) from the Central Florida Phosphate District; pp. 293-312 in X. Wang and L. G. Barnes (eds.), Geology and Vertebrate Paleontology of Western and Southern North America. Contributions in Honor of David P. Whistler. Natural History Museum of Los Angeles County, Science Series 41.

Werdelin, L., and M. E. Lewis. 2000. Carnivora from the South Turkwel hominid site, northern Kenya. Journal of Paleontology 74:1173-1180.

Werdelin, L., and M. E. Lewis. 2002. Species identification in *Megantereon*: a reply to Palmqvist. Journal of Paleontology 76:931-933.

Werdelin, L., and M. E. Lewis. 2013. The Carnivora, Koobi Fora Research Project, Volume 7. California Academy of Sciences, San Francisco, California, 333 pp.

Werdelin, L., N. Yamaguchi, W. E. Johnson, and S. J. O'Brien. 2010. Phylogeny and evolution of cats (Felidae); pp. 59-82 in D. M. Macdonald and A. Loveridge (eds.), Biology and Conservation of Wild Felids. Oxford University Press, Oxford.

3 *Smilodon fatalis* from Talara, Peru: Sex, Age, Mass, and Histology

KEVIN L. SEYMOUR, ASHLEY R. REYNOLDS,
AND CHARLES S. CHURCHER

Introduction

The unique and impressive collection of *Smilodon fatalis* fossils from Rancho La Brea (RLB), California was described in detail by Merriam and Stock (1932). Since that time, several studies have looked at parts of this collection to study different aspects of the taxonomy and biology of this species. Berta (1985) proposed that the wide variation seen in samples in North and South American warranted that all Late Pleistocene samples be assigned to one variable species, *Smilodon populator*. Kurtén and Werdelin (1990) proposed that, based on several morphological features, the name *S. populator* be restricted to South American material and that North American material should be referred to *S. fatalis*. They identified the sample from Talara, Peru, however, as an anomaly, representing a late incursion of *S. fatalis* into north-western South America, and we follow this taxonomic treatment here.

This study provides a summary of the 1,948 fossils of *Smilodon fatalis* from Talara, Peru in the Royal Ontario Museum collection. All bones of the skeleton are present except for the clavicle, xiphisternum, and some hyoid elements. A basihyoid bone is described for the first time, and a proposed caudal series of 13+ bones is presented. Compared to the minimum number of individuals (MNI) of 20 based on postcranial material, there is an MNI of 24 based

on the right dentary. The limb and metapodial elements average shorter in length than the Rancho La Brea *S. fatalis*. The dentaries were aged based on tooth wear, resulting in a count of one juvenile, six young adults, 16 adults, and one old adult, giving a similar proportion as is present in the famous Rancho La Brea collection in California. Osteohistological samples were made for four *S. fatalis* specimens, representing the first multielement histological analysis of any fossil felid. The specimens were aged as follows: one 2 years, two 4+years, and one 7+years old. Tooth-bearing elements were sexed using osteometrics, resulting in two females and one male based on skulls, and eight females and two males based on dentaries. The average carnassial length for these sexed specimens demonstrates sexual dimorphism in this collection. Body mass estimates based on three humeral measures average 244.1 kg, while those based on three femoral measures average 198.4 kg, and on two tibial measures average 154.1 kg. The combination of sexual size dimorphism in the dentition, sexual shape dimorphism in the skulls, a skewing of the sample toward females, and a delayed limb maturation compared to the dentition all suggest some form of sociality in this species.

Many studies confirm the importance of sexual dimorphism in carnivores (e.g., Weckerly, 1998). Because sexual size dimorphism often is associated with competition between males for access to females, there has been some interest in the possibility of discerning sexual dimorphism in *Smilodon fatalis* as a way of understanding more about the social organization of this extinct species. Van Valkenburgh and Sacco (2002) examined variation in skull length, as well as canine and lower molar measures. They found that the possible sexual dimorphism was lower than that of living or fossil lions, and more like that of solitary felids. Meachen-Samuels and Binder (2010) state that, because growth could continue well into adulthood (see Stamps, 1993), specimens need to be aged before they can be sexed. Using the percent pulp cavity closure on the lower canine, they found little to no sexual size dimorphism in the RLB *S. fatalis* population.

Christiansen and Harris (2012), however, found that although there might not be much sexual size dimorphism in the RLB *Smilodon fatalis* sample, there was as much shape dimorphism between the sexes as there is in living pantherine cats. They examined many RLB *S. fatalis* skulls, divided their sample into a large morph and a small morph, and deleted the overlap specimens between them to ensure that both sexes were represented in their sample. Then, using morphometric analyses, they discerned that *S. fatalis* varied in the same way that male and female pantherines varied, and hence the large morph was most likely male and the smaller morph most likely female. Once the skulls were sexed, they then sought out basic proportions that could separate the two sexes, and presented 8 skull and 2 mandible proportions that were reasonably successful in this regard.

Age variation potentially confounds the elucidation of sexual variation, as Stamps (1993) and Meachen-Samuels and Binder (2010) described. Age variation in *Smilodon* has been explored using several different methods. Miller (1968) studied suture fusion and tooth wear of 2,100 skulls from RLB. He found that the proportion of juvenile material varied from one site to another: between 11% and 25% depending on the pit. Since tooth wear was found to be a less accurate method than radiographic analyses in determining the age of *Panthera leo* specimens of known age (Smuts et al., 1978), Meachen-Samuels and Binder (2010) radiographically examined pulp cavity closure in the lower canine of *S. fatalis* as an estimate

for aging the RLB specimens. They found that at least 80% of the specimens showed a closed pulp cavity, indicating either that many specimens were of advanced age or (more likely) that the pulp cavity filled more quickly than in pantherine species, making this method less useful for clarifying the relative age of individuals. Seymour (2015) tallied the proportion of juvenile limb bones in the *Smilodon* sample from Talara and found that 41% of the material represented juveniles; however, this is based on a much smaller sample size than that of Miller (1968). This methodology essentially breaks the sample into two groups, limiting its utility in aging specimens.

Osteohistological examination is another technique for aging fossil taxa, specifically analyzing the presence of cyclical growth marks within the cortices of skeletal elements (Woodward et al., 2013; Kolb et al., 2015). In a study on extant ruminants, Köhler et al. (2012) demonstrated that mammals deposit lines of arrested growth (LAGs) on an annual cycle and therefore that LAGs are well-suited for age estimation. In addition, because a LAG represents the size and shape of the element's circumference when the growth mark was formed, these can be used to estimate body mass throughout an animal's lifetime (Lee et al., 2013). Thus, to lay the groundwork for future studies of growth in *Smilodon fatalis*, some exploratory histological sections were made of four limb bones (two femora, and one tibia and one humerus) from the Talara *S. fatalis* sample, which allows some general inferences on life history.

Body mass estimates for *Smilodon* are another criterion useful for better understanding size variation in this species. Christiansen and Harris (2005) used 36 osteological measurements to arrive at an estimated body mass of between 160 kg and 280 kg in the RLB *S. fatalis* sample. Body mass estimation techniques based on more universal scaling relationships, such as those of Campione and Evans (2012) would likely produce more refined results but require associated forelimb and hind limb material, something that is lacking in the Talara and RLB tar pit material, apart from a single specimen from RLB (Cox and Jefferson, 1988).

Understanding the dynamics of sex, age, and dimorphism allows us to make inferences regarding the social system of extinct taxa. Indeed, the question of whether *Smilodon fatalis* was a social species, like present-day *Panthera leo*, has been a topic of debate within the literature. Gonyea (1976) and Akersten (1985), suggested that *S. fatalis* may have been social based on the sheer numbers of individuals found in the RLB fossil deposit as well as the high incidence of severe pathologies in the RLB collection. Other opinions have been varied on whether the species was social and, if so, what type of social system it may have exhibited. Van Valkenburgh and Sacco (2002) concluded it was unlikely that *S. fatalis* lived in polygynous groups such as those of *Panthera leo*, and if it was social it would have lived in monogamous pairs. McCall et al. (2003) asserted that pathology does not necessarily indicate some form of care by group members and that the species' small brain size is not consistent with the increased encephalization normally seen in social animals. Carbone et al. (2009) drew parallels between the prevalence of social and nonsocial species in tar seeps and in audio playbacks of herbivore distress calls in the African savannah, suggesting that the high incidence of *S. fatalis* at RLB would only make sense if it were social. However, Kiffner (2009) disagreed, pointing out several differences between audio playbacks and a true carnivore trap scenario, in addition to noting that the ecosystem in which *S. fatalis* lived may not have been comparable to that of the present-day African savannah. Sociality was considered plausible by both Binder and Van Valkenburgh (2010) and Meachen-Samuels and Binder (2010). The latter noted, however, that *Smilodon* was likely not as polygynous as extant cats and could have been either promiscuous or monogamous. Christiansen and Harris (2012) agreed that *Smilodon* was not likely to have lived in polygynous groups, but suggested that it could have been either solitary and polygynous or have lived in unisexual groups.

To date no description of the Talara *Smilodon fatalis* sample has been published, and this contribution is meant to fill that gap. Herein, we analyze the collection of 1,948 fossils of the sabertooth cat *S. fatalis*, from Talara, Peru, and investigate sexual dimorphism, age, and body mass in this collection, using some of the methods described above, and give comments on the possible sociality of the Talara population.

Locality

Over 27,000 identified fossil bones were excavated by the Royal Ontario Museum near Talara, Peru in 1958. A map of the location was provided in Campbell (1979), and the history of the collection at this site was outlined in Lemon and Churcher (1961). This site represents a tar seep deposit with 79.5% of the over 17,000 Late Pleistocene mammalian fossils, representing the remains of eight species of Carnivora, thereby resembling the RLB tar pits as a carnivore trap (Seymour, 2015; Lindsey and Seymour, 2015). The Talara site dates to the Late Pleistocene, with radiocarbon dates published by Churcher (1966) as follows: $13,616 \pm 600$ BP and $13,790 \pm 535$ BP on small chewed sticks, as well as $14,150 \pm 564$ BP and $14,418 \pm 500$ BP on large pieces of wood associated with fossil bones. The Talara site is therefore at least partially contemporaneous with RLB. Radiocarbon dating technology has improved greatly since the 1960s, but two recent attempts to obtain radiocarbon dates directly from Talara bones were unsuccessful.

ABBREVIATIONS

C, cervical vertebra; **Ca,** caudal vertebra; **F,** female; **L,** lumbar vertebra; **m,** lower molar; **M,** male; **MNI,** minimum number of individuals; **NISP,** number of individual specimens; **P,** upper premolar; **RLB,** Rancho La Brea site, California; **ROM,** Royal Ontario Museum, Ontario, Canada; **T,** thoracic vertebra.

Materials and Methods

Description

All fossil material was examined to confirm taxonomic and element identification. The seminal work of Merriam and Stock (1932) was consulted to confirm the identifications. In most cases the *Smilodon* elements were sufficiently different from those of the other carnivores preserved at this site that misidentification of the species was not a serious issue. A possible exception is the presence of the large felid *Panthera onca*, which forms a rare part of the Talara fauna (NISP 93, MNI = 3) (Seymour, 2015). Three labeled comparative skeletons of pantherines were also consulted

(ROM R6804 *Panthera tigris* female, ROM R1055 *P. pardus* unsexed, ROM R6704 *P. onca* female), to separate *Smilodon* from *Panthera* and to identify phalanges. One of the authors (C. S. Churcher) performed the preliminary identifications, and another (K. L. Seymour) separated out the jaguar specimens and confirmed those identifications (Seymour, 1983). Identifications were later confirmed (Seymour, 2015), specifically looking for misidentifications between these two taxa. Specimens were counted for each element where possible to establish NISP and MNI. During this count 75 (mostly carpal) records were found to be missing from the database and were added, hence giving a larger total than was reported by Seymour (2015). All specimens are conserved in the vertebrate palaeontology collections of the Department of Natural History at the Royal Ontario Museum in Toronto, Canada. All measurements were made using dial calipers to the nearest 0.1 mm by C. S. Churcher except the few measurements needed for body mass estimates, which were taken by A. R. Reynolds to closely match the measurements of Christiansen and Harris (2005); circumferential measurements were taken by A. R. Reynolds to the nearest 1 mm with a measuring tape. Measurements in the style of Merriam and Stock

(1932) for virtually all elements were taken by C. S. Churcher and are on file at ROM; only select summaries of measurements are given here as was deemed appropriate.

Aging of Dentigerous Elements

For tooth-bearing elements (skulls, maxillae, or dentaries), we follow the age categories as described in Miller (1968). If maxillae or dentaries were known to be associated, they were counted as one individual; otherwise left side and right side specimens were each tallied separately and the higher number was used for MNI. Premaxillae were not included, following Miller (1968).

Histology

For the examination of osteohistology, transverse thin sections were created for four specimens at the minimum circumference of the diaphysis, which varies from specimen to specimen but occurs at the approximate locations indicated in Figure 3.1. Age classes based on epiphyseal fusion were assigned as either juvenile (no epiphyseal fusion) or adult (one or more epiphyses fused). The fragmentary nature of specimens precluded us from assigning subadult

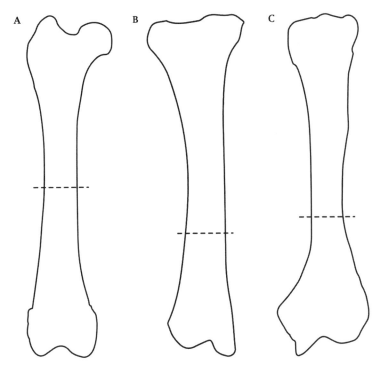

Figure 3.1. Diagram of a *Smilodon fatalis* right femur (A), right tibia (B), and right humerus (C) in anterior view showing the approximate location of thin sectioning. Images adapted from Merriam and Stock (1932).

status where partial epiphyseal fusion was present. Specimens were selected based on their size, age class, and preservation (e.g., state of completeness or extent of tar staining). Each element was measured based on criteria for body mass estimation as laid out in Christiansen and Harris (2005) and photographed and 3D modeled using AgiSoft Photo Scan photogrammetry software.

Prior to sectioning, each specimen was coated and/or impregnated with epoxy resin, either Technovit 5071 (Heraeus Kulzer GmbH, Wehrheim, Germany) or Castolite AC (Eager Polymers, Chicago, Illinois). Where Castolite was used, a thick section was removed from the remaining bone using an IsoMet 1000 precision saw (Buehler, Lake Bluff, Illinois) equipped with a diamond wafering blade. Where Technovit was used, the specimen was sawed transversely through the area coated with resin. Following this first cut, all specimens were wet-ground on the side to be mounted using a CrystalMaster Pro 12 grinding wheel (Crystalite Corporation, Lewis Center, Ohio) and no. 600 silicon carbide grit (Carborundum Company, Niagara Falls, New York). Each specimen was mounted to a frosted, 1.5 mm-thick plexiglass slide using either resin (Polymer Solutions Inc., Mississauga, Ontario) or CA40 cyanoacrylate (3M Scotch-Weld, London, Ontario) as an adhesive. A final cut was made using the wafering blade, leaving an approximately 0.7 mm section of the specimen attached to the slide. This section was then thinned using a Hillquist Thin Section Machine (Hillquist Inc., Denver, Colorado) and silicon carbide grit nos. 600 and 400 (Kingsley North Inc., Norway, Michigan) until the desired thickness was achieved.

Each slide was examined under regular transmitted light and cross-polarized light using a Nikon Multizoom AZ100 microscope (Nikon Instruments Inc., Melville, New York) and immersion in baby oil to increase refraction index and therefore image clarity. All specimens were imaged using a Nikon Digital Sight DS-Fi1 microscope camera head (Nikon Instruments Inc., Melville, New York) at $4 \times$ total magnification and 1280×960 resolution. Images were taken across the entire thin section with 40% overlap and stitched together using NIS-Elements Basic Research 3.0 (Nikon Inc.). All further adjustments to images—for example editing of brightness or contrast and addition of text, scale bars, and symbols,—were completed in Adobe Photoshop CC (Adobe Systems Inc., San Jose, California).

Sexing of Specimens

Christiansen and Harris (2012) gave 8 cranial and 2 mandibular proportions that best separated the sexes of RLB *Smilodon fatalis* fossils (their figs. 6 and 7). We were unable to utilize one of these (because the location of their "anterior edge of infraorbital foramen" utilized in figure 6C was not defined nor was it described in their supplemental data). We used their other 9 measures to estimate the sex of Talara specimens. Once the specimens were sexed, average condylobasal, P4, and m1 lengths were calculated for each sex. The means from these three measures were used to estimate degree of sexual dimorphism by calculating male-to-female ratios, as in Gittleman and Van Valkenburgh (1997), Van Valkenburgh and Sacco (2002), and Cullen et al. (2014).

Body Mass Estimates

In terms of calculating an estimated body mass, Christiansen and Harris (2005, their Figs. 3 and 4) highlighted six forelimb measures and six hind limb measures that yielded the highest correlations in their regressions for the estimation of body mass. Of these twelve, we found eight measures (three for the humerus, three for the femur, and two for the tibia) that were most easily reproducible given the descriptions in their paper, and we used these to give estimates for the body mass of the Talara *Smilodon fatalis*. The two or three estimated masses for each of the limb bones were then averaged to give a mean estimated mass for each bone. These were subsequently averaged to give an overall average mass estimate for each of the three limb bones.

Results

Description and Discussion of Material
Cranial Material: Skulls

There are only two reasonably complete skulls of *Smilodon fatalis* in the Talara collection, ROM 2116 and 3816 (Fig. 3.2). This small sample has hampered comparative analyses with the RLB sample. A third, less complete and partially disarticulated skull

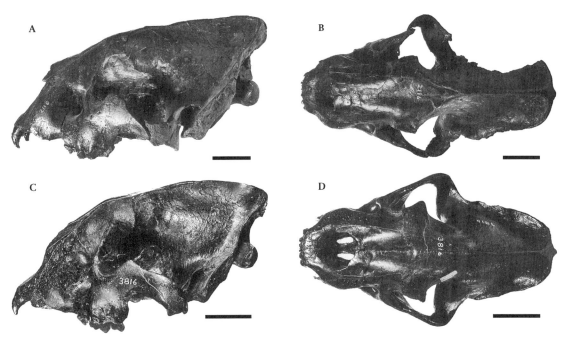

Figure 3.2. *Smilodon fatalis* skulls from Talara, Peru, in lateral (A, C) and dorsal (B, D) views. A, B: ROM 2116; C, D: ROM 3816. Skulls are scaled to be approximately the same length for ease of comparison. Scale bars are 5 cm.

Table 3.1. Counts of skull and dentary elements for *Smilodon fatalis* from Talara, Peru, arranged by numbers of individual specimens and minimum numbers of individuals

Element	NISP Complete	NISP Incomplete	MNI
Skull	2	9	9
Upper canine	5	15	11
P3	19	3	12
P4	12	10	15
Dentary	17	22	22
p4	15	10	19
m1	19	15	17
Juvenile dentary	1	2	2

Abbreviations: NISP = number of individual specimens; MNI = minimum number of individuals

Note: NISP counts are taken regardless of side. MNI counts take into account left and right sides of specimens and therefore can be smaller than NISP.

(ROM 2146) could also be included in most analyses. All, however, show the critical features outlined by Kurtén and Werdelin (1990) as diagnostic for *S. fatalis*: the alignment of the mastoid with the lambdoid plane, and the nasals not raised as in *S. populator*. In addition, there are a number of other partial skulls in the Talara collection; all six additional posterior portions show the mastoid aligned to the lambdoidal crest, and both additional anterior portions exhibit the relatively depressed nasals. This indicates that there is only one species of *Smilodon* at this site (*S. fatalis*), as suggested by Kurtén and Werdelin (1990). Counts for skull material are in Table 3.1.

Cranial Material: Dentaries

After the dentaries were measured (and before this publication), a number of ROM specimens were lent to another institution for radiographic analysis; unfortunately, four were stolen from their locked laboratory and never recovered. However, in case they turn up elsewhere, their ROM numbers are 2018 A and 2230 as well as the maxilla 2121 and squamosal 5325 B. Consequently, the number of Talara specimens available for future study is diminished. Nevertheless, measurements for these specimens are included here.

Seymour (2015) found an MNI of 20 for the Talara *Smilodon* material, based on the astragalus. Unfortunately, he did not examine the mandibular material, where we found 22 adults and two juveniles using the right side, resulting in an MNI of 24 for this collection (Table 3.1).

Cranial Material: Individual Teeth

There are 142 separate teeth and tooth fragments of *Smilodon fatalis* in the Talara collection. These teeth were examined to determine whether their inclusion would affect the MNI count of the mandibular material mentioned above; these isolated teeth had no effect on the total MNI. Even though these elements were identified to tooth position and side, they were not utilized in any of the analyses presented here.

Cranial Material: Hyoids

Two possible *Smilodon* basihyoids have been identified (ROM 42927 and 42928). A possible stylohyoid (ROM 42929) was also tentatively identified but could not be located at the time of writing. There is a dearth of described *Smilodon* hyoid elements due to lack of articulated specimens associated with cranial material. The single possible *Smilodon* hyoid element presented in Merriam in Stock (1932:Fig. 12a) has since been identified as that of a proboscidean (C. Shaw, pers. comm., 2014). The identification of these two elements as *Smilodon* is based on overall similarity to those of pantherines (Pocock, 1916; Weissengruber et al., 2002), but they are more robust (Fig. 3.3). One end is more elongated on both sides, and these may have formed the connection to either the ceratohyoid or the thyrohyoid. In felids the dorsal portion of the basihyoid, connecting to the ceratohyoid, has a small projection or knob, and there is no projection ventrally for the thyrohyoid bone. On this basis the projection would be connecting on the dorsal side with the ceratohyoid. However, the roughened edges of these connections look more like cartilaginous surfaces, as occurs with the thyrohyoid, which is cartilaginous in many felids (Peters and Hast, 1984), and on this basis the projection would be pointing ventrally.

Postcranial Material: Vertebrae

All vertebrae are represented in this collection except for any possible caudals posterior to caudal 13. Table 3.2 lists the number of elements present for each vertebra, separated into relatively complete specimens and relatively incomplete specimens (a somewhat arbitrary division), for a total of 302 specimens. The most commonly preserved and identified elements are the axis and atlas, perhaps because of their distinctive nature.

The vertebral formula seems to be as in other felids except for the shortened tail: 7 cervicals, 13 thoracics, 7 lumbars, and 13+caudals. This is about the same formula surmised by Merriam and Stock (1932), although they noted that some individuals may have had only six lumbars, and they did not give a count for the number of caudals. Méndez-Alzola (1941) counted 14 thoracics and six lumbars in *Smilodon bonaerensis* (now considered to be *S. populator*), but this may only represent a labeling issue for L1, as the total count for T + L is the same, although he did also report 14 pairs of ribs. It is unknown whether this differing vertebral count is a diagnostic character for *S. populator* or the individual described by Méndez-Alzola (1941) was an unusual individual.

The caudal series is of interest because there has been no complete, articulated caudal series reported for *Smilodon*, and hence the shortening of the tail

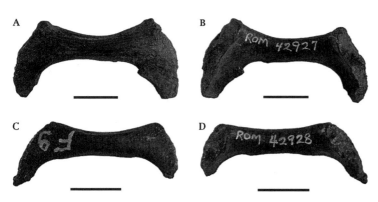

Figure 3.3. Proposed *Smilodon fatalis* basihyoids from Talara, Peru in anterior (A, C) and posterior (B, D) views, with dorsal to the top. A, B: ROM 42927; C, D: ROM 42928. Scale bars are 1 cm.

Table 3.2. Counts of vertebral elements for *Smilodon fatalis* from Talara, Peru, arranged by numbers of individual specimens

Element	NISP Complete	NISP Incomplete
Atlas	5	9
Axis	3	14
C3	2	12
C4	3	10
C5	2	9
C6	3	10
C7	5	4
T1	3	8
T2	0	3
T3	2	5
T4	1	3
T5	1	2
T6	1	6
T7	1	3
T8	0	6
T9	1	6
T10	2	2
T11	1	2
T12	1	6
T13	1	8
L1	2	7
L2	1	13
L3	1	6
L4	0	7
L5	3	9
L6	1	12
L7	2	8
Sacrum	5	6
Ca1	3	2
Ca2	2	1
Ca3	3	1
Ca4	3	0
Ca5	4	2
Ca6	4	1
Ca7	6	0
Ca8	6	0
Ca9	4	0
Ca10	4	0
Ca11	4	0
Ca12	2	0
Ca13	1	0
Total	99	203

Abbreviations: NISP: numbers of individual specimens; C = cervical; T = thoracic; L = lumbar; Ca = caudal.

reported by Merriam and Stock (1932) remains hypothetical. As at RLB, the Talara caudal vertebrae are not associated and all were identified as individual elements. For this reason, perhaps, Merriam and Stock (1932) do not give a count for a total number of

caudals for *S. fatalis* from RLB. Here, we propose a total of 13+caudal vertebrae in *S. fatalis* (Fig. 3.4). We matched caudal 1 to the reduced posterior articulation on the sacrum, and then sized the rest of the series from caudal 1. The total proposed here includes six with a more or less enclosed neural canal and seven posterior to this. The proposed Talara caudal series ends with a vertebra that still has a distal articular surface, implying at least one additional vertebra to complete the series. In comparison, *Panthera pardus* (ROM R1055) has eight caudals with an enclosed neural canal and 16+posterior to those; *Lynx rufus* (ROM R471) has five and 10 respectively, making the caudal series of *Smilodon* similar to, or even more abbreviated than, that of *Lynx rufus*.

Merriam and Stock (1932) only illustrated a selection of caudals rather than a series (their figs. 60 and 61), but noted that for all *Smilodon* caudals there is a shortening and reduction of the elements, making them relatively smaller but also presenting a different morphology from *Panthera*. However, in their Plate 25, a complete mounted skeleton is presented that appears to have five anterior caudals and eight posterior caudals, very similar to our proposal of six and seven, respectively. In addition, a *Smilodon* skeleton mounted at RLB on display at the ROM (ROM 20) also has 6 anterior and 7 posterior caudals, with the last caudal vertebra having an articular facet, giving a count of 13+caudals. Méndez-Alzola (1941) mentioned that for the specimen he studied, the caudal vertebrae were smaller, confirmed by a reduced posterior articulation on the sacrum, although only 6 caudals in total were preserved with the type specimen he described. However, due to their simplicity there is the possibility of misidentification of *Panthera* distal caudals as those of *Smilodon*, as no *Panthera* caudals have yet been identified from Talara.

Postcranial Material: Ribs and Sternal Elements

Ninety-three rib fragments have been tentatively identified as belonging to *Smilodon fatalis*. However, the sole reference for identification is Plate 20 in Merriam and Stock (1932), which illustrates only ribs I and II, and X through XIII. Consequently, most of these elements must remain tentatively identified;

Figure 3.4. Proposed series of caudal vertebrae of *Smilodon fatalis* from Talara, Peru, in dorsal view with anterior to the left. ROM catalogue numbers from left (caudal 1) to right (caudal 13): 4702 C, 4703 B, 4704 A, 4705 B, 4706 E, 4707 C, 4709 A, 4708 E, 4710 B, 4711 C, 4712 C,4713 A, 4714 A. Scale bar is 5 cm.

perhaps only about 15 of these are relatively securely identified, since they can be reasonably matched with one of the illustrations in Merriam and Stock (1932).

There are two manubria and eight mesosternal elements, but no xiphisternal elements identified from Talara. This element may possibly still be present and unrecognized in the several drawers of unidentified and unsorted smaller material, although a search specifically for this element was not successful.

Postcranial Material: Forelimb and Girdle

Table 3.3 lists the identification and counts for forelimb material preserved from Talara; in total, there are 509 forelimb elements. All bones of the forelimb are represented in this collection except clavicles, a relatively small element. The MNIs account for juvenile as well as left and right elements. The phalanges were determined to digit and side using two pantherine skeletons (*Panthera pardus* and *P. onca*) for guidance. Many of these phalanges should be considered as tentatively identified because they are very similar in appearance, although their distribution does not look skewed toward any particular digit. The 98 unguals and 111 sesamoids were not determined to foot or digit, and remain unassigned.

Table 3.4 gives the average lengths of the major forelimb elements from Talara and RLB as presented in Merriam and Stock (1932). These authors always gave only a sample of RLB *Smilodon* limb measures, attempting to represent the range of size. If it is a representative sample (and there is some evidence to suggest their measures are skewed to the large side; C. Shaw, pers. comm. 2016), it shows that on average the Talara cat is a little smaller but still within the range of measurements of the RLB *Smilodon*.

Table 3.3. Counts of forelimb and girdle elements for *Smilodon fatalis* from Talara, Peru, arranged by numbers of individual specimens and minimum numbers of individuals

Element	NISP Complete	NISP Incomplete	MNI
Scapula	2	9	6
Humerus	3	30	13
Ulna	5	26	13
Radius	7	24	12
Scapholunar	14	4	9
Magnum	13	2	8
Unciform	11	1	8
Cuneiform	6	0	5
Trapezoid	11	2	10
Trapezium	6	0	4
Pisiform	15	3	13
MC I	22	1	13
MC II	18	9	12
MC III	19	7	12
MC IV	10	7	8
MC V	15	3	11
Digit I, phalanx 1	24	2	13
Digit II, phalanx 1	22	3	13
Digit II, phalanx 2	28	0	18
Digit III, phalanx 1	21	1	12
Digit III, phalanx 2	25	1	14
Digit IV, phalanx 1	25	0	13
Digit IV, phalanx 2	20	0	11
Digit V, phalanx 1	12	1	9
Digit V, phalanx 2	17	2	10
Total	371	138	
Manus or pes unguals	24	74	?
Manus or pes Sesamoids	111	0	?

Abbreviations: NISP = numbers of individual specimens; MNI = minimum numbers of individuals

Note: NISP counts are taken regardless of side. MNI counts take into account left and right sides of specimens, and therefore can be smaller than NISP.

Table 3.4. Lengths of forelimb and girdle elements for *Smilodon fatalis* from Talara, Peru, compared to those from Rancho La Brea, California

	Talara			RLB		
Element	N	Range	Mean Length	N	Range	Mean Length
Scapula	2	281.0–311.4	296.2	10	266–358	320.6
Humerus	5	319.3–363.1	338.5	10	309–385	345.0
Ulna	5	294.5–334.0	311.5	10	287–372	336.1
Radius	8	228.0–259.0	241.6	10	235–295	265.7
MC I	22	35.3–39.6	37.0	4	27.5–39.6	36.3
MC II	19	76.8–96.2	85.5	6	76.4–105.5	90.5
MC III	19	91.3–106.0	97.3	6	83.0–109.6	96.0
MC IV	11	86.3–99.3	92.2	6	79.4–107.4	92.7
MC V	15	69.7–78.0	73.6	6	61.6–86.6	74.4

Abbreviations: RLB = Rancho La Brea; N = sample size

Note: RLB data taken from Merriam and Stock (1932). Note that most Talara elements are on average smaller than those from RLB.

Postcranial Material: Hind Limb and Girdle

Table 3.5 gives the counts for hind limb material preserved from Talara; in total, there are 515 elements represented. All bones of the hind limb are represented in this collection. The MNIs account for juvenile as well as left and right elements. As with the manual phalanges, the pedal phalanges were determined to digit and side using two pan-therine skeletons (*Panthera pardus* and *P. onca*) for reference. Despite this, many of these phalanges should be considered as tentatively identified because they are very similar, although their distribution does not look skewed toward any particular digit.

Both elements of metatarsal I are fused to the entocuneiform, as is true for just under half the specimens discussed by Merriam and Stock (1932) from RLB. However they noted that metatarsal I was always directed away from the principal axis of the entocuneiform. This is not true for either of the Talara specimens, which are relatively smaller than the RLB specimens and approximately in line with the principal axis of the entocuneiform.

Table 3.6 gives the lengths of the major hind limb elements from Talara and RLB as presented by Merriam and Stock (1932). As mentioned above for the forelimb, these authors gave only a sample of RLB *Smilodon* limb measures, not a random selection of individuals. If theirs is a representative sample (see comment above in forelimb section), it shows that on average the Talara cat is a little smaller but still

within the range of measurements from RLB. Only the tibia reaches the average of the RLB sample and this may be due to sample size. The MNI of 20 for the astragalus provided the largest MNI for any element for *S. fatalis* from Talara (Seymour, 2015) until the dentaries were counted in this study, producing an MNI of 24.

Aging of Dentigerous Elements

Based on the skulls and maxillae, there are no juvenile specimens, five young adults, eight adults, and no old adults, giving an MNI of 13. Using the dentaries, however, there are two juveniles, seven young adults, 14 adults, and one old adult, for an MNI of 24, almost twice as large as for the upper dentition. Using the dentary sample, this can be summarized as 8% juvenile, 29% young adult, 58% adult, and 4% old adult. These proportions are similar to those in the RLB sample studied by Miller (1968), which consisted of 16.6% juveniles, 25.7% young adults, and 56.5% adults, with 1.2% undetermined (old adults were lumped with adults), except that the Talara sample has a lower proportion of juvenile specimens.

Histology

The following histological descriptions follow the terminology of Francillon-Vieillot et al. (1990) and the descriptive approach of Werning (2012) in that collagen orientation, vascular orientation/arrangement, osteonal development, and osteocyte orientation/

Table 3.5. Counts of hind limb and girdle elements for *Smilodon fatalis* from Talara, Peru, arranged by numbers of individual specimens and minimum numbers of individuals

Element	NISP Complete	NISP Incomplete	MNI
Innominate	1	26	16
Femur	5	29	12
Tibia	4	19	10
Fibula	3	19	9
Patella	10	0	8
Calcaneum	17	10	11
Astragalus	25	6	20
Navicular	11	5	7
Cuboid	17	0	9
Ectocuneiform	12	0	6
Mesocuneiform	2	0	1
Entocuneiform	4	0	3
MT I (fused to entocuneiform)	2	0	1
MT II	18	13	14
MT III	18	8	12
MT IV	20	7	15
MT V	25	7	14
Digit II, phalanx 1	22	0	12
Digit II, phalanx 2	24	0	12
Digit III, phalanx 1	19	0	12
Digit III, phalanx 2	23	0	13
Digit IV, phalanx 1	23	0	12
Digit IV, phalanx 2	22	0	13
Digit V, phalanx 1	18	4	11
Digit V, phalanx 2	16	1	9
Unguals (see Manus)			
Sesamoids (see Manus)			
Total	361	154	

Abbreviations: NISP = numbers of individual specimens; MNI = minimum numbers of individuals

Note: NISP counts are taken regardless of side. MNI counts take into account left and right sides of specimens and therefore can be smaller than NISP.

arrangement are discussed separately rather than using tissue-level descriptors. For each specimen, the overall size and shape are described first, followed by the above characteristics of the medullary cavity, endosteal surface, cortex, and finally the presence and condition of any growth marks: annuli, lines of arrested growth (LAGs), and external fundamental system (EFS). Here, we define an EFS as a subperiosteal region of tightly packed lamellar bone that is avascular and acellular (Huttenlocker et al., 2013). The diameter of each sample is reported with the anteroposterior (AP) measurement listed first and the lateromedial (LM) measurement second (e.g., $AP \times LM$). A summary of each specimen, including element, age class, and age estimation based on number of LAGs, may be found in Table 3.7.

Histology of Juvenile Femur ROM 4720

The diaphysis of ROM 4720 (Fig. 3.5) is ovoid with a circumference of 84 mm and diameters of 23.7×28.9 mm. The medullary cavity, however, is circular and resembles that of the adult femur (see below). Trabeculae formed from avascular lamellar tissue extend inward from the endosteal surface of the cortex. An inner circumferential layer is present and is slightly vascularized with longitudinal canals. The primary tissue is woven-fibered throughout the entire cortex. Many secondary osteons are present throughout the inner and mid-cortices in addition to large resorption cavities. Where primary tissue is visible, it is heavily vascularized with longitudinal canals that occasionally anastomose with no directional preference. All vascular canals through

Table 3.6. Lengths of hind limb and girdle elements for *Smilodon fatalis* from Talara, Peru, compared to those from Rancho La Brea, California

Element	Talara			RLB		
	N	Range	Mean Length	N	Range	Mean Length
Femur	7	318.0–358.8	340.8	10	317–408	377.1
Tibia	4	275–281	278.8	10	239–305	273.6
Fibula	3	217.2–222.8	219.5	10	212.7–284.7	248.2
MT II	21	76.4–89.7	82.9	6	73.7–96.5	84.7
MT III	21	87.9–104.8	97	6	85.5–112.6	98.8
MT IV	21	89.7–103.8	97.4	6	84.6–113.8	98.9
MT V	26	71.4–92.3	81.2	6	70.8–94.8	82.5

Abbreviations: RLB = Rancho La Brea; N = sample size

Note: RLB data taken from Merriam and Stock (1932). Note that most Talara elements average smaller than those from RLB.

the inner to middle cortex are associated with osteons, but simple vascular canals are present in all but the posterior outer cortex, which has vascular canals associated with primary osteons. The nutrient canal is located within the posterior mid-cortex; it is lined by lamellar bone on its posterior edge. Osteocyte lacunar density is consistently high within all regions of the cortex, although osteocyte lacunae occur more frequently in woven-fibered bone than in the lamellar endosteal tissue.

Up to two LAGs are visible and are indicated by arrows (in Fig. 3.5) in the outer lamellar layer of ROM 4720. The first occurs at the base of the outer cortex in a region dense with vacuities, so can therefore only be traced intermittently. The second is close to the periosteal surface and can be traced around the entire circumference of the shaft, with a few interruptions due to localized secondary osteons. Based on these LAGs, this individual may be inferred to have been at least two years old at the time of death.

Table 3.7. List of skeletal elements of *Smilodon fatalis* from Talara, Peru, sectioned (Figures 3.5 to 3.8) and their deduced ages in years based on LAG counts

Specimen	Element	Age Class	Deduced Age (years)
ROM 4720	Femur	Juvenile	2+
ROM 3026	Femur	Adult	7+
ROM 4737	Tibia	Adult	4+
ROM 2255	Humerus	Adult	4+

Histology of Adult Femur ROM 3026

The adult femur of *Smilodon fatalis* (Fig. 3.6) is circular in cross-section, although it is slightly wider lateromedially than anteroposteriorly (circumference: 91 mm; diameters: 27.3 × 29.5 mm). The medullary cavity is circular and located in the center of the shaft, with a thin layer of coarse cancellous bone and thick trabeculae along the endosteal margin. An inner circumferential layer is present. The inner to middle cortex is comprised of a woven-fibered matrix, while the outer cortex is formed by lamellar bone. Secondary osteons are present throughout the cortex but decrease in density toward the outer cortex. Longitudinal vascular canals are observed throughout the entire cortex. Anastomosing canals are more common than in ROM 4720 and generally occur in a random arrangement between two to three vascular canals except in the posteromedial outer cortex, where they are predominantly circumferential. Vascularity is dense throughout the inner and middle cortex but is markedly lower in the inner and outer lamellar layers. Almost all vascular canals are associated with osteons, except for many of those present in the outer lamellae. The nutrient canal is visible in the posterior mid-cortex and is lined with lamellar bone on its posterior, medial, and lateral edges. Osteocyte lacunar density generally remains consistent throughout the entire sample.

One growth mark (LAG) is present in the mid-cortex on the posteromedial side but is truncated by secondary osteons and cannot be traced circumferentially. Six LAGs are present in the lamellae of the

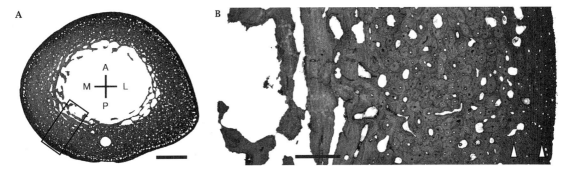

Figure 3.5. Section of femur ROM 4720 of *Smilodon fatalis* from Talara, Peru. A: full section; scale bar is 5 mm. B: magnified portion (enclosed in a box in A), rotated such that interior is at left and exterior is at right. White arrows represent locations of lines of arrested growth (LAGs); scale bar is 1 mm. Abbreviations: A = anterior, P = posterior, L = lateral, M = medial.

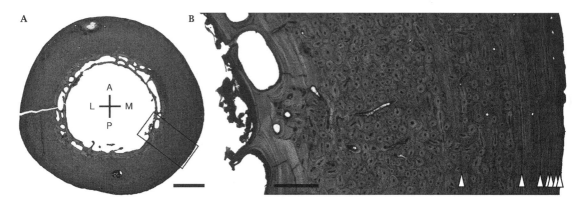

Figure 3.6. Section of femur ROM 3026 of *Smilodon fatalis* from Talara, Peru. A: full section; scale bar is 5 mm. B: magnified portion (enclosed in a box in A), rotated such that interior is at left and exterior is at right. White arrows represent locations of lines of arrested growth (LAGs); scale bar is 1 mm. Abbreviations: A = anterior, P = posterior, L = lateral, M = medial.

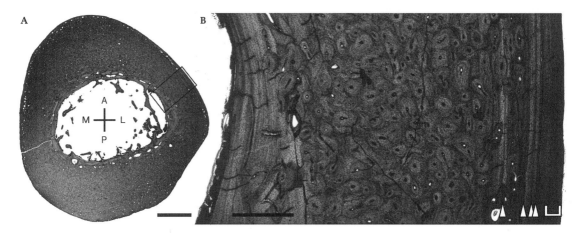

Figure 3.7. Section of tibia ROM 4737 of *Smilodon fatalis* from Talara, Peru. A: full section; scale bar is 5 mm. B: magnified portion (enclosed in a box in A), rotated such that interior is at left and exterior is at right. White arrows represent locations of lines of arrested growth (LAGs); EFS is indicated by bracketed area; scale bar is 1 mm. Abbreviations: A = anterior, P = posterior, L = lateral, M = medial.

posteromedial outer cortex and can be traced along approximately 60% to 70% of the bone's circumference with some interruptions by secondary osteons. The LAG spacing is greater posteromedially than in the anterolateral cortex. This LAG count throughout the cortex suggests that the individual was at least seven years of age at time of death, but it is possible that the innermost LAGs have been masked by remodeling, suggesting that it might have been older than this. An EFS is not present, but LAG spacing indicates a gradual slowing of growth.

Histology of Adult Tibia ROM 4737

The diaphysis of the adult tibia (Fig. 3.7) is a rounded triangle in cross-section, with a circumference of 91 mm and diameters of 30.3 × 27.3 mm. The medullary cavity is oval and is widest anteroposteriorly. Trabeculae extend from the endosteal surface into the medullary cavity; an inner circumferential layer is present. A few Haversian canals are found in the endosteal lamellae, and the remainder of the cortex is extensively remodeled. Where primary tissue is visible between secondary osteons, it appears to follow the same pattern as the adult femur, with woven-fibered bone in the inner and mid-cortex but lamellar bone in the outer cortex. Both primary bone types are observed most readily within the anterolateral cortex. Vascularity is expressed as longitudinal canals that anastomose with a similar frequency to the adult femur (ROM 3026). These anastomosing canals tend to be radially arranged in the medial inner cortex. Vascular canals occur in high densities throughout the inner and middle cortex, but reduce drastically in density in the outer lamellae. Osteocyte lacunae occur with high density throughout the entire cortex and are evenly distributed between osteons and interstitial tissue.

Four LAGs are visible in the outer region of lamellar bone. The third from the inside is partially split into a double LAG. An EFS is present. None of the visible growth marks can be followed outside the anterolateral region of the cortex due to remodeling. Thus, this individual was a minimum of four years old at time of death, but would likely have been older as LAGs have likely been obscured by remodeling and/or medullary expansion.

Histology of Adult Humerus ROM 2255

The humerus (Fig. 3.8) is ovoid in cross-section and is certainly the largest of the sampled limb bones, with a circumference of 115 mm and diameters of 38.4×34.3 mm. The medullary cavity is oval and centrally positioned. Lamellar bone with a moderate amount of remodeling forms the cancellous endosteal surface and the inner circumferential layer. The entire cortex consists primarily of secondary osteons, but where primary tissues are visible they show as woven-fibered bone in the inner cortex and lamellar bone in the mid- to outer cortex. Intriguingly, a series of folding lamellae are visible in the anterior inner to middle cortex, which appear more like endosteal

rather than subperiosteal lamellae in that they are more finely fibered than the latter. Vascularity in this specimen is almost exclusively present as longitudinal vascular canals, but reticular anastomosing canals can be seen prominently at the posterolateral tip of the cortex. The nutrient canal is present in the posterior mid-cortex and is not accompanied by any lamellar bone. Vascular density is decreased in lamellar bone. Osteocyte lacunar density is generally consistent throughout the cortex with some small patches of increased density in the posterior cortex.

One LAG is present in the anterolateral mid-cortex but is almost completely obliterated by secondary osteons and therefore cannot be seen outside this region. Three additional LAGs can be seen in the lateral outer cortex; however, they are faint and are thus difficult to discern. As with the tibia, the extent of remodeling suggests that this individual would have been older than the four years indicated by the visible LAGs. While it is difficult to say due to the faint nature of the visible growth marks, an EFS does not appear to be present in this specimen.

Sexing of Specimens

Table 3.8 presents the hypothesized sex of each of the three skulls using seven cranial proportions elucidated by Christiansen and Harris (2012). Six of the seven proportions suggest that ROM 3816 was female, and only one suggests it was male. The results for the other two skulls are more equivocal, with four suggesting male and two suggesting female and

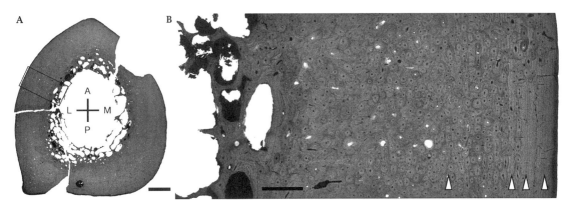

Figure 3.8. Section of humerus ROM 2255 of *Smilodon fatalis* from Talara, Peru. A: full section; scale bar is 5 mm. B: magnified portion (enclosed in a box in A), rotated such that interior is at left and exterior is at right. White arrows represent locations of lines of arrested growth (LAGs); scale bar is 1 mm. Abbreviations: A = anterior, P = posterior, L = lateral, M = medial.

Table 3.8. Sexes of three Talara *Smilodon fatalis* skulls based on ratios of seven measures compared to the condylobasal length

ROM #	CBL	A	A/CBL	A Sex	B	B/CBL	B Sex	D	D/CBL	D Sex	E	E/CBL	E Sex	F	F/CBL	F Sex
2116	344	150	0.436	F	89.9	0.261	overlap	60.2	0.175	M	148.2	0.431	M	20.5	0.060	F
2146	299	115.6	0.387	overlap	70.9	0.237	F	51.2	0.171	M	-	-	-	16.2	0.054	Overlap
3816	283	121.5	0.429	F	69.8	0.247	F	49.1	0.173	M	133.1	0.470	overlap	17.3	0.061	F

ROM #	G	G/CBL	G Sex	H	H/CBL	H Sex	Total M	Total Overlap	Total F	Deduced Sex	P4 length
2116	44.5	0.129	M	58.8	0.171	M	4	1	2	M	44.5
2146	39.7	0.133	overlap	62	0.207	F	1	3	2	F	39.7
3816	39.5	0.140	F	63.5	0.224	F	1	1	5	F	38.9

Abbreviations: CBL = condylobasal length; A = distance from anterior rim of preglenoid process to posterior edge of occipital condyle; B = mastoid height; D = width across the incisor arcade; E = palatal width across carnassial notch of UP4; F = UP3 crown length; G = UP4 crown length; H = postorbital constriction width.

Note: The Talara ratio was plotted onto the appropriate plot in figure 6 in Christiansen and Harris (2012); for ease of comparison the letters for their characters from their figure 6 are used here. If the number fell into the range of one sex only, it was classified as either male (M) or female (F). If it fell into the range of overlap between the two sexes, it was classified as overlap, or equivocal. The summary tallies the number of deduced sexes for each specimen and gives an overall conclusion on the sex of each of the three specimens, as well as the P4 carnassial length for each. Note that the male specimen is larger than the two female specimens.

Table 3.9. Sexes of 10 Talara *Smilodon fatalis* dentaries based on ratios of p4 and m1 crown lengths compared to dentary length as described by Christiansen and Harris (2012)

ROM #	Dentary Length	p4 Length	p4/ Dentary	p4 Sex	m1 Length	m1/ Dentary	m1 Sex	Deduced Sex	m1 Length (F)	m1 Length (M)
2117	189.8	25.6	0.135	F	26.2	0.138	F	F	26.2	
2125	229.0	27.3	0.119	M	30.1	0.131	overlap	M		30.1
2146	198.0	27.0	0.136	F	30.3	0.153	F	F	30.3	
2228	228.0	26.8	0.118	M	29.8	0.131	overlap	M		29.8
2229	189.7	25.2	0.133	F	27.1	0.143	F	F	27.1	
2230	186.3	28.9	0.155	F	27.2	0.146	F	F	27.2	
2233	205.1	25.7	0.125	F	26.0	0.127	overlap	F	26.0	
2245	187.0	25.3	0.135	F	26.9	0.144	F	F	26.9	
2258	210.0	28.4	0.135	F	30.1	0.143	F	F	29.0	
3055	208.7	26.0	0.125	F	27.7	0.133	overlap	F	27.7	
								Mean	27.6	30.0

Note: The Talara ratio was plotted onto the appropriate plot of figure 7 in Christiansen and Harris (2012). If the number fell into the range of one sex only, it was classified as either male (M) or female (F). If it fell into the range of overlap between the two sexes, it was classified as overlap or equivocal. The summary gives the deduced sex based on p4 as there were a number of overlaps in the m1 ratios, although none of the m1 deductions disagreed with those of p4. The summary also gives the m1 carnassial length for each. Note that the two male specimens average larger than the eight female specimens.

one falling in the zone of overlap for ROM 2116, and one suggesting male, two suggesting female, and three in the zone of overlap for ROM 2146. Specimen ROM 2146 has an associated mandible, however, and both mandibular measures suggest female.

Ten complete or almost complete dentaries from the Talara collection could be measured for the ratios used by Christiansen and Harris (2012) to sex their RLB specimens. Table 3.9 presents the hypothesized sex for each of these dentaries. Using the p4, eight are female and two are male, while using the m1, six are female, two are probably female, and two are probably male. Using both teeth together we can conclude that eight dentaries were probably female and two were probably male.

Sexual Dimorphism

Once the skulls and dentaries were sexed, the average condylobasal, P4, and m1 lengths were calculated for males and females. Mean condylobasal length was 291 mm (N = 2) for females and 344 mm for the single male, with a male-to-female ratio of 1.18. The female P4 averaged 39.3 mm (N = 2) and the sole male P4 was 44.5 mm, with a male-to-female ratio of 1.13. For the lower carnassial, the females average 27.6 mm (N = 8) and the males average 30.0 mm (N = 2), producing a male-to-female ratio of 1.09.

Body Mass Estimates

Table 3.10 gives the body mass estimates for the Talara *Smilodon fatalis*. The average mass using the three humeral measures is 244.1 kg. The average mass is 198.4 kg using the three femoral measures, and 154.1 kg using the two tibial measures. The hind limb estimates are considerably lower than the forelimb estimates. Christiansen and Harris (2005) found a range of estimates from 160 kg to 280 kg for the RLB *S. fatalis;* both the humerus and femur estimates for the Talara specimens fall in this range but the estimates based on the tibia fall just below the RLB range.

Discussion

Almost all studies of possible sexual dimorphism or size variation in *Smilodon fatalis* have utilized the skull (e.g., Van Valkenburgh and Sacco, 2002; Meachen-Samuels and Binder, 2010), and the small sample from Talara hampers comparison of the Talara sample with that of RLB. C. S. Churcher supplied basic measurements and photos of the Talara *S. fatalis* to Kurtén and Werdelin (1990:fig. 4), who concluded that although the Talara sample represented *S. fatalis,* two of the three skulls were unusually elongated, more resembling *S. populator* (their Fig. 4). However, all other characters discussed

Table 3.10. Estimated body mass in kg of *Smilodon fatalis* from Talara, Peru, using the variables and regression statistics as presented in Tables 2 and 3 of Christiansen and Harris (2005)

	HDAP	Estimated Mass in kg	HLC	Estimated Mass in kg	HHLM	Estimated Mass in kg	Mean Individual Mass Estimate
ROM 2145	-	-	-	-	52.9	147.5	147.5
ROM 2239	39.9	231.2	117	226.5	-	-	228.9
ROM 2240	40.7	243.1	124	261.0	-	-	252.0
ROM 2255	38.4	209.9	115	217.2	64.7	233.4	220.2
ROM 2350	31.1	123.2	107	182.2	-	-	152.7
ROM 2873	38.8	215.5	117	226.5	-	-	221.0
ROM 2934	46.7	344.1	140	350.7	66.8	251.0	315.3
ROM 2935	48.9	386.5	145	382.0	69.2	272.0	346.9
ROM 2956	49.2	392.6	132	303.9	-	-	348.2
ROM 3034	-	-	-	-	65.3	238.3	238.3
ROM 3992	42.4	269.6	123	255.9	-	-	262.7
ROM 3993	35.7	174.6	112	203.7	-	-	189.1
ROM 3994	39.4	224.0	125	266.1	59.3	191.3	227.1
ROM 3998	38.8	215.5	117	226.5	64.8	234.2	225.4
ROM 4002	37.7	200.4	110	194.9	-	-	197.6
ROM 4163	47.3	355.4	139	344.7	71.9	296.8	332.3
						Total Mean Mass	244.1

	FAW	Estimated Mass in kg	FLCAP	Estimated Mass in kg	FMCAP	Estimated Mass in kg	
ROM 2332	73.6	226.7	58.8	254.3	54.6	247.1	242.7
ROM 4717	63.5	155.6	54.1	209.5	52.4	224.7	196.6
ROM 2023	63.6	156.2	52.9	198.8	52.4	224.7	193.2
ROM 2243	69.8	198.0	49.1	167.2	53.4	234.7	200.0
ROM 2909	67.0	178.4	50.5	178.5	51.2	212.9	189.9
ROM 2908	72.1	215.1	51.1	183.4	49.6	197.9	198.8
ROM 2561	74.2	231.4	49.4	169.6	47.2	176.4	192.5
ROM 2250	68.3	187.3	48.4	161.7	46.7	172.1	173.7
						Total Mean Mass	198.4

	TC	Estimated Mass in kg	TDLM	Estimated Mass in kg			
ROM 2238	79.0	114.2	23.7	88.9			101.6
ROM 4731	80.0	118.0	24.8	99.8			108.9
ROM 4730	89.0	155.2	28.6	143.6			149.4
ROM 4737	91.0	164.3	27.3	127.5			145.9
ROM 2430	99.0	204.1	31.6	185.2			194.6
ROM 3070	99.0	204.1	30.6	170.6			187.4
ROM 29663	100.0	209.4	30.7	172.1			190.7
						Total Mean Mass	154.1

Abbreviations: HDAP=humerus diaphysis measured anterior posteriorly; HLC=humerus least circumference; HHLM=humerus head measured lateral medially; FAW=femur articular width; FLCAP=femur lateral condyle measured anterior posteriorly; FMCAP=femur medial condyle measured anterior posteriorly; TC=tibia least circumference; TDLM=tibia diaphysis measured lateral medially.

Note: The body mass of each animal represented by a bone was estimated using more than one variable if possible. These estimates were next averaged to give a total average estimate for each bone. All the bone averages were then averaged to give an overall body mass estimate for all humeri, femora, and tibiae separately.

by them aligned these skulls with *S. fatalis* and in this we agree. It should be noted, however, that their figure 4 is open to other interpretations. If the single smallest of their *S. populator* skulls were removed from their sample, the regression lines for both *S. fatalis* and *S. populator* would be nearly parallel. If a larger sample of *S. populator* were graphed and a cluster the size of the *S. fatalis* sample were available to sample, the regression lines for the two ellipses could very well be subparallel, and not represent a different growth pattern for *S. populator*. Regardless of the growth pattern, they do demonstrate that the *S. populator* population tends to have a narrower skull for its length, but only just outside the 95% confidence ellipse for *S. fatalis*. Because of this we prefer to rely on their other characters for identification rather than on overall skull proportions, which was also their conclusion.

The postcranial skeleton of *Smilodon* was well-studied and illustrated by Merriam and Stock (1932) and to some extent most authors have assumed that there was little more to learn. The Talara sample is an example of this: it was collected and prepared in the late 1950s and early 1960s but remained essentially unstudied for 50 years. However, some issues of skeletal anatomy still need clarification, such as the number of caudal vertebrae, and the Talara sample adds to our knowledge in this regard.

Smilodon limb proportions were explored by Berta (1985), and she found that the limb proportions of *S. populator* and *S. fatalis* were similar, and that there was a decrease in size from the Middle to Late Pleistocene. Because the RLB and Talara samples are essentially coeval, it is uncertain whether the slightly smaller overall size of the Talara population is significant, given the smaller sample size available from Talara and the possibility of skewed measurements reported in Merriam and Stock (1932). It also might be an example of Bergmann's rule however, reflecting smaller overall stature in more tropical locations. Other than limb proportions, the postcranial skeleton has not been explored in detail for possible specific discrimination between populations of *Smilodon*. This may be due to the large sample at RLB, which provides a window into ample variation and can make small variations seem insignificant.

Aging of Dentigerous Elements

In terms of overall age, the Talara sample seems to compare well with the RLB sample except that it has a smaller proportion of juvenile individuals. This could easily be a sample size issue since the Talara sample (MNI = 24) is so much smaller than the RLB sample (MNI = about 2,500 [Seymour, 2015]), or 2,100 in the survey by Miller (1968).

Discussion of Histology

Based on the thin-sectioned elements of four individuals, several general statements can be made regarding the osteohistology of *Smilodon fatalis* from Talara, Peru. In summary, the medullary cavity is large and circular or oval in all specimens. The endosteal surface is characterized by some degree of cancellous bone composed of lamellar bone tissue in all samples. The inner to mid-cortex is composed of woven-fibered bone in all specimens, while the outer cortex contains lamellar bone in the adults only; the outer cortex is composed of woven-fibered bone in the juvenile. The primary type of vascular canal orientation is longitudinal. When visible, the nutrient canal may or may not be accompanied by lamellar bone. Vascularity is denser in the woven-fibered bone of these specimens when compared to lamellar tissue. Secondary osteons were present in all specimens, but with varying prevalence. Osteocyte lacunar density is high in all specimens with an even arrangement. Growth marks, represented by LAGs, are present in all individuals.

All elements sectioned show some degree of secondary remodeling. This is expected in mature individuals but is also notable in the juvenile specimen. As noted by Enlow and Brown (1958), secondary remodeling is not unusual in mammalian tissue. Straehl et al. (2013) have proposed that remodeling may be more extensive in larger taxa compared to smaller relatives due to increased loading or prolonged lifespan, both of which are associated with larger body size. Our results do not disagree with this observation, but further sampling is needed to test whether this pattern holds true for all felids. It has also been suggested that in large, fast-growing animals, secondary osteons may be more prevalent in relatively small bones compared to larger elements (Padian et al., 2015). This may contribute to the degree of secondary remodeling present in the juvenile from Talara, as in general the *Smilodon fatalis* femur is relatively small in comparison with the robust forelimb elements.

This study represents the first multielement histological analysis of a fossil carnivoran, and is the first paleohistological study on an extinct felid. The histological features seen in *Smilodon fatalis* are consistent with those of extant felids as presented by Singh et al. (1974), except in contrast we found distinct outer lamellae in all adult specimens; this difference could possibly be explained by the young age of the specimens they examined. Interestingly, most carnivoran vascular canals described in Enlow and Brown (1958) were not longitudinally oriented as those found in *S. fatalis*; those of the domestic cat (*Felis catus*) were described as reticular. Unfortunately, they did not describe the vascular canal orientation of the lion (*Panthera leo*) femur they did sample, which would have been a better comparator to *S. fatalis* at least in terms of body size.

Age Assessment Based on Histology

Based on the preserved lines of arrested growth (LAGs) in the sampled elements, these individuals ranged between at least two and more than seven years old. It is probable that obscuring of older LAGs in the three adult individuals may lead to underestimations of the true age. Given a larger sample size such as those available at RLB, it is possible to retrocalculate the missing LAGs in older individuals; this work is in progress by A. R. Reynolds.

The qualitative histological characteristics of the Talara skeletal elements can be used to make some qualitative statements about the growth patterns of the taxon. According to Amprino's rule, the densely vascularized woven-fibered bone of the inner and mid-cortex is indicative of relatively high growth rates, whereas the lamellar tissue in the outer cortex represents slower growth (Amprino, 1947; Castanet et al., 2001; de Margerie et al., 2002, 2004). When we combine this observation with the occurrence of growth marks, we can make preliminary inferences regarding the timing of these shifts in growth rate. Since both LAGs in the juvenile femur (ROM 4720) occur within a woven-fibered, densely vascularized

tissue (both characteristics of rapid growth), it is reasonable to conclude that *Smilodon fatalis* grew rapidly at least into its third year of life. Furthermore, although seven lines of arrested growth are visible in the cortex of the adult femur (ROM 3026), the lack of an EFS suggests this animal was still actively growing and not yet skeletally mature. The age at which the long bones reach the end of their centrifugal periosteal growth is not known for any extant felid, but some data on other metrics of skeletal maturation are known for lions (*Panthera leo*). In a study on the anatomy of free-ranging *P. leo*, Kirberger et al. (2005a, 2005b) found that all forelimb and hind limb epiphyses were completely fused by 5.5 and 4.5 years, respectively. This agrees with previous work by Smuts et al. (1980), which showed that female and male lions reached maximum shoulder height, and therefore would have completed epiphyseal fusion of the forelimb, by the ages of four and six years, respectively. Additionally, their data showed that asymptotic body size for female and male lions is reached at approximately ages five and eight years, respectively. If the formation of the EFS coincides with attainment of asymptotic body size, then our data suggests *S. fatalis* took approximately the same amount of time to reach skeletal maturity as male *P. leo*. However, if the EFS more closely approximates the timing of epiphyseal fusion, then it is possible that *S. fatalis* could have had an extended growth period compared to living cats.

While the histological data herein allow us to begin addressing the questions regarding longevity and growth in *Smilodon* and how these may affect morphological analyses and assessments of sexual dimorphism, they are only a first step toward more rigorous and definitive analyses. Unfortunately, the Talara sample is too small for such an undertaking, so it is recommended that further work utilize the much larger RLB sample.

Sexing of Specimens

Besides the possibility that two Talara skulls are somewhat laterally distorted (ROM 3816 does not appear to be crushed but ROM 2116 may be), the proportions that Christiansen and Harris (2012) utilized to sex the RLB skulls are reasonably successful for the Talara material. The skull

ROM 3816 is most likely to have been female; ROM 2146 was probably also female and ROM 2116 was probably male. The sexing of the dentaries is a bit more definitive, resulting in eight females and two males. That this suite of proportions does not provide a definitive sex for all the Talara specimens is not surprising, given that in the Christiansen and Harris (2012) analysis, they eliminated the intermediate-sized specimens (31 or 39% of the skulls and 18 or 29% of the dentaries) to ensure that their sample represented unmixed samples of the two sexes. Even after this elimination, their Figures 6 and 7 still show some overlap between the sexes, depending on the proportion. Thus, it is not surprising that the Talara sample should include some of those problematic intermediate-sized specimens.

Sexual Dimorphism

Once the dentigerous elements were sexed, an average tooth measurement for the sexes shows that males were larger than females, with varying magnitudes depending on the element measured. The sexual dimorphism ratio of 1.18 for condylobasal length is larger than the 1.06 previously reported for *Smilodon fatalis* (Van Valkenburgh and Sacco, 2002), and is in fact larger than the mean for extant felids, reported as 1.10 (Gittleman and Van Valkenburgh, 1997) or 1.12 (Van Valkenburgh and Sacco, 2002). Similarly, our calculated ratio of 1.13 for the upper carnassial is larger than the mean of 1.06 reported for felids in Gittleman and Van Valkenburgh (1997). Our ratio of 1.09 for the m1, is larger than, but closer to, the 1.07 reported for both extant felids and *S. fatalis* in Van Valkenburgh and Sacco (2002). Although our sample sizes are small, particularly for males, the assignment of sex based on differences in morphology rather than absolute size may change our understanding of sexual size dimorphism in *S. fatalis*.

Kurtén and Werdelin (1990) noted that the two narrower Talara skulls (ROM 2116 and 2716) are anomalously narrow compared with *Smilodon fatalis* from RLB, and are more similar in overall proportion to *S. populator* in this respect. They suggested that the Talara sample might represent a different growth pattern than *S. fatalis* from RLB. Here we determine one of the narrower skulls to be male (ROM 2116)

and the other to be probably female (ROM 2146), whereas the third skull (ROM 3816), which is of a similar proportion to RLB skulls, is female, so sexual dimorphism does not account for the difference. All three skulls also have adult age tooth wear (not young adult) as defined by Miller (1968); therefore, it seems that these skulls had already grown into their adult shape, and so age does not account for this difference. Perhaps this difference is just variation in the population, since an analogous situation was found by Kurtén (1964) in a sample of *Ursus spelaeus*.

Kurtén and Werdelin (1990) obtained basic length measurements for the metapodials from CSC and determined that there was a possible bimodal distribution of these elements, at least in MT III. We found this same bimodality in the other metatarsals and to a lesser extent the metacarpals, suggesting sexual dimorphism (these data are not presented here). Although both right and left side elements were pooled, possibly duplicating individuals and skewing the distribution, the plots by either side show the same basic pattern as presented by Kurtén and Werdelin (1990). However, once again there is not a large sample size. Lacking a method to sex the metapodials means that using metapodials to determine sexual dimorphism must remain hypothetical.

Body Mass Estimates

Smilodon fatalis is known to have a relatively massive forelimb for functional reasons (Gonyea, 1976), which may explain why body mass estimates based on humeral measures are so much greater than those for femoral and tibial measures. Nevertheless, the average mass estimates based on both the humerus and femur fall into the range proposed by Christiansen and Harris (2005). The estimates based on the tibia fall just below their range, however. This agrees with the overall observation presented in Tables 3.5 and 3.7 that the Talara *S. fatalis* averages smaller in size than *S. fatalis* from RLB.

Comments on Sociality

The possible sociality of *Smilodon* has been disputed, as noted earlier. The Talara collection can also contribute to this debate despite its relatively small sample size. Several lines of evidence presented here suggest *S. fatalis* was social and may have hunted cooperatively: the skull and dentition are sexually dimorphic in size, and the skulls are sexually dimorphic in shape (although this is also true of many solitary felids [Gittleman and Van Valkenburgh, 1997]). Yet these data, when combined with the fact that a larger proportion of females to males was collected from Talara also suggests sociality where females hunted cooperatively (Kitchener, 1991). An alternative explanation noted by Van Valkenburgh and Sacco (2002) is that females may have fed together regardless of whether the species was primarily solitary or had functional prides or coalitions, as this is also observed in some living cat species (Kitchener, 1991). While the dental data shows very few juvenile specimens, the aging based on limbs resulted in a figure of 41% juveniles in the Talara collection (Seymour, 2015). If one can reasonably assume that to some extent the limbs are associated with the cranial material in this deposit (which is reasonable in an entrapment scenario), this suggests, along with the histological data presented here, that the limb bones may have had delayed epiphyseal fusion compared to the acquisition of the permanent dentition. This therefore also implies delayed maturation and extended maternal or familial care, as suggested by Van Valkenburgh and Sacco (2002). Earlier maturation of the dentition compared to the limbs in *Smilodon* could explain why the pulp cavity closure method was found not to be as useful for aging *Smilodon* as it is for pantherines. Some 80% of the *S. fatalis* specimens examined by Meachen-Samuels and Binder (2010) showed significant pulp cavity closure.

The Talara collection may represent the second largest sample of *S. fatalis* (Seymour, 2015), with a an MNI of at least 24. These data from the Talara *Smilodon* fossils contribute to the understanding of the distribution, biology and behavior of *S. fatalis*, and further studies comparing the Talara and RLB samples could yield interesting results regarding regional differences between *Smilodon* populations.

ACKNOWLEDGMENTS

First and foremost, the authors thank A. G. Edmund for collecting the Talara fauna in 1958, and A. G. Edmund, Rudy Zimmerman, and Gord Gyrmov (as well as C. S. Churcher) for the messy preparation of the tar pit material. In addition, C. S. Churcher soaked many specimens in Varsol to remove additional asphalt. Jackie Heath, Marili Moore, Pamela Purves, and Kay Sunahara assisted in the inventory of and database creation for the ROM specimens from Talara.

The authors are grateful to John Glover of the Faculty of Arts and Science, University of Toronto, for the excellent photography of the *Smilodon* skulls.

Last but not least, the authors thank the members of the Evans, Caron, and Reisz labs for their consultations regarding the histology as well as for proofreading.

REFERENCES

Akersten, W. A. 1985. Canine function in *Smilodon* (Mammalia; Felidae; Machairodontinae). Natural History Museum of Los Angeles County, Contributions in Science 356:1-22.

Amprino, R. 1947. La structure du tissu osseux envisagée comme expression de différences dans la vitesse de l'accroissement. Archives de Biologie 58:315-330.

Berta, A. 1985. The status of *Smilodon* in North and South America. Natural History Museum of Los Angeles County, Contributions in Science 370:1-15.

Binder, W. J., and B. Van Valkenburgh. 2010. A comparison of tooth wear and breakage in Rancho La Brea saber-tooth cats and dire wolves across time. Journal of Vertebrate Paleontology 30:255-261.

Campbell, K. E. Jr. 1979. The non-passerine Pleistocene avifauna of the Talara tar seeps, Northwestern Peru. Royal Ontario Museum Life Sciences Contributions 118:1-203.

Campione, N. E., and D. C. Evans. 2012. A universal scaling relationship between body mass and proximal limb bone dimensions in quadrupedal terrestrial tetrapods. BMC Biology 10:60.

Carbone, C., T. Maddox, P. J. Funston, M. G. L. Mills, G. F. Grether, and B. Van Valkenburgh. 2009. Parallels between playbacks and Pleistocene tar seeps suggest sociality in an extinct sabretooth cat, *Smilodon*. Biology Letters 5:81-85.

Castanet, J., J. Cubo, and E. de Margerie. 2001. Signification de l'histodiversité osseuse: le message de l'os. Biosystema 19:133-147.

Christiansen, P. and J. M. Harris. 2005. Body size of *Smilodon* (Mammalia:Felidae). Journal of Morphology 266:369-384.

Christiansen, P., and J. M. Harris. 2012. Variation in craniomandibular morphology and sexual dimorphism in pantherines and the sabercat *Smilodon fatalis*. PLoS One 7(10):e48352

Churcher, C. S. 1966. The insect fauna from the Talara tar-seeps, Peru. Canadian Journal of Zoology 44:985-993.

Cox, S. M., and G. T. Jefferson. 1988. The first individual skeleton of *Smilodon* from Rancho la Brea. Current Research in the Pleistocene 5:66-67.

Cullen, T. M., D. Fraser, N. Rybczynski, and C. Schröder-Adams. 2014. Early evolution of sexual dimorphism and polygyny in Pinnipedia. Evolution 68:1469-1484.

de Margerie, E., J. Cubo, and J. Castanet. 2002. Bone typology and growth rate: testing and quantifying "Amprino's rule" in the mallard (*Anas platyrhynchos*). Comptes Rendus Biologies 325:221-230.

de Margerie, E., J.-P. Robin, D. Verrier, J. Cubo, R. Groscolas, and J. Castanet. 2004. Assessing a relationship between bone microstructure and growth rate: a fluorescent labelling study in the king penguin chick (*Aptenodytes patagonicus*). Journal of Experimental Biology 207:869-879.

Enlow, D. H., and S. O. Brown. 1958. A comparative histological study of fossil and recent bone tissues, part III. Texas Journal of Science 10:187-230.

Francillon-Vieillot, H., V. de Buffrénil, J. Castanet, J. Géraudie, F. J. Meunier, J. Y. Sire, L. Zylberberg, and A. de Ricqlès. 1990. Microstructure and mineralization of vertebrate skeletal tissues; pp. 471-671 in J. G. Carter (ed.), Skeletal Biomineralization: Patterns, Processes and Evolutionary Trends. Van Nostrand Reinhold, New York.

Gittleman, J. L., and B. Van Valkenburgh. 1997. Sexual dimorphism in the canines and skulls of carnivores: effects of size, phylogeny, and behavioural ecology. Journal of Zoology 242: 97-117.

Gonyea, W. J. 1976. Behavioral implications of saber-toothed felid morphology. Paleobiology 2:332-342.

Huttenlocker, A. K., H. N. Woodward, and B. K. Hall. 2013. The biology of bone; pp. 13-34 in K. Padian and E.-T. Lamm (eds.), Bone Histology of Fossil Tetrapods: Advancing Methods, Analysis, and Interpretation. University of California Press, Berkeley, California.

Kiffner, C. 2009. Coincidence or evidence: was the sabretooth cat *Smilodon* social? Biology Letters 5:561-562.

Kirberger, R. M., W. M. du Plessis, and P. H. Turner. 2005a. Radiologic anatomy of the normal appendicular skeleton of the lion (*Panthera leo*). Part 1: thoracic limb. Journal of Zoo and Wildlife Medicine 36:21-28.

Kirberger, R. M., W. M. du Plessis, and P. H. Turner. 2005b. Radiologic anatomy of the normal appendicular skeleton of the lion (*Panthera leo*). Part 2: pelvic limb. Journal of Zoo and Wildlife Medicine 36:29-35.

Kitchener, A. 1991. The Natural History of the Wild Cats. Comstock, Ithaca, New York, 280 pp.

Köhler, M., N. Marín-Moratalla, X. Jordana, and R. Aanes. 2012. Seasonal bone growth and physiology in endotherms shed light on dinosaur physiology. Nature 487:358-361.

Kolb, C., T. M. Scheyer, K. Veitschegger, A. M. Forasiepi, E. Amson, A. A. Van der Geer, L. W. Van den Hoek Ostende, S. Hayashi, and M. R. Sánchez-Villagra. 2015. Mammalian bone palaeohistology: a survey and new data with an emphasis on island forms. PeerJ 3:e1358.

Kurtén, B. 1964. The evolution of the polar bear, *Ursus maritimus* Phipps. Acta Zoologica Fennica 108:1-30.

Kurtén, B., and L. Werdelin. 1990. Relationships between North and South American *Smilodon*. Journal of Vertebrate Paleontology 10:158-169.

Lee, A. H., A. K. Huttenlocker, K. Padian, and H. N. Woodward. 2013. Analysis of growth rates; pp. 217-252 in K. Padian and E.-T. Lamm (eds.), Bone Histology of Fossil Tetrapods: Advancing Methods, Analysis, and Interpretation. University of California Press, Berkeley, California.

Lemon, R. R. H., and C. S. Churcher. 1961. Pleistocene geology and paleontology of the Talara region, northwest Peru. American Journal of Science 259:410-429.

Lindsey, E. L., and K. L. Seymour. 2015. "Tar Pits" of the western neotropics: paleoecolology, taphonomy, and mammalian biogeography, pp. 111-123 in J. M. Harris (ed.), La Brea and Beyond: The Paleontology of Asphalt-Preserved Biotas. Natural History Museum of Los Angeles County, Science Series 42.

McCall, S., V. Naples, and L. Martin. 2003. Assessing behavior in extinct animals: was *Smilodon* social? Brain, Behavior and Evolution 61:159-164.

Meachen-Samuels, J. A., and W. J. Binder. 2010. Sexual dimorphism and ontogenetic growth in the American lion and sabertoothed cat from Rancho La Brea. Journal of Zoology 280:271-279.

Méndez-Alzola, R. 1941. El *Smilodon bonaërensis* (Muñiz): estudio osteológico y osteométrico del gran tigre fósil de la pampa comparado con otros félidos actuals y fósiles. Anales del Museo Argentino de Ciencias Naturales "Bernardino Rivadavia," Ciencias Zoológicas 40:135-252.

Merriam, J. C., and C. Stock. 1932. The Felidae of Rancho La Brea. Carnegie Institution of Washington Publication No. 422:1-231.

Miller, G. J. 1968. On the age distribution of *Smilodon californicus* Bovard from Rancho La Brea. Natural History Museum of Los Angeles County, Contributions in Science 131:1-17.

Padian, K., S. Werning, and J. R. Horner. 2015. A hypothesis of differential secondary bone formation in dinosaurs. Comptes Rendus Palevol 15:40-48. doi: 10.1016/j.crpv.2015.03.002.

Peters, G., and M. H. Hast. 1984. Hyoid structure, laryngeal anatomy, and vocalization in felids (Mammalia, Carnivora, Felidae). Zeitschrift für Säugetierkunde 59:87-104.

Pocock, R. I. 1916. On the hyoidean apparatus of the lion (*F. leo*) and related species of Felidae. Annals and Magazine of Natural History, 18:222-229.

Seymour, K. L. 1983. The Felinae (Mammalia: Felidae) from the Late Pleistocene tar seeps at Talara, Peru, with a critical examination of the fossil and recent felines from North and South America. MSc thesis, Department of Geology, University of Toronto, Toronto, Canada, 230 pp.

Seymour, K. L. 2015. Perusing Talara: Overview of the Late Pleistocene fossils from the tar seeps of Peru; pp. 97-109 in J. M. Harris (ed.), La Brea and Beyond: The Paleontology of Asphalt-Preserved Biotas, Natural History Museum of Los Angeles County, Science Series 42.

Singh, I. J., E. A. Tonna, and C. P. Gandel. 1974. A comparative histological study of mammalian bone. Journal of Morphology 144:421-437.

Smuts, G. L., J. L. Anderson, and J. C. Austin. 1978. Age determination of the African lion (*Panthera leo*). Journal of Zoology 185:115-146.

Smuts, G. L., G. A. Robinson, and I. J. Whyte. 1980. Comparative growth of wild male and female lions (*Panthera leo*). Journal of Zoology 190:365-373.

Stamps, J. A. 1993. Sexual size dimorphism in species with asymptotic growth after maturity. Biological Journal of the Linnean Society 50:123-145.

Straehl, F. R., T. M. Scheyer, A. M. Forasiepi, R. D. MacPhee, and M. R. Sánchez-Villagra. 2013. Evolutionary patterns of bone histology and bone compactness in Xenarthran mammal long bones. PLoS One 8:e69275.

Van Valkenburgh, B., and T. Sacco. 2002. Sexual dimorphism, social behavior, and intrasexual competition in large Pleistocene carnivorans. Journal of Vertebrate Paleontology 22:164-169.

Weckerly, F. W. 1998. Sexual-size dimorphism: influence of body mass and mating systems in the most dimorphic mammals. Journal of Mammalogy 79:33-52.

Weissengruber, G. E., G. Forstenpointer, G. Peters, A. Kübber-Heiss, and W. T. Fitch. 2002. Hyoid apparatus and pharynx in the lion (*Panthera leo*), jaguar (*Panthera onca*), tiger (*Panthera tigris*), cheetah (*Acinonyx jubatus*) and domestic cat (*Felis silvestris* f. *catus*). Journal of Anatomy 201:195-209.

Werning, S. 2012. The ontogenetic osteohistology of *Tenontosaurus tilletti*. PLoS One 7: e33539.

Woodward, H. N., K. Padian, and A. H. Lee. 2013. Skeletochronology; pp. 195-216 in K. Padian and E.-T. Lamm (eds.), Bone Histology of Fossil Tetrapods: Advancing Methods, Analysis, and Interpretation. University of California Press, Berkeley, California.

4 The Sabertooth Cat, *Smilodon populator* (Carnivora, Felidae), from Cueva del Milodón, Chile

H. GREGORY MCDONALD AND LARS WERDELIN

Introduction

Given an invitation to name a single sabertooth cat, most people, amateur and professional alike, will choose *Smilodon*. In large part due to the mass occurrence of *Smilodon* at Rancho La Brea in California (Merriam and Stock, 1932) its detailed anatomy, as well as its general appearance, has made it almost iconic of the term 'sabertooth cat.' For example, Akersten (1985) developed his model of canine function in sabertooths, the so-called canine shear-bite, mainly on the anatomy of the North American species, *Smilodon fatalis*. And when the ecology of sabertooths is discussed in general terms, most authors have *Smilodon* in mind. Nevertheless, it is important to realize that *Smilodon* in reality represents an extreme form in one lineage of sabertooth cat, and that other lineages, represented by genera such as *Machairodus, Dinofelis, Homotherium,* and *Xenosmilus,* had quite different evolutionary histories and ecologies (Antón, 2013).

Despite it being extremely well known for an extinct predator, there is also much we do not yet know about the biology of *Smilodon*, beginning with the question of how many species should be recognized within the genus. A further question of considerable interest is the biogeography of Late Pleistocene members of the genus and the significance of these biogeographic patterns to the paleoecology of North and South American species of *Smilodon*. How different and how similar were these two species, especially given their different evolutionary histories on two different continents? Kurtén and Werdelin (1990) demonstrate a biogeographic pattern in which *S. fatalis* is found west of the Andes and the South American species, *S. populator*, east of the mountains. Why didn't *S. fatalis* disperse east of the Andes as well? Recently, *Smilodon populator* was reported from two localities in Venezuela, the Inciarte Tar Pits in Zulia State, and Zumbador Cave (10° 51′N) in Falcon State (Rincón, 2003, 2006), indicating that the distribution of *Smilodon populator* extended the entire length of the South American continent on the east side of the Andes. It is possible that the existence of a previously established larger species of *Smilodon* on the eastern side of the Andes may have resulted in the competitive exclusion of the smaller *S. fatalis* and prevented the later and smaller species from dispersing eastward, thus restricting its distribution to the western side of the Andes, which did not have the larger *S. populator*. This chapter, describing specimens of *Smilodon* from Cueva del Milodón, Magallanes region, southern Chile, represents a contribution to these issues.

Materials and Methods

Smilodon Material from Cueva del Milodón

The collections, originally housed in the Zoological Museum at the University of Amsterdam (ZMA), contain at least two individuals of *Smilodon*, both adults. One individual is represented by a partial skeleton including the following: ZMA 20,031, right mandible with i3, c, p3, p4; ZMA 20,033, distal left scapula; ZMA 20,044, distal right humerus; ZMA 20,035, left ulna; ZMA 20,034, left radius; ZMA 20,100, right radius; ZMA 20,039, right femur; ZMA 20,037, left femur; ZMA 20,042, left tibia; ZMA 20,036, left and right pelvis and sacrum; ZMA 20,043, distal left fibula; ZMA 20,038, axis, cervicals 3,4,5,6 and additional cervical (second individual); ZMA 20,045, five anterior lumbar vertebrae. In addition to the cervicals mentioned above, the second individual is represented by ZMA 20,041, a left radius with exostoses. The collection has been transferred to the Naturalis Biodiversity Center in Leiden, but the specimens still retain the original catalog numbers of the Zoological Museum, University of Amsterdam.

ABBREVIATIONS

DMNS, Denver Museum of Nature and Science, Denver; **EPN,** Escuela Politécnica Nacional, Quito, Ecuador; **FM,** Field Museum of Natural History, Chicago; **IMNH,** Idaho Museum of Natural History, Pocatello; **ISM,** Illinois State Museum, Springfield; **LACM,** Natural History Museum of Los Angeles County, Los Angeles; **MNHN,** Museum National d'Histoire Naturelle, Paris, France; **MMP,** Museo Municipal de Ciencias Naturales de Mar del Plata, Argentina; **MZUC,** Museum of Zoology, University of Cambridge, UK; **NHMUK,** Natural History Museum, London, UK; **PUC-MG,** Secao de Paleontologia do Museu de Ciencias Naturais da Pontifica Universidad Catolica, Minas Gerais, Brazil; **ROM,** Royal Ontario Museum, Toronto, Canada; **SBCM,** San Bernardino County Museum, Riverside; **SMU,** Shuler Museum, Southern Methodist University, Dallas; **TMM,** Texas Memorial Museum, Austin; **UF,** Florida Museum of Natural History, Gainesville; **UUVP,** University of Utah Vertebrate Paleontology Collection, Salt Lake City; **ZMA,** Zoological Museum,

University of Amsterdam, The Netherlands; **ZMUC,** Zoological Museum, University of Copenhagen, Denmark.

Whenever possible specimens were examined and measured with dial calipers to the nearest tenth of a millimeter. In some instances, access to the specimens was not possible and measurements from the literature have been included in our data sets. For those specimens we have given the citation in the list of material from each locality. The latitude for each locality has been determined to the nearest half degree. For many of the specimens from Argentina, specific locality information is not available other than indicating Buenos Aires Province, and for the purposes of this study, average latitude for the province has been used for all of these specimens. Distribution of the localities in South and North America used for comparison are shown in Figure 4.1.

Sample of South American *Smilodon* Used for Comparison

Buenos Aires Province, Argentina, Latitude 34.5° S—NHMUK 21000 complete mandible; NHMUK 43236 right mandible; NHMUK 43235 left mandible; ZMUC Lausen #2, ZMUC Lausen #12, ZMUC Lausen #52 left mandible; ZMUC Lausen #4, ZMUC Lausen #9 right mandible; MNHN PAM 607 left mandible; FM P14294 left mandible; FM P14271 right mandible; NHMUK 37682 left humerus; ZMUC Lausen #2 left humerus; ZMUC Lausen #11, ZMUC Lausen #54 right humerus; FM P14294, FM P14271 left humerus; ZMUC Lausen #8 right radius; ZMUC Lausen #7 left radius; MNHN PAM 629 right radius; FM P14294 left radius; ZMUC Lausen #2, ZMUC Lausen #7 left ulna; ZMUC Lausen #3, ZMUC Lausen #61 right ulna; FM P14294 left ulna; ZMUC Lausen #2, ZMUC Lausen #7 right femur; MNHN PAM 566, MNHN PAM 611 right femur; MZUC K5381 left femur; FM P14294 left femur; ZMUC Lausen #2 right tibia; MNHN PAM 610, MNHN PAM 565 right tibia; MNHN PAM 557 left tibia; FM P14294 right tibia; FM P14271 right fibula.
Mar del Plata, Argentina, Latitude 38° S—MMP 652 humerus, MMP 5-I femur, MMP 465-I femur, MMP

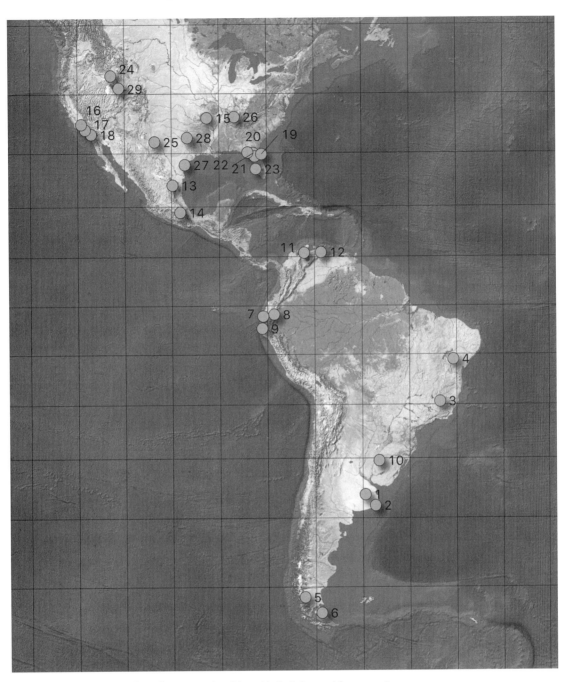

Figure 4.1. Map of South and North American localities with *Smilodon* used for comparison.

495b-Dj tibia, MMP 796g-I tibia. Measurements from Churcher (1967).

Lagoa Santa, Brazil, Latitude 17° S—ZMUC 5350, ZMUC 2024, right mandible; ZMUC 5332 left mandible; MNHN BRD 21 right mandible; NHMUK 18972 right humerus; ZMUC 2003,

ZMUC 2001 left humerus; NHMUK 32992 right radius; ZMUC 2004, 3685 left radius, NHMUK 18972 right ulna; ZMUC 2005 left ulna; ZMUC 2006 left femur; MNHN BRD 22 right femur; NHMUK 18972 right tibia; ZMUC 2002 right tibia.

Tocas das Onca, Brazil, Latitude 11° S—PUC-MG 7157 right ulna, PUC-MG 7159 right tibia, PUC-MG 7161 right femur, PUC-MG 7160 left femur. Measurements from Cartelle and Abuhid (1989).

La Carolina, Ecuador, Latitude 2° S—ROM 5100 left mandible, ROM 5101 left mandible; ROM 5106; right humerus. Measurements from Hoffstetter (1952).

Punín, Chimborazo, Ecuador, 1.5° S—EPN V1894 right tibia, measurements from Hoffstetter (1952).

Talara, Peru, Latitude 4.5° S—ROM 2118, ROM 2146, ROM 2245, ROM 2562, ROM 2563, ROM 2230 left mandible; ROM 2117, ROM 2228, ROM 2229, ROM 2258, ROM 2429, ROM 3055, ROM 4419 right mandible; ROM 2934, ROM 2873, ROM 3993, ROM 3994, ROM 3998, ROM 4007, ROM 3996, ROM 3997 right humerus; ROM 2956, ROM 2239, ROM 2350, ROM 2240, ROM 4163 left humerus; ROM 2520, ROM 2906, 4124, 2904 right radius; ROM 2112, 2251, 2905, 4111 left radius; ROM 2899, ROM 4059, ROM 2900, ROM 2243, ROM 2241, ROM 3053 left ulna; ROM 2901, 2902, 4123 right ulna; ROM 2243, ROM 2023, ROM 4717, ROM 2332 right femur; ROM 2561, 2908, 2909 left femur; ROM 2430, ROM 3033, ROM 4730, ROM 4737 left tibia; ROM 2238, ROM 3070, ROM 4731, ROM 4732 right tibia; ROM 4767, ROM 4768, ROM 4775, ROM 4776 right fibula; ROM 2236 left fibula.

Rio Negro, Uruguay, Latitude 32.5° S—NHMUK 39400 right femur.

Sample of North American *Smilodon* Used for Comparison

Hurricane River Cave, Arkansas, Latitude 36° N—ISM 702.102 left radius; ISM 702.103 left ulna; ISM 702.96 right femur, ISM 702.98 right tibia; ISM 702.101 left fibula.

Maricopa, California, Latitude 35° N—LACM 17792 left mandible, LACM 46830 complete mandible, LACM 21996 right radius, LACM 22597 left ulna, LACM 42077 right ulna, LACM 67389 left ulna, LACM 17775 left tibia.

Rancho La Brea, California, Latitude 34° N—FM P12417 left humerus; FM P 12417 left ulna; FM P12417, FM P12356 left femur; FM P12449 right femur; FM P 12357 right tibia and measurements from Merriam and Stock (1932).

Winchester Meadows, California, Latitude 33° N—SBCM A2715-0485 left radius.

Arredondo, Florida, Latitude 29.5° N—UF 2562 left humerus, left radius, right ulna, right femur, left tibia of one individual.

Aucilla River, Florida, Latitude 30.5° N—UF 14894 mandible.

Devil's Den, Florida, Latitude 29.5° N—UF 9023 right tibia.

Ichetucknee River, Florida, Latitude 29.5° N—UF 4114 mandible; UF 4115 mandible.

Warm Mineral Springs, Florida, Latitude 27° N—UF 23866 left radius; UF 23867 left tibia.

American Falls Reservoir, Idaho, Latitude 43° N—IMNH 28196 left mandible; IMNH 16026 right humerus; IMNH 108, IMNH 2888 left femur; IMNH 1623, IMNH 39092 right tibia; IMNH 28894 left tibia.

Pecos River Valley, New Mexico, Latitude 32.5°—DMNS 1229 mandible. Morgan and Lucas (2001).

First American Bank Site, Tennessee, Latitude 36° N—radius, tibia. Measurements from Guilday (1977).

Ingleside, Texas, Latitude 28° N—TMM 30967-198 right mandible, TMM 30967-1713 left radius. Measurements from Lundelius (1972).

Moore Pit, Texas, Latitude 33° N—SMU 60105 right mandible.

Pemberton Hill (=Moore Pit), Latitude 33° N—SMU 60030 right mandible (Holotype *Smilodon trinitensis*).

Wood Pit, Latitude 33° N (=Moore Pit)—SMU 60227 left mandible.

Silver Creek Site, Utah, Latitude 41° N—UUVP 7315 right humerus.

San Josecito Cave, Nuevo León, Mexico, Latitude 24° N—LACM 3938 right mandible; LACM 8600 right humerus, radius, and ulna; LACM 8601 left humerus and ulna; LACM 8608, LACM 8606 right tibia.

Valsequillo, Puebla, Mexico, Latitude 19° N—right mandible cited in Kurtén (1967).

Age of Cueva del Milodón Specimen

A portion of the distal end of the left fibula (ZMA 20,043) was submitted to Dr. Thomas Stafford for accelerator mass spectrometry dating. The resulting radiocarbon date for the *Smilodon* is 11,330±60 radiocarbon years (CAMS 33969) using XAD filtration. Using OxCal 4.2 and the IntCal 13 calibration curve, this gives a range of 13,290 to 13,075 cal BP.

History of Studies of Cueva del Milodón Fauna

Location of Cave

Since its initial discovery in 1895, Cueva del Milodón has been referred to in the literature by a variety of names. These include Eberhardt Cave or Cueva Eberhardt (Lönnberg, 1900; Studer, 1905); Mylodon Cave (Salmi, 1955; Saxon, 1979) or Cueva del Milodón (Moore, 1978; Rau and Yañez, 1980), Last Hope Cave (Sutcliffe, 1985) and Glossotherium Cave (Nordenskiöld, 1900). The cave is located in XII Region, Magallanes of Chile, approximately 24 km northeast of Puerto Natales at 51°35′S lat, 72°38′W long. Elevation is approximately 150 m above sea level. The cave is formed in a conglomerate (Wellman, 1972).

History of Collecting

Hauthal et al. (1899) state that Cueva del Milodón was discovered in January 1895 by Herman Eberhardt while he was exploring his property and that skin fragments of the sloth, *Mylodon listai* (now considered a junior synonym of *Mylodon darwinii*) were collected from near the entrance at the time of the cave's discovery. The following year, Eberhardt gave a fragment of skin to Otto Nordenskjöld, the Antarctic explorer, who was visiting the area. This piece of skin and some other specimens were later described by Lönnberg (1899). In 1897 F. P. Moreno of the La Plata Museum, while a member of the commission to settle the boundary dispute between Chile and Argentina, found a large piece of sloth skin tacked to a tree in the vicinity of the cave. Moreno presented the skin to Arthur Smith Woodward of the British Museum (Natural History) (now Natural History Museum, London) and a description of the specimen was published in the Proceedings of the Zoological Society of London (Moreno, 1899). Prior to Moreno's publication, F. Ameghino obtained some dermal ossicles from the cave, which he described as a new type of extinct sloth, *Neomylodon listai* (Ameghino, 1898).

In March of 1899 Erland Nordenskiöld, the cousin of Otto, conducted the first excavations in the cave. This collection was subsequently deposited in the Malmö Museum, Sweden, and was described by Lönnberg (1900). Rudolf Hauthal of the La Plata Museum excavated in the cave in April of the same year. Results of this excavation included a second description of the ground sloth as *Grypotherium domesticum* (Hauthal et al., 1899). That same year E. Nordenskiöld published three papers on the cave, in French, German, and Swedish (Nordenskiöld, 1899a, 1899b, 1899c) and another in 1900. Woodward (1900) published a second description of material from the cave based in part on material collected by Hauthal and deposited in the British Museum (Natural History). Santiago Roth (1902) described additional remains from the Hauthal excavation in the cave, including *Felis listai*, *Canis avus*, and *Onohippidium saldiasi*.

On November 4, 1901 the British Museum (Natural History) received a letter from Enrique Hansen with a photograph of a piece of sloth skin, offering the piece for sale. Hansen later sold the piece and other material to Walter Neuman. In April, 1902 Neuman brought the collection to London to show to the Zoological Society. Woodward offered to buy the collection from Neuman for the British Museum of Natural History, but they could not agree on a price. The collection was subsequently sold by Neuman to the Humboldt Museum in Berlin. These skin fragments were used by Branco (1906) in one of the first studies using x-rays to study fossils. In late April, Hansen again wrote the British Museum (Natural History) informing them that a second collection from the cave was available. After negotiations, this second collection, composed of one piece of skin and 23 bones, was sold to the museum in late May 1902.

In September, 1902, Albert Conrad and John Berg excavated in the cave, and in February 1903 Conrad conducted a second excavation. The specimens from these excavations were given to Captain Charles A. Milward, then British vice-consul at Punta Arenas. In January, 1904 Woodward contacted Graham Milward (brother of Charles) regarding the purchase of this collection.

Studer (1905) described specimens of sloth, horse, and humans from the cave in the collections of the Zoological Museum at the University of Zurich acquired from Santiago Roth. The cricetid rodents in the fossil fauna of the cave are described in Rau and Yañez (1980).

Saxon (1979) conducted excavations in the cave in 1976 and 1979 and provided some of the first radiocarbon dates for the site. All material from these excavations is deposited in Instituto de la Patagonia, Punta Arenas, Chile. Based on his excavations, Saxon (1979) concluded that humans and members of the extinct fauna were contemporaneous with the sloth, which survived in the region until as late as 5,000 years BP. A summary of radiocarbon dates from southern Patagonia and Tierra del Fuego, including those from Cueva del Milodón, was published by Ortiz-Troncoso (1980). Borrero et al. (1991) examined the site and challenged Saxon's chronology, concluding that the sloth had become extinct in the area at the end of the Pleistocene. Tonni et al (2013) provided an updated review of the radiocarbon dates from the cave. A more comprehensive overview of the literature pertaining to Cueva del Milodón is provided by Prieto (2013).

History of the Collection of the Zoological Museum of Amsterdam *Smilodon*

The collection from Cueva del Milodón was originally housed in the Instituut voor Taxonomische Zoologie in Amsterdam but has since been transferred to the Naturalis Biodiversity Center in Leiden. This material was obtained by Jan Herman Kruimel during a collecting trip to Patagonia for the Zoological Museum, University of Amsterdam from December 4, 1908 until the first half of June, 1909. An overview of this trip is given by van Bree (1989) and Mol et al. (2003). The present collection is a combination of two separate collections. One collection was obtained by Kruimel from a Mr. H. Hansen, a Dane, and the other from the Englishman Charles A. Milward, who had previously sold two collections from Cueva del Milodón to the British Museum (Natural History). Following his return from Patagonia, Kruimel visited the British Museum (Natural History) and consulted with Arthur Smith Woodward on the identification of the specimens in

preparation of a publication on the collection. Unfortunately, his untimely death in 1916 prevented the completion of this paper.

Description of *Smilodon* Material

Craniodental Material

The complete right ramus of the mandible (ZMA 20,031) is preserved and has separated from the left ramus at the mandibular symphysis (Fig. 4.2). Although the tooth itself is not preserved, there are two alveoli for the third premolar. Examination of the third premolar in other South American specimens indicates that size and degree of development varies considerably and there is no consistent morphology. It is evident that the p3 in the Cueva del Milodón specimen has two separate and distinct roots; this is also seen in other specimens of *Smilodon populator*, such as NHMUK 43236. In other individuals, such as MNHN BRD 1 from Brazil, the two roots are confluent and the crown has a simple, single cusp.

A small number of individuals of *Smilodon* from North America also retain a third premolar in the mandible. Merriam and Stock (1932:51) stated that the percentage of mandibles in the collections from Rancho La Brea with p3s was 6% (41 of 678). Tejada-Flores (pers. comm. 1985) reexamined specimens from Rancho La Brea in the Page Museum and found 41 mandibles with the p3 out of 1,711 specimens examined or 2.39% of the population. Since no other North American localities have produced as large a sample of *Smilodon* as Rancho La Brea, it is not possible to make any direct comparisons. Out of the Late Pleistocene sample used in this study only four specimens had a lower third premolar.

These include an individual from Maricopa, California, one from Ichetucknee, Florida, and two from Dallas County, Texas gravel pits. Slaughter (1963) figured the Texas specimens, one of which retains the lower third premolar with a single root (SMU 60227), while another has a single alveolus for the lower third premolar (SMU 60105). Given this small sample (N = 10) of *Smilodon fatalis* mandibles from sites other than Rancho La Brea, the presence of four specimens with a lower third premolar gives a disproportionately high 40% compared to the Rancho La Brea sample.

A

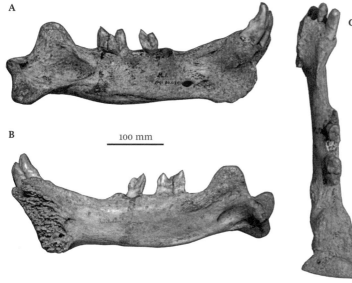

B

100 mm

C

Figure 4.2. *Smilodon populator* right mandible, ZMA 20,031. A = lateral (buccal) view; B = medial (lingual) view; C = occlusal view.

The sample from South America is larger than that from North America (excluding Rancho La Brea), with a total of 31 specimens. Of these 31 specimens, 5 (16.1%) retained the lower third premolar. If these specimens are separated into two groups, east or west of the Andes, the percentage of specimens with p3 in west (Ecuador and Peru, considered to be *Smilodon fatalis*) is 11.8% (N = 17; 2 with) and in east (Argentina, Brazil, and Chile, assigned to *Smilodon populator*), 21.4% (N = 14; 3 with).

The crown of the lower canine is larger in South American *Smilodon* east of the Andes than it is in the population west of the Andes or the North American specimens (Fig. 4.3A). As in a number of the other graphs, the sample from Ecuador and Peru overlaps the North American sample in these dimensions. The width of the lower canine in the eastern South American *Smilodon* sample (*S. populator*) also tends to be broader relative to the mesiodistal length of the tooth than the specimens from Peru and Ecuador referred to *Smilodon fatalis*, a pattern that may be related to differences in crown-to-tip length of the canines.

Berta (1985) noted that the diastema in South American *Smilodon* is relatively shorter than in North American specimens. Plotting diastema length from canine to lower fourth premolar against total length of the mandible (Fig. 4.3B) supports her observation that, despite the greater length of the mandible in *S.*

populator, there is no similar increase in the length of the diastema. In both species, despite differences in size and the observed changes in size with latitude, the relationship of diastema length to mandibular length remains the same (Fig. 4.3C).

The length of the diastema from the distal margin of the canine to the mesial margin of the lower fourth premolar was plotted against alveolar length of the lower fourth premolar and first molar (Fig. 4.3D). In addition to *Smilodon* east of the Andes being larger than *Smilodon* from Ecuador and Peru or North America, there is also a difference in relative proportions, and while there is some overlap between the largest *S. fatalis* and the smallest *S. populator,* in *S. populator* the length of the lower fourth premolar and first molar is slightly greater relative to the length of the diastema than in *S. fatalis*.

The mandible of *Smilodon* east of the Andes tends to be more massive relative to its depth than that of *Smilodon* west of the Andes or in North America. Plotting the dorsoventral depth of the diastema against diastema length (Fig. 4.3E) shows a similar separation between *Smilodon* east of the Andes and the others, although as before the samples grade into each other and there is no marked separation.

Depth of the mandible at the midpoint of the diastema plotted against depth of the horizontal ramus posterior to the lower first molar (Fig. 4.3F) indicates that, while *S. populator* is larger than *S.*

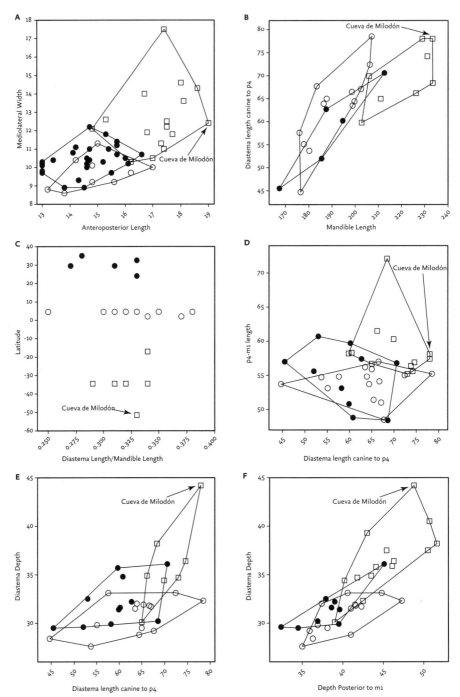

Figure 4.3. Scatter diagrams of selected dimensions of North and South American *Smilodon*. Solid circles = *Smilodon fatalis*, North America; open circles = *Smilodon fatalis*, Peru and Ecuador; squares = *Smilodon populator*. A = anteroposterior length against mediolateral width of lower canine; B = mandible length against length of diastema from distal margin of lower canine to mesial margin of lower fourth premolar; C = length of diastema as a fraction of total mandibular length plotted against latitude; D = length of diastema from distal margin of lower canine to mesial margin of lower fourth premolar against tooth row length of lower fourth premolar and first molar; E = diastema length from distal margin of lower canine to mesial margin of lower fourth premolar against dorsoventral depth of diastema at midpoint; F = depth of mandible posterior to lower first molar against diastema depth at midpoint.

fatalis, larger individuals of *S. populator* do not have a greater depth of the mandible behind the lower first molar despite their larger size. As seen before, *Smilodon* from Peru and Ecuador overlaps with the sample from North America.

As a general observation, South American *Smilodon* east of the Andes consistently tend to be larger than individuals in North America, although there is some overlap between larger individuals of *S. fatalis* and smaller individuals of *S. populator*. There is no exact symmetry relative to the equator with regard to size, and populations from similar northern and southern latitudes are not similar in size. The two samples that can be directly compared are those from Rancho La Brea and Buenos Aires Province. Guilday (1977) calculated means and observed ranges of various dimensions of *Smilodon* from Rancho La Brea utilizing measurements in Merriam and Stock (1932). A comparison of these measurements with those of *Smilodon* from Buenos Aires Province shows that for some dimensions *Smilodon* from Buenos Aires Province are larger, but in other dimensions they are similar. In some cases, this results in differences in body proportions; for example, the femur in both samples is similar in length but in the sample from Argentina, the tibia is shorter than in the California sample.

Postcranial Material

The postcranial skeletal material of the Cueva del Milodón *Smilodon* is that of mature individuals. All of the epiphyses are fused to the diaphyses and in none of the specimens is the epiphyseal line visible. The precise circumstances of the find are not known (see above), and as noted below for the radius, material of at least two individuals is present. The majority of the material likely does belong to one individual, however, as evidenced by the general similarity in size and preservation, and the absence of duplication among the elements (except for the second radius).

A portion of the vertebral column is preserved and includes the axis and cervicals 3 through 6 (ZMA 20,038), the last thoracic and lumbars 1 through 4 (ZMA 20,045), and the sacrum (ZMA 20,038). Measurements are provided in Table 4.1. Osteophytes are present on vertebrae three and four of the lumbar series (Fig. 4.4). Similar pathologies of spondylosis deformans or ankylosing spondylosis of the lumbar vertebrae in *Smilodon fatalis* have been described for individuals from Rancho La Brea (Moodie, 1973; Bjorkengren et al., 1987).

The two left radii (ZMA 20,034 and ZMA 20,041) are quite different in their lengths: 259.6 and 224.9 mm, respectively (Fig. 4.5). The difference in length may be related to the presence of exostoses on

A

100 mm

B

Figure 4.4. *Smilodon populator.*
A = cervical vertebrae ZMA 20,038;
B = last thoracic and first four lumbar vertebrae, ZMA 20,045 with osteophytes on lumbar vertebrae three and four.

Table 4.1. Measurements of *Smilodon populator* from Cueva del Milodón, Chile (in mm.)

Mandible (ZMA 20,031)

Maximum length of mandible from mandibular symphysis to posterior margin of condyle - 233.4

Length of mandible from mandibular symphysis to posterior margin of angular process - 229.6

Length of mandible from posterior margin of canine to posterior margin of condyle - 219.7

Length of mandible from posterior margin of canine to posterior margin of angular process - 220.8

Alveolar length of tooth row from lower third premolar to first molar - 70.6

Alveolar length of tooth from lower fourth premolar to first molar - 58.1

Alveolar length of first molar - 28.9

Height of coronoid process - 70.2

Depth of mandibular ramus posterior to the lower first molar - 48.8

Depth of midpoint of diastema - 44.2

Diastema length from distal margin of canine to mesial margin of alveolus of lower third premolar - 64.8

Diastema length from distal margin of canine to mesial margin of alveolus of lower fourth premolar - 78

Height of mandibular symphysis - 78.9

Distance from posterior margin of alveolus of lower first molar to posterior margin of condyle - 85.5

Lower dentition in mandible (ZMA 20,031)

Mesiodistal length of crown of lower canine - 17.4

Labiolingual width of crown of lower canine - 11.0

Mesiodistal length of crown of lower fourth premolar - ?

Labiolingual width of anterior cusplet of lower fourth premolar - 11.4

Labiolingual width of principal cusp of lower fourth premolar - 11.8

Mesiodistal length of crown of lower first molar - 28.3

Labiolingual width of paraconid of lower first molar - 13.5

Labiolingual width of midpoint of lower first molar - 13.3

Mesiodistal length of blade of paraconid - 10.9

Mesiodistal length of blade of protoconid - 17.9

Axis (ZMA 20,038)

Length of centrum including dens - 102.8

Mediolateral width of anterior of centrum - 88.0

Length of neural spine - 130.0

Dorsoventral height of anterior articular surface - 39.9

Mediolateral width of posterior centrum - 46.9

Dorsoventral height of posterior centrum - 26.9

Mediolateral width across postzygapophyses - 61.4

Cervical vertebrae 3-6

Length of centrum - 57.8; 54.5; 52.5; 53.5

Mediolateral width of anterior of centrum - 38.9; 40.1; 40.1; 38.6

Dorsoventral height of anterior articular surface - 29.6; 27.2; 32.1; 32.3

Length of neural spine - 47.7; 44.2; 38.2; 26.8

Mediolateral width of posterior centrum - 47.4; 48.4; 47.6; 46.4

Dorsoventral height of posterior centrum - 31.4; 33.2; 35.1; 37.1

Mediolateral width across postzygapophyses - 80.2; 86.4; 93.2; 89.1

Thoracic and lumbar vertebrae (ZMA 20,045)

Length of centrum - 41.7; 42.0; 45.2; 48.5; 49.3

Mediolateral width of anterior of centrum - 50.0; 49.4; 45.3; 43.7; 47.9

Dorsoventral height of anterior articular surface - 33.3; 33.4; 34.3; 34.4; 36.6

Table 4.1. (continued)

Mediolateral width of posterior centrum - 52.5; 53.4; 53.8; 55.2; 55.5
Dorsoventral height of posterior centrum - 32.7; 33.9; 34.3; 35.8; 39.8
Mediolateral width across prezygapophyses - 62.7; 66.3; 64.9; 58.7; ?
Mediolateral width across postzygapophyses - 38.3; 42.6; 46.0; 44.1; 39.8
Height from base of centrum to top of neural spine - 96.8; 102.3; 107.8; ?; ?

Sacrum (ZMA 20,038)
Length of centrum - 113.4
Width across iliac processes - 106.5
Mediolateral width across anterior centrum - 50.0
Dorsoventral height of anterior centrum - 27.7
Width of anterior zygapophysis - 63.0
Height of iliac articulation - 78.6

Distal left scapula (ZMA 20,033)
Anteroposterior length of glenoid - 91.5
Mediolateral width of glenoid - 57.6

Distal right humerus (ZMA 20,044)
Transverse diameter of midshaft - 37
Anteroposterior diameter of midshaft - 55.8
Anteroposterior diameter of shaft above entepicondylar foramen - 43.9
Transverse diameter of distal end - 123.3
Transverse diameter of distal articular surface - 93.9
Anteroposterior width of medial side of distal end - 69.7

Left radius (ZMA 20,100 and 20,041)
Length between articular surfaces - 259.6; 224.9
Anteroposterior length of proximal end (long axis) - 52.5; 55.5
Mediolateral width of proximal end (short axis) - 41.2; 42.7
Anteroposterior width of midshaft - 38.1; 35.2
Mediolateral width of midshaft - 23.8; 19.7
Anteroposterior length of distal end - 65.5; 61.7
Mediolateral width of distal end - 47.7; 49.7

Left ulna (ZMA 20,100 and 20,035)
Maximum length - 361.6
Length of the olecranon process - 78.5
Length from the radial notch to the distal end - 274.1
Height of the anconeal process - 71.6
Height of coronoid process - 83.7
Anteroposterior length of distal end - 46.5
Mediolateral width of distal end - 29.8

Left pelvis (ZMA 20,036)
Total length - 388.4
Length of ilium - 203.2
Dorsoventral height of ilium - 82.8
Anteroposterior diameter of acetabulum - 55.3
Dorsoventral diameter of acetabulum - 54.5
Dorsoventral width of neck of ilium - 64.1
Dorsoventral width of neck of ischium - 51.8

Left (ZMA 20,037) and right (ZMA 20,039) femur
Greatest length from greater trochanter to lateral condyle - 396.7; 405.5
Length from head to medial condyle - 396.2; 401.3
Transverse diameter of distal end - 107.7; 108.2
Anteroposterior diameter of greater trochanter - 54.2; 58.5
Anteroposterior diameter of head - 51.1; 50.8

(continued)

Table 4.1. (continued)

Minimum diameter of neck - 35.4; 35.0
Anteroposterior diameter of shaft - 36.8; 36.7
Mediolateral diameter of shaft - 40.3; 45.3
Transverse diameter of distal end across epicondyles - 81.8; 83.0
Transverse diameter across distal condyles - 85.0; 87.2
Transverse diameter of lateral condyle - 31.4; 33.1
Transverse diameter of medial condyle - 35.8; 37.3
Anteroposterior thickness of distal end on medial side - 82.4; 83.6

Left tibia (ZMA 20,042)
Length - 278.7
Mediolateral diameter of proximal end - 89.3
Anteroposterior diameter of proximal end on lateral side - 70.1
Anteroposterior diameter of shaft - 44.7
Mediolateral diameter of shaft - 35.4
Anteroposterior diameter of distal end - 41.3
Mediolateral width of distal end - 57.2

Left fibula (ZMA 20,043)
Anteroposterior length of distal end - 30.6
Mediolateral width of distal end - 30.6

ZMA 20,041, suggesting a pathological condition, or it may simply reflect two differently sized individuals. A large exostosis is located toward the proximal end of the posterior edge and probably formed as an ossification within the interosseous membrane. There is also extra bone deposition around the distal edge of the head of the radius, but this does not extend onto the articular surface.

Preserved bones of the forelimb include a distal right humerus (ZMA 20,044), left ulna (ZMA 20,035), and left radius (ZMA 20,034), all considered to be part of a single individual, and a second left radius with exostosis (ZMA 20,041), also from an adult individual (Fig. 4.5).

The hind limb is represented by a right pelvis (ZMA 20,036), left femur (ZMA 20,037), and left tibia (ZMA 20,042), considered to be part of the same individual as the mandible and forelimb (Fig. 4.6). Anatomically they do not present any significant differences from other specimens of *S. populator*.

Discussion

Taxonomy

There are three alternatives with regard to the relationship of Late Pleistocene *Smilodon* in North and South America. The traditional view is that Late Pleistocene *Smilodon* in North and South America are separate species. Although various names have been applied to the North American species, *Smilodon fatalis* (with *S. floridanus, S. californicus, S. nebraskensis, S. trinitiensis,* and others as junior synonyms; see Kurtén and Anderson, 1980) is currently utilized. *Smilodon populator,* the type species of the genus, has been utilized for the Late Pleistocene South American species, as previously discussed.

Paula Couto (1955) proposed that all Late Pleistocene *Smilodon* in North and South America were the same species, *Smilodon populator,* with two subspecies, *Smilodon populator populator* in South America and *Smilodon populator californicus* (=*S. p. fatalis*) in North America. This interpretation of the relationship of *Smilodon* in North and South America was supported by Berta (1985).

Contrary to Paula Couto and Berta's interpretation of the relationships of North and South American *Smilodon,* Kurtén and Werdelin (1990) suggested that *Smilodon* on the east side of Andes be referred to *Smilodon populator* and that *Smilodon* on the west side of the Andes is conspecific with the North American *S. fatalis. Smilodon* on the western side of the Andes would then represent a Late Pleistocene migration into South America along with other North American species such as *Canis dirus,* which is also known only from the western side of the Andes. This

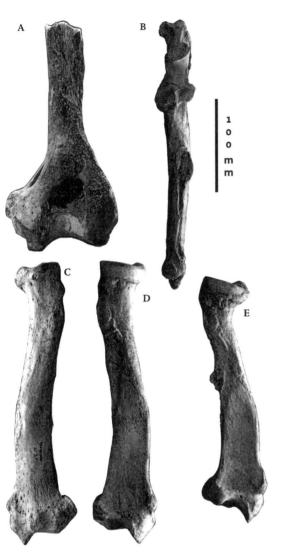

Figure 4.5. *Smilodon populator*. A = distal right humerus, posterior view ZMA 20,044; B = left ulna, anterior view, ZMA 20,035; C–D = left radius, lateral view, medial view ZMA 20,034; E = left radius with exostosis, medial view ZMA 20,041.

interpretation of two species of Late Pleistocene *Smilodon* in South America dispersing into the continent at different times parallels Webb's (1978) suggestion of two alternative savanna corridors, his high and low road scenario, by which the North American species entered South America: the Andean savanna route (high road) and the eastern savanna route (low road). *Smilodon* in Ecuador and Peru, therefore, would have entered via the Andean savanna route, while the ancestor of *Smilodon populator* would have utilized the eastern savanna

route. Support for an earlier dispersal of *Smilodon* into South America is provided by the presence of *Smilodon gracilis* from the Early to Middle Pleistocene El Breal de Orocual asphalt deposits in Monagas state, northeastern Venezuela (Rincón et al., 2011). Consequently, *Smilodon populator* on the eastern side of the Andes evolved from this earlier migration into South America, with subsequent speciation. There is a questionable record of *Smilodon* from 'Uquian' (Marplatan or Early Ensenadan) deposits in Argentina (Berta, 1985), but *S. populator* is well documented beginning in the Ensenadan of Bolivia and Argentina. For a discussion of putative pre-Ensenadan records of *Smilodon*, see Prevosti and Pomi (2007).

Unfortunately, the specimen from Cueva del Milodón does not preserve the skeletal elements, skull and metapodials, utilized by Kurtén and Werdelin to distinguish *Smilodon populator* from *Smilodon fatalis*, thus raising some questions as to its allocation to either species. A characteristic utilized by Kurtén and Werdelin (1990) to distinguish the two species available in the partial skeleton is hind limb proportions. Kurtén and Werdelin note that in *S. populator* the proximal limb elements are long and the distal segments are shortened. In addition, the forelimb is somewhat elongated relative to the hind limb. A plot of relative length of limb segments (Fig. 4.7) shows that the Cueva del Milodón *Smilodon* does not fit the pattern of *Smilodon fatalis*. We here follow Kurtén and Werdelin (1990) in recognizing two species of *Smilodon* in the Late Pleistocene of South America. The specimen from Cueva del Milodón is referred to *Smilodon populator* and is part of the southernmost record of this earlier established population east of the Andes. It does not appear to be a southern extension of the Ecuadorean and Peruvian population along the western coast of South America.

The southernmost record of *Smilodon* utilized by Berta (1985) is from Mar del Plata in Argentina at 38° S. However, since Berta's paper, Prieto et al. (1991) and Borrero et al. (1997) have reported on remains of *Smilodon* from the Lago Sofia 4 cave, which lies only a kilometer or so north of Cueva del Milodón. This, and the material reported here, thus represents the southernmost documented records of *Smilodon* and extends the range of *Smilodon* in South America by more than 1,500 km.

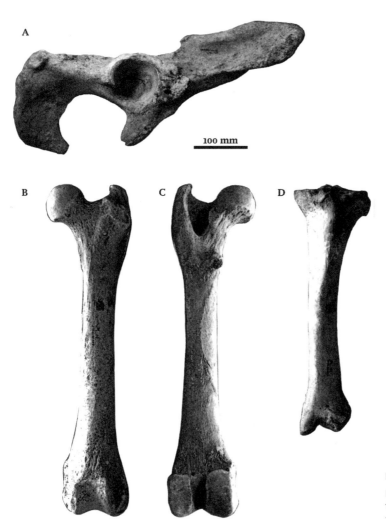

Figure 4.6. *Smilodon populator.* A = right pelvis, lateral view ZMA 20,036; B–C = left femur, anterior view, posterior view ZMA 20,037; D = left tibia, anterior view 20,042.

Bergmann's Rule

Kurtén and Werdelin (1990:159) suggested that *Smilodon* might conform to Bergmann's rule. They note however, that if this were the case, the large size of lower-latitude South American *Smilodon* relative to higher-latitude North American *Smilodon* would be exceptional. The discovery of *Smilodon* from Cueva del Milodón provides an opportunity to test this hypothesis, since at 51°35′ S lat, this partial skeleton is the best specimen from the highest latitude in either hemisphere at which any species of *Smilodon* has been recorded.

The size of *Smilodon* from Cueva del Milodón was compared with *Smilodon* from other South American localities to determine if any relationship existed

between size of individuals and latitude. Such a pattern has been identified in the puma, *Puma concolor*, whose latitudinal range, like that of *Smilodon*, extends across both hemispheres (Kurtén, 1973). It is therefore not unreasonable to expect a similar pattern in *Smilodon*.

Documentation of conformation to Bergmann's rule in *Smilodon* is best accomplished by having all specimens as close in geological age as possible, preferably Late Pleistocene. Although *Smilodon* is known from a number of localities, the overall sample size for each analysis is small, because the same bone may not be available from each locality. Due to the small amount of *Smilodon* material available from sites covering a wide range of latitudes, departure from the ideal was required and some older sites

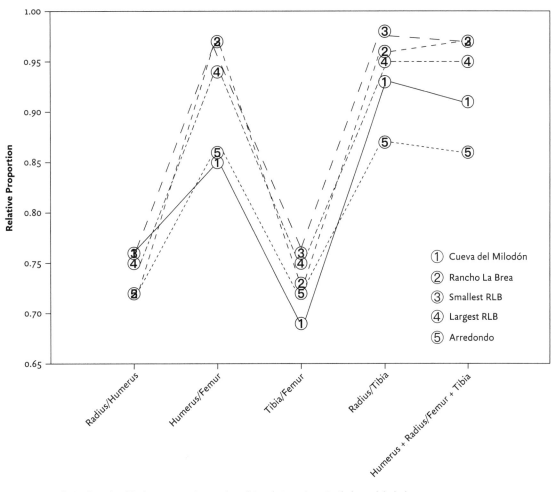

Figure 4.7. Relative length of limb segments in North and South American *Smilodon* as labeled.

in North America were incorporated into the sample. These include American Falls Reservoir, Idaho, Silver Creek, Utah (Miller, 1976), and Ingleside, Texas (Lundelius, 1972), all considered to be Sangamonian in age.

Kurtén (1965) documented a size increase in North American *Smilodon* through the Pleistocene. A comparison of the body mass of the older *S. gracilis* and younger *S. fatalis* by Christiansen and Harris (2005) estimated the body mass of the former to between 55 kg and 100 kg and of the latter to between 160 kg and 280 kg. Shaw and Tejada-Flores (1985) demonstrated an increase in size over time in *Smilodon* at Rancho La Brea, but Meachen et al. (2014) documented a more complex pattern for *Smilodon fatalis* from Rancho La Brea based on the mandible. Their study shows oscillations in mandibular morphology between a small, ancestral-type morph, a larger more derived morph, and an intermediate morph, from different pits at Rancho La Brea.

The general increase in size of *Smilodon* through time and incorporation of older samples into the analysis should weaken the argument for conformation to Bergmann's rule, since individuals from older sites should be smaller than Late Pleistocene individuals for any given latitude. Although considerably older than individuals from more southern latitudes, the specimen from American Falls Reservoir, the northernmost record for North American *Smilodon*, is larger than individuals from younger but more southern latitudes. It might be

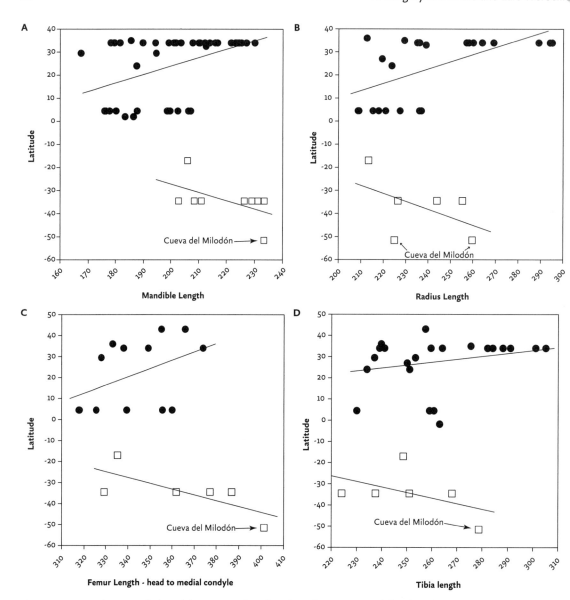

Figure 4.8. Scatter diagrams of selected dimensions of North and South American *Smilodon* plotted against latitude. Circles = *S. fatalis*; squares = *S. populator*. A = length of mandible; B = length of radius; C = length of femur; D = length of tibia.

expected that Late Pleistocene *Smilodon* in Idaho would be even larger. Thus, we do not feel that introducing *Smilodon* of the Sangamonian into the sample obscures the latitudinal patterns present in the data.

Utilizing the maximum length of the mandible and the lengths of limb bones as indicators of size, the small sample size only indicates that there is a general trend of larger individuals at higher latitudes (Fig. 4.8). In most plots, the *Smilodon* from Cueva del

Milodón is larger and more robust than other South American specimens. The range of latitude in the North American sample was more limited, 24° N to 43° N (San Josecito Cave, Nuevo Leon, Mexico to American Falls Reservoir, Idaho), than was available for South America (2° S to 51.5° S). If the samples from Ecuador and Peru are considered to be *Smilodon fatalis* and are included as part of the North American sample, the pattern for that species becomes stronger.

While the samples available are limited in both number of specimens and number of localities (with the *S. fatalis* sample heavily dominated by Rancho La Brea and Talara), and do not provide strong evidence, there is an indication for all variables examined that *Smilodon* conforms to Bergmann's rule. The discovery of additional specimens is obviously needed to confirm the validity of this pattern.

As can be seen in all the graphs, individuals of *Smilodon* from La Carolina, Ecuador and Talara, Peru, the most equatorial populations, are smaller than individuals from the more temperate populations in either North or South American. Whether *Smilodon* on the western side of the Andes is considered to be more closely related to the North American species, as suggested by Kurtén and Werdelin (1990) and herein, or to the South American species as suggested by Berta (1985), cannot be determined from these data. As can be seen in the scatter diagrams, specimens from Ecuador and Peru overlap closely with the North American sample. Yet, if the North American sample is not included in the analysis, then based on size alone, *Smilodon* from Ecuador and Peru could easily be interpreted as a continuation of the trend in the South American sample. As noted by Berta (1985), specific distinctions of wide-ranging carnivores based on size alone are doubtful, and resolution of species identification of *Smilodon* from Ecuador and Peru should be obtained through use of discrete morphological features and not by size-related criteria. While Kurtén and Werdelin (1990) utilized such morphological criteria to distinguish the North and South American species of *Smilodon*, it cannot be ruled out that at least some of the differences, such as those in skull proportions, which they attribute to changes in allometric growth patterns, may reflect size differences due to conformation to Bergmann's rule.

At the same time, the argument presented by Kurtén and Werdelin (1990) regarding the different alignments of the occipital and mastoid planes in *S. fatalis* and *S. populator* have been considerably strengthened by the realization that the morphology of *Megantereon* as reconstructed by them based on the Senèze *Megantereon* figured by Piveteau (1961) is erroneous due to incorrect restoration of the specimen (Antón and Werdelin, 1998). In fact,

Megantereon had an occipital and mastoid morphology matching that seen in *S. fatalis* and the more primitive *S. gracilis*, with this representing the plesiomorphic condition in the clade. The morphology seen in *S. populator* is thus derived and this has significant consequences for posture and perhaps hunting mode in the species. This is, in our opinion, a very strong argument for specific distinction and for attributing the western South American *Smilodon* to *S. fatalis*, since in occipital and mastoid morphology it is identical to *Smilodon* from North America.

Biogeography and Paleoecology

Prior to the discovery and description of *Smilodon* remains from the Magallanes region in Chile (Canto et al., 1991; Prieto et al., 1991; Borrero et al., 1997; this chapter), the latitudinal limits of *Smilodon* in North and South America were not very different—43°N and 38°S, with the difference mainly ascribable to a lack of suitable sites in Argentina south of Mar del Plata. However, now that several finds at 51°S are known besides those from Cueva del Milodón, including Cueva de Medio and Cueva Lago Sofía 4 (Labarca et al., 2008; Prieto et al., 2010) and the recent discovery of *Smilodon* from Tres Arroyos Site 1 on the island of Tierra del Fuego at 53°21′ S (Prevosti et al., 2013), this latitudinal difference is now significantly greater. This difference might also be simply due to the lack of sites in the northern United States and Canada, given the presence of continental glaciers during the Wisconsinan. However, the presence in permafrost localities in Alaska of *Homotherium* and *Panthera* but not *Smilodon* (Alan Cooper, pers. comm. 2014) may indicate that perhaps the environmental conditions suitable for those large predators were less suitable for *Smilodon*. Perhaps conditions north of the known northern limit of *Smilodon* were generally such that this sabertooth was outcompeted by the related, but paleoecologically quite different, *Homotherium* (Martin, 1980; Martin et al., 2000).

Homotherium is currently known only from northern South America at El Breal de Orocual, Venezuela, which is considered to be Plio-Pleistocene in age. However, although Rincón et al. (2011) dismissed a specimen from Uruguay described by Mones and Rinderknecht (2004), it is certainly a Homotheriini although not the genus *Xenosmilus* as proposed by those authors. Given the rarity of Homotheriini in South America it is not possible at this time to determine its geographic range in southern South America or its chronologic range into the Late Pleistocene on that continent, or whether its relative rarity may have allowed *Smilodon* to extend its range to higher latitudes in South America than in the northern hemisphere. In the southern tip of South America, the only other large cat was the jaguar, *Panthera onca mesembrina* (Cabrera, 1934). These possibilities should be explored further, as they may strongly impact our understanding of the paleoecology, environmental requirements, and niche separation of *Smilodon*

from *Homotherium*, as well as with other large cats such as the jaguar, where they overlap in their distribution.

Smilodon is one of four large predators present in the Late Pleistocene fauna preserved in Cueva del Milodón (Table 4.2), the other three being short-faced bear (*Arctotherium* sp.), jaguar, *Panthera onca,* and puma, *Puma concolor.* While the jaguar at the cave represents the large end of the size range for the species (Seymour, 1993), *S. populator* would still have been significantly larger, with an estimated body mass between 220 kg and 360 kg (Christiansen and Harris, 2005), making it comparable in size to the bear, *Arctotherium,* which had an estimated body mass of 400 kg (Soibelzon and Tartarini, 2009).

Table 4.2. Faunal list for Cueva del Milodón Order Xenarthra

Family Mylodontidae
    Mylodon darwinii (= *Grypotherium domesticum, Neomylodon listai, Mylodon listai*)
Order Rodentia
  Family Cricetidae
    Reithrodon physodes
    Akodon sp.
    Phyllotis sp.
    Lagostomus sp.
    Ctenomys magellanicus
Order Carnivora
  Family Felidae
    Panthera onca (= *Yemisch listai, Felis listai, Felis onca mesembrina*)
    Puma concolor
    Smilodon populator
  Family Ursidae
    Arctotherium sp.
  Family Canidae
    Dusicyon avus
    Canis familiaris (= *Canis magellanicus*; introduced by European shipwrecks)
  Family Mustelidae
    Lyncodon patagonicus
Order Perissodactyla
  Family Equidae
    Equus curvidens
    Onohippidium saldiasi
    Hippidium principale
Order Artiodactyla
  Family Camelidae
    Lama guanicoe
  Family Cervidae
    Hippocamelus bisulcus
Order Litopterna
  Family Macraucheniidae
    Macrauchenia sp

Potential prey of *Smilodon* preserved in the cave include the ground sloth, *Mylodon darwinii,* the horses *Equus curvidens, Onohippidium saldiasi,* and *Hippidium principale,* a camelid, *Lama guanicoe,* a deer, *Hippocamelus bisulcus,* and a litoptern, *Macrauchenia* sp. Presumably the ability of the four large predators to coexist reflects their varying specializations with regard to available prey.

Prevosti and Martin (2013) proposed that for the Pleistocene mammalian fauna of Patagonia the large hypercarnivores, *Smilodon* and *Panthera onca,* could prey on most large mammals, while morphology suggests that the short-faced bear was mainly an omnivore that may have scavenged and occasionally hunted medium to large mammals like camelids and horses. Results of a study based on stable isotopes (δ13C and δ15N) by Bocherens et al. (2016), including many of the same taxa from the Pampas of Argentina, are congruent with these interpretations, although they indicate that *Arctotherium* was highly carnivorous, possibly due to having scavenging habits. The carbon and nitrogen stable isotopic values of collagen were measured for skeletal remains from *Smilodon* in the Pampas region of Argentina, ranging in age from 25 to 10 kyr BP. By comparison with similar values obtained on coeval predators, including *Panthera onca,* and potential prey, such as giant ground sloths, glyptodontids, *Macrauchenia, Toxodon,* equids, cervids, and rodents, established that *Smilodon* consumed essentially large prey from open landscapes, such as *Macrauchenia* and giant ground sloths during the last 15,000 years of the Late Pleistocene in the Pampas region. The stable isotopes indicate that *P. o. mesembrina* ate larger proportions of *Hippidion* and *Lama gracilis,* and taphonomic studies show that it may have gnawed bones of *Mylodon, Hippidion,* and camelids. This suggests that these taxa were common prey, and agrees with the ecomorphological and stable isotope interpretations.

Akersten (1985) argued that the enlarged canines of *Smilodon* were utilized for killing juveniles of large, thick-skinned herbivores. Although juvenile proboscideans are commonly utilized as examples, based on the association of juvenile mammoths and *Homotherium* in Friesenhahn Cave (Meade, 1961; Rawn-Schatzinger, 1992; Marean and Ehrhardt, 1995),

ground sloths and other animals of radii of curvature of neck or trunk within certain limits (Andersson et al., 2011) can be considered anatomical and ecological equivalents as potential prey for saber-tooth cats. Considering the list of associated fauna from Cueva del Milodón, it is not unreasonable to speculate that *Smilodon* in the region was feeding predominantly on the young of *Mylodon darwinii.* Examination of some of the collections (Zoological Museum University of Amsterdam, Natural History Museum County of Los Angeles, British Museum [Natural History], Museo de La Plata, Paleontological Institute Museum, University of Zurich, Museo nacional de Historia Natural, Santiago; Malmö Museum) with material from Cueva del Milodón indicates a minimum of 23 individuals of *M. darwinii* based on a count of right mandibles. This sample includes juveniles as well as adults. Given the lack of provenance for most of the collections, it is not possible to determine whether the presence of some of the juveniles in the cave resulted from hunting and subsequent denning activities of *Smilodon* or whether the cave was utilized by female *Mylodon* and juveniles for other purposes, such as a denning site, as such behavior has been proposed for other sloths (Akersten and McDonald, 1991).

Analysis of the plants found in the dung of *Mylodon darwinii* from the cave indicates that habitat in the vicinity of the cave during the Pleistocene was cool, wet sedge-grasslands (Moore, 1978). None of the plant taxa identified in the dung were characteristic of taxa confined to woodland habitat, while all were characteristic of moist, exposed communities. Evidence of trees and shrubs appears above the sloth dung layer and indicates that development of forested habitats in the region was a post-Pleistocene event.

This evidence for an open habitat fits Kurtén and Werdelin's (1990) analysis that the differences in limb proportions (heavier build and longer forelimbs), and hence posture, that distinguish *Smilodon populator* from *Smilodon fatalis* suggest that *S. populator* might have moved with a canter typical of plains animals to a greater degree than did *S. fatalis.* In posture, if not limb length, *S. populator* was more like the clearly open-habitat adapted *Homotherium* than was *S. fatalis.* It is tempting to speculate that, whereas in North

America, *S. fatalis* and *Homotherium* had different habitat requirements and hence different distributions, in South America, *S. populator* may have adapted to more open habitats in the absence of *Homotherium*.

With this said, we should note that the reason for the rarity of *Homotherium* from southern South America is not wholly clear. It may be due to simple chance, or it may, as seems plausible, be due to the absence of suitable habitat for *Homotherium* between northern and southern South America that prevented its dispersal farther south. The types and distribution of Pleistocene habitats are poorly documented or unknown, but likely included significant amounts of tropical forest, as is the case today. This is a habitat type that *Smilodon* and pantherine cats could well have utilized, but it may have excluded *Homotherium*, thus limiting its dispersal into the southernmost part of South America.

This hypothesized adaptation of *S. populator* to an open habitat might also have been expressed in other features, such as pelage color and pattern. Although the fossil record does not often preserve this information, given the unique preservation of soft tissues of other members of the Pleistocene fauna, including skin in Cueva del Milodón, it is not unreasonable to hope that perhaps somewhere in Cueva del Milodón or another cave in the region, skin of *Smilodon populator* is preserved and will someday be found to provide added insight into the ecology of this extinct species.

The causes of the extinction of *Smilodon populator* in South America, as with its counterpart in North America, are still debated. Villavicencio et al. (2015) proposed that extinction of the megafauna in the Última Esperanza region of South America was the result of multiple causes, from both human impacts and climate-vegetation change (see also Metcalf et al., 2016). They note that the balance of factors was taxon specific and competition between humans and large carnivores seems to be the most plausible cause for the extinction of the latter. Based on the radiocarbon chronology of Villavicencio et al. (2015), humans coexisted with extinct horses, extinct camels, and mylodont sloths for several thousand years, which rules out a scenario of blitzkrieg overkill of mega-

fauna by humans. The transition of the vegetation from cold grasslands to *Nothofagus* forests corresponds with the disappearance of *Onohippidion saldiasi* and *Lama* cf. *owenii*, while the subsequent full establishment of *Nothofagus* forests and an increasing fire frequency coincided with the disappearance of the mylodonts. A climate-driven reduction in open environments plausibly reduced herbivore populations, making them susceptible to local extinction, and if *Smilodon populator* was more open-habitat adapted than its North American counterpart, as discussed above, the loss of both prey species and preferred habitat very likely also was a major contributor to the extinction of *Smilodon populator* in the region.

ACKNOWLEDGMENTS

The authors would like to thank Peter van Bree for providing initial access to the Kruimel collection and permission to describe the *Smilodon* material, and for his gracious hospitality during H. G. McDonald's visit. The authors were introduced to the collections at the Zoological Museum by Dick Mol, who also made their trip productive in many ways and has continued to facilitate this project. Dr. Earl Saxon provided H. G. McDonald with E. Nordenskiöld's 1900 paper, translated for him by Dr. L. N. Holmberg.

The authors would also like to express their thanks to the following people who allowed access to specimens in their care: W. A. Akersten, Idaho Museum of Natural History; T. Jefferson and C. A. Shaw, Page Museum; R. W. Graham, Illinois State Museum; W. E. Miller, Brigham Young University Geology Museum; S. D. Webb and G. S. Morgan, Florida Museum of Natural History; K. Seymour, Royal Ontario Museum; W. F. Simpson, Field Museum of Natural History; J. J. Hooker, Natural History Museum, London; K. A. Sorenson and K. Rosenlund, Zoological Museum University of Copenhagen; C. de Muizon, Museum National d'Histoire Naturelle, Paris; A. Friday, Museum of Zoology, University of Cambridge; and A. Cooper, University of Adelaide, who provided information on molecular work on *Smilodon* and *Homotherium*. L. Werdelin is supported by grants from the Swedish Research Council.

REFERENCES

Akersten, W. A. 1985. Canine function in *Smilodon* (Mammalia, Felidae, Machairodontinae). Natural History Museum of Los Angeles County, Contributions in Science, 356:1-22.

Akersten, W. A., and H. G. McDonald. 1991. *Nothrotheriops* (Xenarthra) from the Pleistocene of Oklahoma and paleogeography of the genus. Southwestern Naturalist 36:178-185.

Ameghino, F. 1898. Première notice sur le *Neomylodon listai*, un représentant vivant des anciens édentés gravigrades fossiles de l'Argentine. Imprenta La Libertad, La Plata, Argentina, 8 pp.

Andersson, K., D. Norman, and L. Werdelin. 2011. Sabertoothed carnivores and the killing of large prey. PLoS One 6(10):e24971.

Antón, M. 2013. Sabertooth. Indiana University Press, Bloomington, Indiana, 243 pp.

Antón, M., and L. Werdelin. 1998. Too well restored? The case of the *Megantereon* skull from Senèze. Lethaia 31:158-160.

Berta, A. 1985. The status of *Smilodon* in North and South America. Natural History Museum of Los Angeles County, Contributions in Science 370:1-15.

Bjorkengren, A. G., D. J. Sartoris, S. Shermis, and D. Resnick. 1987. Patterns of paravertebral ossification in the prehistoric saber-toothed cat. American Journal of Roentgenology 148:779-782.

Bocherens, H., M. Cotte, R. Bonini, D. Scian, P. Straccia, L. Soibelzon, and F. J. Prevosti. 2016. Paleobiology of sabretooth cat *Smilodon populator* in the Pampean region (Buenos Aires Province, Argentina) around the last glacial maximum: insights from carbon and nitrogen stable isotopes in bone collagen. Palaeogeography, Palaeoclimatology, Palaeoecology 449:463-474.

Borrero, L. A., J. L. Lanata, and P. Cárdenas. 1991. Re-estudiando cuevas: nuevas excavaciones en Ultima Esperanza, Magallanes. Anales del Instituto de la Patagonia 20:101-110.

Borrero, L. A., F. M. Martin, and Y. A. Prieto. 1997. La cueva Lago Sofia 4, Ultima Esperanza, Chile: una madriguera de felino del Pleistoceno tardío. Anales del Instituto de la Patagonia, serie Ciencias Humanas 25:103-122.

Branco, W. 1906. Die Anwendung der Röntgenstrahlen in der Paläontologie. Abhandlungen der Königlich Preussischen Akademie der Wissenschaften, Berlin. Physikalische Abhandlungen 2:1-55.

Bree, P. J. H. van. 1989. Een Amsterdammer ter walvisvaart in 1909: de reis van J. H. Kruimel naar Zuid Chili. Tijdschrift voor zeegeschiedenis 8:45-66.

Cabrera, A. 1934. Los yaguares vivientes y extinguidos de la América austral. Notas Preliminares del Museo de La Plata 2:9-39.

Canto, J. 1991. Posible presencia de una variedad de *Smilodon* en el Pleistoceno tardío de Magallanes. Anales del Instituto de la Patagonia, ser. Ciencias Sociales 20:96-99.

Cartelle, C., and V. S. Abuhid. 1989. Novos espécimenes Brasileiros de *Smilodon populator* Lund, 1842 (Carnivora, Machairodontinae): morfologia e conclusoes taxonomicas. Anais do XI Congresso Brasileiro de Paleontologia 1:607-620.

Christiansen, P., and J. M. Harris. 2005. Body size of *Smilodon* (Mammalia, Felidae). Journal of Morphology 266:369-384.

Churcher, C. S. 1967. *Smilodon neogaeus* en las barrancas costeras de Mar del Plata, provincia de Buenos Aires. Publicaciones del Museo Municipal de Ciencias Naturales de Mar del Plata 1:245-262.

Guilday, J. E. 1977. Sabertooth cat, *Smilodon floridanus* (Leidy), and associated fauna from a Tennessee cave (40 Dv 40), the First American Bank site. Journal of the Tennessee Academy of Science 52:84-94.

Hauthal, R., S. Roth, and R. Lehmann-Nitsche. 1899. El mamífero misterioso de la Patagonia "*Grypotherium domesticum.*" Revista del Museo de La Plata 9:409-474

Hoffstetter, R. 1952. Les mammifères Pléistocènes de la République de L'Equateur. Mémoires de la Société géologique de France 66:1-391.

Kurtén, B. 1965. The Pleistocene Felidae of Florida. Bulletin of the Florida State Museum, Biological Sciences 9:215-273.

Kurtén, B. 1967. Präriewolf und Säbelzahntiger aus dem Pleistozän des Valsequillo, Mexiko. Quartär 18:173-178.

Kurtén, B. 1973. Geographic variation in size in puma (*Felis concolor*). Commentationes Biologicae 63:1-8.

Kurtén, B., and E. Anderson. 1980. Pleistocene Mammals of North America. Columbia University Press, New York, 422 pp.

Kurtén, B., and L. Werdelin. 1990. Relationships between North and South American *Smilodon*. Journal of Vertebrate Paleontology 10:158-169.

Labarca, R. E., A. Prieto, and V. Sierpe 2008. Sobre la presencia de *Smilodon populator* Lund (Felidae, Machairodontinae) en el Pleistocene tardío de la Patagonia meridional chilenea. I Simposio—Paleontológica en Chile, Libro de Actas 1:131-135.

Lönnberg, E. 1899. On some remains of *Neomylodon listai* Ameghino, brought home by the Swedish expedition to Tierra del Fuego, 1895-1897. Svenska Expeditionen till Magellansländerna 2:149-170.

Lönnberg, E. 1900. On a remarkable piece of skin from Cueva Eberhardt, Last Hope Inlet, Patagonia. Proceedings of the Zoological Society of London 1900:379-384.

Lundelius, E. L., Jr. 1972. Fossil Vertebrates from the Late Pleistocene Ingleside Fauna, San Patricio County, Texas. Bureau of Economic Geology, Report of Investigations No. 77, 74 pp.

Marean, C. W., and C. L. Ehrhardt. 1995. Paleoanthropological and paleoecological implications of the taphonomy of a sabertooth's den. Journal of Human Evolution 29:515-547.

Martin, L. D. 1980. Functional morphology and the evolution of cats. Transactions of the Nebraska Academy of Science 8:141-154.

Martin, L. D., J. P. Babiarz, V. L. Naples, and J. Hearst. 2000. Three ways to be a saber- toothed cat. Naturwissenschaften 87:41-44.

Meachen, J. A., F. R. O'Keefe, and R. W. Sadleir. 2014. Evolution in the sabre-tooth cat, Smilodon fatalis, in response to Pleistocene climate change. Journal of Evolutionary Biology 27:714-723.

Meade, G. 1961. The saber-toothed cat Dinobastis serus. Bulletin of the Texas Memorial Museum 2:23-60.

Merriam, J. C., and C. Stock. 1932. The Felidae of Rancho La Brea. Carnegie Institute of Washington Publication No. 422:1-232.

Metcalf, J. L., C. Turney, R. Barnett, F. Martin, S. C. Bray, J. T. Vilstrup, L. Orlando, R. Salas-Gismondi, D. Loponte, M. Medina, M. De Nigris, T. Civalero, P. M. Fernández, A. Gasco, V. Duran, K. L. Seymour, C. Otaola, A. Gil, R. Paunero, F. J. Prevosti, C. J. A. Bradshaw, J. C. Wheeler, L. Borrero, J. J. Austin, and A. Cooper. 2016. Synergistic roles of climate warming and human occupation in Patagonian megafaunal extinctions during the last deglaciation. Scientific Advances 2:e1501682.

Miller, W. E. 1976. Late Pleistocene vertebrates of the Silver Creek Local Fauna from north central Utah. Great Basin Naturalist 36:387-424.

Mol, D., P. J. H. van Bree, and H. G. McDonald. 2003. De Amsterdamse collectie fossielen uit de Grot van Ultima Esperanza (Patagonië, Zuid-Chili). Grondboor & Hamer 2003:26-36.

Mones, A., and A. Rinderknecht. 2004. The first South American Homotheriini (Mammalia, Carnivora, Felidae). Comunicaciones Paleontológicas del Museo Nacional de Historia Natural y Antropología 35:201-212.

Moodie, R. L. 1923. Palaeopathology: An Introduction to the Study of Ancient Evidence of Disease. University of Illinois Press, Urbana, Illinois, 567 pp.

Moore, D. M. 1978. Post-glacial vegetation in the South Patagonian territory of the giant ground sloth, Mylodon. Botanical Journal of the Linnean Society 77:177-202.

Moreno, F. P. 1899. On a portion of mammalian skin, named Neomylodon listai, from a cavern near Consuelo Cove, Last Hope Inlet, Patagonia. Proceedings of the Zoological Society of London 1899:144-156.

Morgan, G. S., and S. G. Lucas. 2001. The sabertooth cat Smilodon fatalis (Mammalia, Felidae) from a Pleistocene (Rancholabrean) site in the Pecos River valley of southeastern New Mexico/Southwestern Texas. New Mexico Geology 23:130-133.

Nordenskiöld, E. 1899a. La Grotte du Glossotherium (Neomylodon) en Patagonie. Bulletin de la Société géologique de France 29:1216-1217.

Nordenskiöld, E. 1899b. Neue Untersuchungen über Neomylodon listai. Zoologische Anzeiger 22:335-336.

Nordenskiöld, E. 1899c. Meddelande rörande gräfningar i grottorna vid Ultima Esperanza (Södra Patagonien). Ymer 3:265-266. [Swedish]

Nordenskiöld, E. 1900. Iakttagelser och fynd i Grottor vid Ultima Esperanza i sydvestra Patagonien. Kongliga Svenska Vetenskaps-Akademiens Handlingar 33(3):1-23. [Swedish]

Ortiz-Troncoso, O. R. 1980. Inventory of radiocarbon dates from southern Patagonia and Tierra del Fuego. Journal de la Société des Américanistes 67:185-212.

Paula Couto, C. de. 1955. O "tigre-dentes-de-sabre" do Brasil. Conselho Nacional de Pesquisas 1:1-30.

Piveteau, J. 1961. Les carnivores; pp. 641-820 in J. Piveteau (ed.), Traité de paléontologie, Tome 6, L'origine des mammiferes et les aspects fondamentaux de leur evolution, Vol. 1. Masson et Cie, Paris.

Prevosti, F. J., and F. M. Martin. 2013. Paleoecology of the mammalian predator guild of southern Patagonia during the latest Pleistocene: ecomorphology, stable isotopes, and taphonomy. Quaternary International 305:74-84.

Prevosti, F. J., and L. H. Pomi. 2007. Revisión sistemática y antigüedad de Smilodontidion riggii (Carnivora, Felidae, Machairodontinae). Revista del Museo Argentino de Ciencias Naturales 9:67-77.

Prevosti, F. J., F. M. Martin, and M. Massone. 2013. First record of Smilodon Lund (Felidae, Machairodontinae) in Tierra del Fuego Island (Chile). Ameghiniana 50:605-610.

Prieto, A. (ed.) 2013. Cueva del Milodón, publicaciones desde 1899-1996. Ediciones de la Universidad de Magallanes, Punta Arenas, Chile, 318 pp.

Prieto, A., J. Canto, and X. Prieto. 1991. Cazadores tempranos y tardíos en la Cueva 1 del Lago Sofia. Anales del Instituto de la Patagonia 20:75-99.

Prieto, A., R. Labarca, and V. Sierpe. 2010. New evidence of the sabertooth cat *Smilodon* (Carnivora, Machairodontinae) in the Late Pleistocene of southern Chilean Patagonia. Revista Chilena de Historia Natural 83:299-307.

Rau, J., and J. Yañez. 1980. Cricétidos fósiles de la Cueva del Milodón, Chile. Noticiario Mensual 24:9-10.

Rawn-Schatzinger. V. 1992. The Scimitar Cat *Homotherium serum:* Cope, Osteology, Functional Morphology, and Predatory Behavior. Illinois State Museum Reports of Investigations No. 47, 80 pp.

Rincón, A. D. 2003. Los mamíferos fósiles del Pleistoceno de la Cueva del Zumbador (Fa. 116), Estado Falcón, Venezuela. Boletín Sociedad Venezolana Espeleología 37:18-26.

Rincón, A. D. 2006. A first record of the Pleistocene saber-toothed cat *Smilodon populator* Lund, 1842 (Carnivora, Felidae, Machairodontinae) from Venezuela. Ameghiniana 43:499-501.

Rincón, A. D., F. J. Prevosti, and G. E. Parra. 2011. New saber-toothed cat records (Felidae, Machairodontinae) for the Pleistocene of Venezuela, and the great American biotic interchange. Journal of Vertebrate Paleontology 31:468-478.

Roth, S. 1902. Nuevos restos de mamíferos de la Caverna Eberhardt en Ultima Esperanza. Revista Museo de La Plata 11:1-18.

Salmi, M. 1955. Additional information on the findings in the Mylodon Cave at Ultima Esperanza. Acta Geographica 14:314-333.

Saxon, E. C. 1979. Natural prehistory: the archaeology of Fuego-Patagonian ecology. Quaternaria 21:329-356.

Seymour, K. E. 1993. Size change in North American Quaternary jaguars; pp. 343-372 in R. A. Martin and A. D. Barnosky (eds.), Morphological Change in Quaternary Mammals of North America. Cambridge University Press, Cambridge, UK.

Shaw, C. A., and A. E. Tejada-Flores. 1985. Biomechanical implications of the variation in *Smilodon* ectocuneiforms from Rancho La Brea. Natural History Museum of Los Angeles County, Contributions in Science, 359:1-8.

Slaughter, B. H. 1963. Some observations concerning the genus *Smilodon*, with special reference to *Smilodon fatalis*. Texas Journal of Science 15:68-81.

Soibelzon, L. H., and V. B. Tartarini. 2009. Estimación de la masa corporal de las especies de osos fósiles y actuales (Ursidae, Tremarctinae) de América del Sur. Revista del Museo Argentino de Ciencias Naturales "Bernardino Rivadavia," 11:243-254.

Studer, T. 1905. Ueber neue funde von *Grypotherium listaei* Amegh in der Eberhardtshöhle von Ultima Esperanza. Neue Denkschrifte der allgemeine schweizerischen Gesellschaft für die gesammten Naturwissenschaften 40:1-18.

Sutcliffe, A. J. 1985. On the Track of Ice Age Mammals. Harvard University Press, Cambridge, Massachusetts, 224 pp.

Tonni, E. P., A. A. Carlini, G. J. Scillato Yané, and A. J. Figini. 2013. Cronología radiocarbónica y condiciones climáticas en la "Cueva del Milodón" (sur de Chile) durante el Pleistoceno Tardío. Ameghiniana 40:609-615.

Villavicencio, N. A., E. L. Lindsey, F. M. Martin, L. A. Borrero, P. I. Moreno, C. R. Marshall, and A. D. Barnosky. 2015. Combination of humans, climate, and vegetation change triggered Late Quaternary megafauna extinction in the Última Esperanza region, southern Patagonia, Chile. Ecography 39:125-140.

Webb, S. D. 1978. A history of savanna vertebrates in the New World. Part 2: South America and the great interchange. Annual Review of Ecology and Systematics 9:393-426.

Wellman, R. W. 1972. Origen de la Cueva del Mylodon en Ultima Esperanza. Anales del Instituto de la Patagonia 3:97-101.

Woodward, A. S. 1900. On some remains of *Grypotherium* (*Neomylodon*) *listai* and associated mammals from a cavern near Consuelo Cove, Last Hope Inlet, Patagonia. Proceedings of the Zoological Society of London 1900:64-79.

5 *Smilodon* from South Carolina: Implications for the Taxonomy of the Genus

JOHN P. BABIARZ, H. TODD WHEELER,
JAMES L. KNIGHT, AND LARRY D. MARTIN

Introduction

Until recently, remains of the sabertooth cat *Smilodon fatalis* were uncommon in South Carolina. Such fossils would periodically show up along with other Pleistocene material, but because they were river finds, stratigraphic provenience was never reliable. We report the discovery of at least five individuals, including two crania and two neurocrania, four left hemimandibles and one right hemimandible, along with associated postcranial elements, belonging to *Smilodon fatalis*. A morphologic comparison has been made of these recently discovered South Carolina specimens to other known Smilodontini of North America. This chapter discusses how these new specimens suggest that they are part of a single anagenetic lineage. The presence of a substantial number of these hypercarnivorous felids may have been the result of a carnivore trap or some other natural occurrence that brought all these individuals to a single burial site.

During routine mining operations at a South Carolina commercial limestone quarry, a large, associated late Irvingtonian megafauna, including *Smilodon fatalis,* was discovered in a channel fill deposit. At least five individual specimens of this large dirk-toothed carnivore were uncovered in one small area. Due to the large numbers of the camelids *Hemiauchenia macrocephala* and *Paleolama mirifica* present, the locality, near Harleyville, in Dorchester County, was named the Camelot Local Fauna by the collecting party from the South Carolina State Museum. This site has produced the first substantial sample defining the *Smilodon fatalis* chronocline that in our view ends as *Smilodon californicus* (or *S. f. californicus*) of the asphaltic deposits that functioned as carnivore traps, such as Rancho La Brea (RLB) in Los Angeles, California.

Overview of the Fauna

Along with *Smilodon*, two additional carnivores are known from the site, the cheetah-like felid, *Miracinonyx inexpectatus,* and the wolf, *Canis armbrusteri.* The majority of carnivore remains at the site belong to *Smilodon fatalis,* however. Artiodactyls are represented by the camelids *Hemiauchenia macrocephala* and *Paleolama mirifica,* and a deer, *Odocoileus virginianus* (Knight and Martin, 2005, 2007). *Hemiauchenia macrocephala* has been identified from the Leisey Shell Pit locality, Florida, which dates to the early Irvingtonian (Morgan and Hulbert, 1995). Perissodactyls include the tapir, *Tapirus veroensis,* and one as yet undetermined species of *Equus.* The sloth, *Megalonyx jeffersonii,* is present and is usually considered to be indicative of Late Pleistocene faunas (Hulbert, 2001). If assignment to this species is correct, its occurrence in the Camelot fauna would mean an earlier appearance of the species. The presence of the tapir *T. veroensis,* which is considered to be Middle to Late Pleistocene in age (Baskin and Thomas, 2007) would place an additional time constraint on the fauna. Based on carbon and oxygen isotope analyses of samples taken from the teeth of both carnivores and herbivores found associated at this site, the area appears to be a marginal woodland-grassland habitat (Kohn et al., 2005).

Taxonomic History

Smilodon is perhaps the best-known large carnivore of the Cenozoic and is well established in the literature. Specimens of this lineage have been collected from numerous fossil deposits of varying ages throughout both North America and South America, beginning with *Rhizosmilodon fiteae* from the Late Miocene Upper Bone Valley Formation, Florida (very late Hemphillian) (Wallace and Hulbert, 2013) and terminating with *Smilodon fatalis* (= *S. californicus*) in North America (Merriam and Stock, 1932; Woodard and Marcus, 1973) and *S. populator* in South America (Lund, 1842).

Kurtén and Werdelin (1990) addressed the taxonomy and systematic relationships of the genus. Using osteometric and odontometric analysis, they concluded that *Smilodon populator* was endemic to South America, existing east of the Andes, and that the *Smilodon* population from Talara, Peru was a late intrusion of *Smilodon fatalis* from North America. They recognized only two species as being endemic to North America: *Smilodon gracilis* (early Irvingtonian North American land mammal age [NALMA]) and the wide-ranging late Irvingtonian through Rancholabrean NALMA (Middle to Late Pleistocene) species, *S. fatalis*.

The Camelot *Smilodon fatalis* sample is intermediate in size and other features between the older *S. gracilis* and the more derived specimens from Rancho La Brea (here referred to as *S. californicus*). For example, *S. californicus* lacks a p3, while *S. fatalis* has a single-rooted p3. In *S. gracilis* this tooth is double-rooted. This morphocline/chronocline and size difference, formerly ascribed to variability in *Smilodon*, might be viewed as specific to distinct species in light of the limited individual variation seen within the Camelot sample. The number of individuals from this site provides a new perspective on the variability in this taxon. A reappraisal of the

synonymy of *S. californicus* with the late Irvingtonian, medium-sized *S. fatalis* from Harleyville, South Carolina may be appropriate. Clearly *S. californicus* is valid at some level; the issue is what level this might be. Because we consider the type specimen of *Smilodon floridanus* doubtfully diagnostic, we use *S. californicus* in this chapter to distinguish between the more derived sample attributed to RLB from the late Irvingtonian population (*S. fatalis*). A detailed taxonomic comparison based on osteological features of the skull, mandible, and dentition follows.

ABBREVIATIONS

Abbreviations for skull and mandible measurements follow Kurtén and Werdelin (1990) and Martin and Czerkas (2000). These are as follows: Skull: **BL,** basal length (prosthion-basion); **CBL,** condylobasal length (prosthion-condylion); **PL,** palatal length (prosthion-staphylion); **ZB,** bizygomatic width; **I-I,** width between outer faces of third incisors; **C-C,** rostral width across canines; **P-P,** width between outer (buccal) faces of carnassials; **MB,** bimastoid width; **CB,** bicondylar width; **AP,** anteroposterior diameter, **IOB,** interorbital width; **POP,** width across postorbital processes; **POC,** width of postorbital constriction. **BB,** breadth of braincase.

Abbreviations for the mandible: **JLI1,** length of jaw from anterior face of i1 (including bone) to posterior end of condyle taken at center; **JL,** length of jaw from anterior face of canine to middle posterior edge of condyle (condyloid process); **HPC,** height from upper (posterior) face of condyle to base of angle; **HPCP,** height from upper (posterior) face of condyle to top of coronoid process; **DM1,** depth of ramus behind M^1 carnassial; **DP3,** depth of ramus anterior to P3; **ML,** mediolateral.

Abbreviations for the teeth: Measurements based on Hillson (1986), Slater and Van Valkenburgh (2008), and Naples et al. (2011). **APD,** anteroposterior distance; **LLD,** labial lingual distance.

Museum collections: **AMNH,** American Museum of Natural History, New York, New York; **BIOPSI,** Babiarz Institute of Paleontological Studies, Mesa, Arizona; **FMNH,** Florida Museum of Natural History, Gainesville, Florida; **KUVP,** University of Kansas Vertebrate Collection, Lawrence, Kansas; **LACMHC,** Natural History Museum of Los Angeles

County, Hancock Collection, Los Angeles, California; **SCSM,** South Carolina State Museum, Columbia, South Carolina; **UF,** University of Florida, Gainesville, Florida; **USNM,** United States National Museum of Natural History, Washington, D.C.; **NHMB,** Natural History Museum, Basel, Switzerland. Casts utilized (CB and BC) are from Bone Clones, Canoga Park, California.

Other abbreviations: **RLB,** Rancho La Brea; upper dentition designated by upper case letters, lower dentition by lowercase letters; **AP,** anteroposterior; **ML,** mediolateral.

Materials and Methods

Based on skull and hemimandible elements, the Camelot *Smilodon fatalis* sample represents a minimum of five specimens. One cranium (SCSM 2004.1) was found intact and another three (SCSM 2003.75.155–157) are represented by the neurocranium; in addition, a fifth left maxilla was found associated with SCSM 2003.75.155 in the field jacket. (See appendix for measurement tables.) Hemimandibles representing five individuals were also recovered. Two of these, specimens SCSM 2003.75.12 and SCSM 2003.75.162, are similar in dimensions and degree of dental wear, and may belong to a single individual, but symphyseal fit between them is not definite. Thus, the minimum number of individuals based on hemimandibles is five. Assorted postcranial skeletal elements were found associated with the cranial elements, but neither appendicular nor axial elements were articulated. Consequently, individual elements cannot be assigned to any one individual cranium or mandible. Neither axial nor appendicular elements ascribed to *Smilodon fatalis* are described here; our focus is on the description of the skull, mandible, and dentition.

The Camelot *S. fatalis* specimens were first compared to the holotype of *S. fatalis* (AMNH FM 10395) (Fig. 5.1), then to other homologous elements of *Smilodon*, including *S. gracilis* (UF87238, UF87236, UF84187, UF 206593) from the early Irvingtonian Leisey Shell Pit site, Hillsborough County, Florida, and *S. californicus* specimens KUVP 98124 and LACMHC (2001-249; 2002-L&R-250 and 2002-L-729) all from Pit 67 at Rancho La Brea. The two California *Smilodon* skulls are considered males, based on a CBL

in excess of 300 mm (Christiansen and Harris, 2012). Two replica casts of *Megantereon cultridens*, CB-20 and BC-106, were utilized, in addition to *M. cultridens* (NHMB SE311) from Senèze, France (Christiansen and Adolfssen, 2007). Some additional comparisons were made with extant jaguar (*Panthera onca*). All measurements were taken from original material when possible. Otherwise, casts of original material were measured, and some measurements were extracted from published literature. Measurements were acquired using digital calipers.

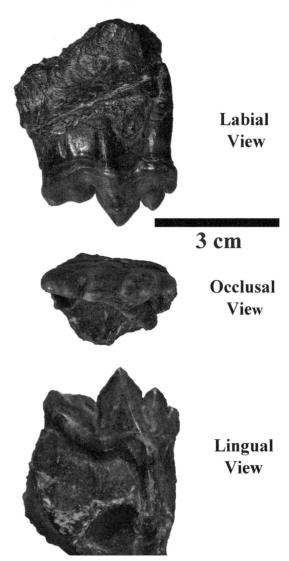

Labial View

3 cm

Occlusal View

Lingual View

Figure 5.1. Holotype of *Trucifelis fatalis* Leidy, 1868 (= *Smilodon*), AMNH FM 10395, American Museum of Natural History. Photograph by Henry Galiano. Licensed from AMNH, New York.

Description of Camelot Material

Skull

The complete *Smilodon* skull (Figs. 5.2 and 5.3) and partial crania (Fig. 5.4) from the Camelot fauna are intermediate in size when compared with *Smilodon gracilis* and the California sample of *S. californicus* (Tables 5.A1 and 5.A2). With a CBL of 247 mm for SCSM 2003.75.154 and 253 mm for SCSM 2004.1.1, the Camelot skulls are smaller than the observed range of 271-344 mm for the Rancho La Brea sample of *Smilodon* (Merriam and Stock, 1932). With an average CBL of close to 307 mm, the California sample is much larger than the earlier and more primitive taxon from Camelot. With a CBL of 237 mm, the

UF 206593 *Smilodon gracilis* is closer in overall size to the Camelot sample. The CBL for *Megantereon cultridens* specimens (CB-20 and BC-106) were 240 mm and 243 mm, respectively.

In anterior view, the nasal cavity is heart shaped in *Smilodon gracilis*. The Camelot skull (Fig. 5.3C) is similar in shape but with a dorsoventral increase in elevation. As expected, the nasal cavity in the RLB sample is larger in this respect by approximately 8%. In anterior view, the combined width of the nasals in *S. gracilis* is greater than the mediolateral width of the nasal aperture. In the Camelot sample, in anterior view, the anterolateral edge of the nasal bones curves laterally, along the premaxilla-nasal suture, yet the ML width of the nasal opening

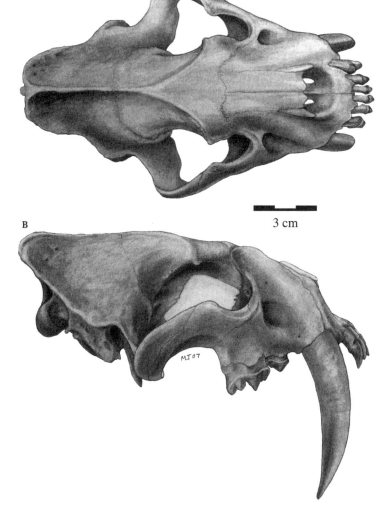

3 cm

Figure 5.2. Restored skull of *S. fatalis*, Camelot sample, SCSM 2004.1. A = dorsal view; B = right lateral view. Drawings by Mary Tanner.

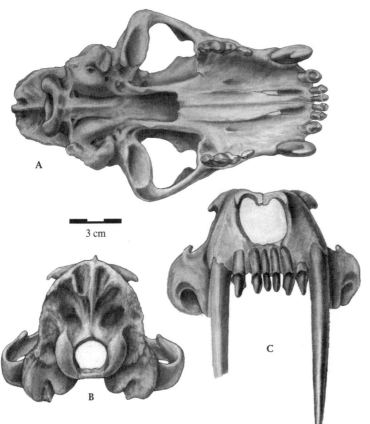

Figure 5.3. Restored skull of *S. fatalis*, Camelot sample, SCSM 2004.1. A = ventral view; B = posterior view; C = anterior view. Drawings by Mary Tanner.

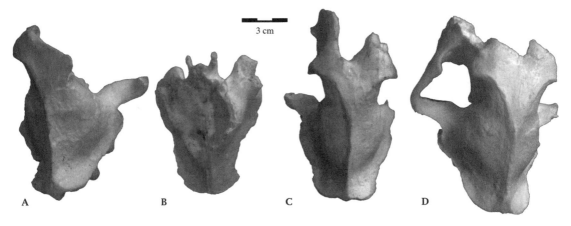

Figure 5.4. Crania of *S. fatalis* Camelot sample. A = SCSM 2003.75.154; B = SCSM 2003.75.155; C = SCSM 2003.75.157; D = SCSM 2004.1. All in dorsal view. Photograph by Brad Archer.

remains slightly narrower than the combined ML width of the nasal bones. In *S. californicus* the nasals, although much larger, no longer curve laterally but have a straight anteroposterior line along the premaxillary-nasal suture. Here, the ML width across the nasal aperture is now greater than the combined width of the nasal bones. This larger nasal cavity may suggest an improved sense of smell in the California sample.

Anteroventral of the infraorbital foramen there is a slight depression in the maxilla, thought to be a muscle attachment area providing additional leverage for mandible closure at high gape (Naples and Martin, 2000). This derived feature is more prominent in the Camelot sample than the KUVP 98124 specimen from RLB. In *Smilodon californicus* the nasolacrimal canal is very large compared to the Camelot sample, and there are several smaller foramina present just caudal and ventral to this canal that are not present in the Camelot sample. The nasolacrimal canal is ventrally bordered by a bony ridge that forms the dorsal aspect of the large interior opening of the infraorbital foramen. In *S. californicus* this horizontal bony ridge runs caudally and slightly ventrally, terminating its ventral progression just anteroposterior to the sphenopalatine foramen. From this ventral point, the bony ridge then continues horizontally and caudally toward the optic foramen and orbital fissure. In lateral view, the anterior opening of the infraorbital foramen in *Smilodon* aligns with the anterior margin of the P4 paracone, whereas in *Panthera*, this anterior opening aligns slightly more rostrally, between the anterior margin of P4 and posterior margin of P3. In the Camelot *Smilodon fatalis*, the tear duct canal is much smaller and the bony ridge runs ventrally from the canal and then curves caudally above the sphenopalatine foramen toward the optic foramen, similar to the California sample in that respect. The sphenopalatine foramen in *S. californicus* is much larger than in the Camelot sample.

The dorsal aspect of the frontal continues posteriorly to the postorbital processes, turns abruptly medially for about 20 mm in both *Smilodon gracilis* and Camelot *S. fatalis* (Fig. 5.2A), compared with 40 mm in the *S. californicus* sample, and then caudally, forming a medial arching crest along the frontal-parietal suture. These two crests combine dorsally and form the sagittal crest of the parietal. The sagittal crest continues posteriorly on a horizontal trajectory, where it bifurcates at the interparietal suture and then forms the lateral edges of the lambdoidal ridge. The varied shapes of the sagittal crest found in the same taxa may reflect the relative age of the individual (Duckler, 1997; García-Perea, 2002). In addition, rugose ridges may develop along the sagittal crest due to stresses created by the action of the *M. temporalis*. The greater the stress, perhaps the greater the possibility that convoluted ridges may form, as seen in *Xenosmilus hodsonae* (Martin et al., 2000).

The posterior aspect of the skull (Fig. 5.3B) is formed by the parietal bone (including the interparietal bone and lambdoidal ridge). This area extends considerably past the occipital condyles in all *Smilodon* species. In *Smilodon gracilis*, this caudal extension of the parietal is about equal in development to that of *Panthera onca*. The parietal continues to extend posteriorly in the Camelot sample, with the greatest extension found in the RLB sample. In lateral view, the sagittal crest of *Smilodon gracilis* is short and barely rises over the parietals, being slightly taller in its posterior extent as it approaches the lambdoidal crest. In the Camelot sample the sagittal crest is visibly taller than in *S. gracilis* but is much shorter when compared to the California sample. Large robust skulls with stout sagittal crests are sometimes referred to as males of the RLB species (Christiansen and Harris, 2012), but with the limited sample from the Camelot site, sex could not be ascertained. A measurement from the apex of the postorbital constriction to the junction of the two medial temporal crests is approximately 25 mm in *S. gracilis*, 30 mm in the Camelot sample, and 35–38 mm in the RLB sample. Interestingly, the distance from the juncture of the parietal crests to the occiput is the same in both *S. gracilis* and the Camelot sample, yet the distance from the parietal crest juncture to the postorbital constriction increases by approximately 5 mm from the one sample to the other. Again, this extension of both the frontal and nasal bones may be the result of a continued development of the sinus cavities within the Camelot population. Although the sinus cavity increase (extension of the nasal-frontal area) in the Camelot sample is somewhat greater when compared to *S. gracilis*, it does not appear to be much greater in the RLB sample. The most noticeable change to the *Smilodon* skull condition, in dorsal view, is the posterior extension of the parietal. In *Smilodon*, posterior cranial enlargement, represented by the larger temporalis musculature, remained a critical factor for survival. As noted (Emerson and Radinsky, 1980; Van Valkenburgh and Ruff, 1987), skull morphology adapted to the increasing gape needed to accommodate hypertrophied canines. To ensure survival, as prey size increased during the Late Pleistocene, effective means to secure such prey animals required adaptive modification.

Over time, morphological adaptations in skull and mandible shape were affected by the increase in upper canine length. In order to reach high gape angles to clear the hypertrophied sabers, modifications in both musculature and positive allometry were required. Some of these features included dorsally arched zygomatic arches, inclined posterior extension of the parietal, adjustments to the size and position of the mastoid process, and a reduction in height of the coronoid processes of the mandible. In addition, with the evolutionary changes to the upper canine, there simultaneously occurred a reduction in the height and width of the lower canines. Tooth size reduction also included upper and lower incisor arcades. This cause and effect were most likely the result of a diminished function in the lower anterior dentition along with the upper incisors. In addition, lengthening of the upper canine resulted in elongation and upward rotation of the face together with ventral displacement of the glenoid process. (Slater and Van Valkenburgh, 2008).

Increase in body size may have been more important than brain size during the evolution of the Smilodontini. The transverse measurement (BB) of the *Smilodon gracilis* cranial vault is approximately 80 mm at its widest point. The Camelot specimen braincase width is also within this range, approximately 83 mm. In the RLB sample, the cranial vault remains close to the same width, about 87 mm for LACMHC 2001-249 and about 97 mm for the larger KUVP 98124, despite these RLB skulls having a ca. 33% increase in overall (CBL) length. Thus, over an

interval in excess of one million years, the breadth of the cranial vault remains almost static, or reasonably close to the same size during this dramatic increase in overall skull size.

The parietals in the adult Camelot sample are totally fused all along their ventral contact with the squamosal, including the dorsal attachment to the frontal and the somewhat vertically oriented attachment to the supraoccipital (lambdoidal crest). The dorsal suture runs almost horizontally on the dorsal aspect of the frontal, forming the sagittal crest, but the ventral connection forms a mild, anteroposteriorly directed curve with the apex of the curve positioned directly above the external auditory meatus. This curved suture is very similar to that of *Panthera onca*. In the RLB sample of *Smilodon*, on the other hand, the suture connecting the squamosal to the parietal is almost parallel to the sagittal crest, forming a somewhat rectangular parietal in lateral view (Merriam and Stock, 1932). The comparative sample of *Smilodon gracilis* was obscured in this area and cannot be compared to the latter sample. As the entire rear portion of the skull of the Senèze specimen of *Megantereon cultridens* (NHMB SE-311) is artificial (Antón and Werdelin, 1998; Christiansen and Adolfssen, 2007), no comparisons posterior to the POP are possible.

In dorsal view (Fig. 5.2A) the posteriormost end of the supraoccipital in *Smilodon* forms the lambdoidal crests, and its ventral extent contacts the mastoid processes on the lateral sides of the skull. The shape of the lambdoidal crest in dorsal view varies, especially with skull size and age. The size of the temporal muscle, along with the constant strain from the muscle's action, forms a variable wave along both lateral edges in lateral view. This shape can be uniform, but in many cases the crests are asymmetrical. This wave is often apparent in the sagittal crest, again from the action of *M. temporalis* (Duckler, 1997). The lambdoidal crests can also vary in thickness. Mediolateral width across this crest is actually somewhat greater in *S. gracilis* than it is in the Camelot sample, but the thickness appears to be identical. In the Camelot sample the posterior aspect of the crests forms a mildly medial curve ending just above the paramastoid. In the KU *Smilodon* skull, the lambdoidal crests have the same curve but are larger than any other *Smilodon* examined and actually have a thin, lateral expansion about midway toward the mastoids. Along the ventral extent and at the termination of the lambdoidal crests, the lateral edges fold horizontally and slightly cup at their base, supporting the temporal muscle. This fold continues rostrally until it meets the malar. The temporal bone expands laterally, forming the infratemporal fossa. Its ventral and horizontal flare contacts the jugal and terminates just below and caudal to the rounded, vertical projection of the jugal below the orbit. Due to the arched and medially compressed zygomatic arch, this fossa is mediolaterally and anteroposteriorly small in the Camelot *Smilodon*; and smaller still in *S. gracilis*. Due to the overall larger skull in the Rancho La Brea sample, this fossa is correspondingly larger than in the Camelot *S. fatalis* sample.

In ventral view (Fig. 5.3A) the temporal (squamosal) forms the glenoid fossa, which articulates with the mandibular condyle to form the temporomandibular joint (TMJ). There is a very tiny postglenoid foramen just medial and dorsal to the glenoid fossa, as described for *Smilodon* (Merriam and Stock, 1932; Berta, 1987; Hulbert, 2001) but this foramen is barely visible in the Camelot sample. This feature is unlike the larger and more lateral postglenoid foramen seen in the American Oligocene nimravids. The anterior aspect of the glenoid fossa forms a short lip that curves ventrally and runs mediolaterally, with its deepest portion beginning with the lateral edge. The reverse is true on the opposite side of the fossa, where the posterior aspect of the lip extends further ventrally than the anterior lip, but with its greatest depth beginning on the medial end and tapering off toward the lateral end of the fossa. In all *Smilodon* specimens, in lateral view, the posterior lip extends ventrally below the mastoid process, with the least depth found in *S. gracilis,* where the lip is almost even with the mastoid process. There is an increase in size of the ventral extension of the posterior glenoid lip below the mastoid process from *S. gracilis* to the Camelot sample, with the greatest depth found in the California sample. This development of the fossa would facilitate a larger gape, necessary to clear the elongated upper canines. The anterior aspect of both the Camelot and California samples is about equal in size and shape, but the

distal lip is much longer mediolaterally and encompasses a larger surface area of contact with the mandibular condyle in the California sample. This additional ventral surface area of the TMJ most likely indicates a greater gape and may have prevented mandibular dislocation during any ventral hyperextension of the mandible. Mediolaterally, in the Camelot *Smilodon* this ventral lip is as deep as in the California sample, but is about 60% shorter.

The characters found in the mastoid processes and bullae of the Camelot sample of *Smilodon fatalis* are distinct from both *Smilodon gracilis* and *Smilodon californicus* regardless of the obvious increase in size over time. In *S. gracilis*, the tympanic bullae are fully ossified as they are in all Felidae, and the anteroposteriorly oriented body of the bullae is more or less parallel to the basioccipital and just slightly lateral to the pterygoid wings in ventral view. The bullae in the Camelot *S. fatalis* sample (Fig 5.3B) show an increase in size and incline ventrally, almost to a level equal to the ventral extent of the mastoid process when compared to *S. gracilis*. The bullae are no longer parallel to each other, as in *S. gracilis*, but extend medially, encroaching on the basioccipital plane. In *S. californicus*, there is the obvious size increase, but the medial bulge and orientation of the bullae increases medially and lies just medial to the pterygoid processes and nearly level with the distal end of the pterygoid wings in ventral view.

The mastoid process (Fig. 5.2B) is a bony knob extending from the lower portion of the temporal bone. Caudally the mastoid process terminates in a bony protuberance, the paroccipital process. The mastoid and paroccipital processes serve as the attachment point for four muscles used for head and mandible manipulation (Antón and Galobart, 1999). In *Smilodon gracilis*, the anteroventral portion of the mastoid process is very rugose, and the entire mastoid is separated from the bulla by a narrow cleft that runs caudally almost to the distal end of the bulla (Berta, 1995). This rugose ventral portion of the mastoid is the attachment point for the *M. brachiocephalicus* (Antón and Galobart, 1999). It is mediolaterally and obliquely grooved and terminates medially in alignment with the medial edge of the postglenoid process. In the Camelot sample, this oblique groove begins to extend medially past the postglenoid alignment but remains anteroposteriorly in alignment with the lateral edge of the bulla. In the Camelot *Smilodon*, the distal end of the mastoid begins to coalesce with the bulla, separated from it by a shorter groove, in which the stylomastoid and tympanohyal foramina are located. In *S. californicus*, this bulla-mastoid coalescence is extensive and results in only the large stylomastoid and tympanohyal foramen separating the two features. In the most derived California sample, in ventral view, this medial excursion of the mastoid begins to fold over the proximal end of the bulla, and it terminates about midway between the medial edge of the postglenoid process and the pterygoid wings. Proximally the mastoid process is only a millimeter or two from making contact with the postglenoid process in the California sample. In the jaguar, the entire mastoid is much smaller than in the Camelot *Smilodon* and compares more closely in size with that of *S. gracilis*, although it is more vertically oriented and caudally about midway from the center of the bulla when viewed laterally.

In *Smilodon gracilis*, just medial to the bulla and along most of its entire anteroposterior length there is a groove that is bordered laterally by the basioccipital. The posterior end of this channel culminates at the large jugular foramen (combined posterior lacerate foramen and anterior condyloid foramen = hypoglossal foramen). The condyles and foramen magnum of *S. gracilis* and *Panthera onca* are about equal in size and shape, although the smooth, articulating surface in *S. gracilis* is slightly broader. This lateral expansion of the articulating surfaces would allow for greater lateral rotation of the head. The entire dorsal surface area of the condyles in the Camelot sample is somewhat broader than in *S. gracilis* and in addition there is a slight vertical increase in the length of the condyle. This would allow the Camelot *Smilodon* greater elevation of the head than in *S. gracilis*, further reflecting the modifications taking place in a larger skull with longer sabers. This lateral expansion along with both dorsal and ventral articular surface length is greater in *S. californicus*, again, a reflection of a large skull with even larger, hypertrophied upper canines. Although the overall size of the condyles in *Smilodon* has increased over time, the opening for the foramen

magnum has remained nearly constant in size, reflecting the minimal increase in size of the cranial vault.

The posterior narial opening in *Smilodon gracilis* is small, being about 18 mm in both width and height. In the Camelot *S. fatalis* sample (Fig. 5.3A) the depth of the posterior narial opening is slightly greater than the width, as measured from the level of the presphenoid. In *S. gracilis*, the anterior border of the narial opening is situated posterior to the posterior border of M1 by about 10 mm. This distance increases to approximately 15 mm in the Camelot sample and eventually measures 22-23 mm in the *S. californicus* sample.

Rostrally from the narial aperture, moderately elevated palatine ridges begin at the anterior end of the posterior palatine canal. They begin to coalesce, forming a triangle as they extend rostrally and ending at the posterior border of the incisive foramina. Christiansen and Adolfssen (2007) state that the palatine ridges are not present in *Megantereon cultridens*. However, although there is some lateral compression of the palate in both CB-20 and BC-106, the palatine ridges are obviously present in this sample of *M. cultridens*. In *Smilodon gracilis*, despite damage to the palate, the posterior end of both ridges seems to originate closer to the anterior border of the narial aperture. The Camelot *Smilodon* shows the palatine ridges to be present, although they are not as robust as in *S. californicus*. In the Camelot sample, the palatine ridges appear as moderately heavy, rounded ridges. As in *S. californicus*, they form close to the narial aperture then gradually become prominent sharp ridges rostrally, again terminating at the posterior borders of the incisive foramina. In *S. fatalis*, lateral to both of these ridges, and beginning at the posterior palatine canal, there is a slight recessed trough, which terminates at the posterior margins of the incisive foramina. This concave groove originates and terminates at identical positions in all *Smilodon*, but the trough is deeply grooved just medial to the canines in *S. californicus* as compared to *S. fatalis*. The area between these two palatine ridges is recessed and forms a concave sulcus that gradually rises as the ridges converge, eventually forming the medial edge of the two incisive foramina. In the California sample, the two ridges are accentuated to the extreme, as compared to the Camelot and *S. gracilis* sample. A third longitudinal ridge actually develops in the California sample, between the two median ridges.

Distal to the narial aperture, the palate terminates caudally with the thin pterygoid wings (processes). These form the lateral margins or walls of the nasopharynx and extend ventrally about to the level of the glenoid process in the Camelot *Smilodon*. In the California sample, the pterygoid processes arch ventrally almost to the level of the postglenoid process. This area cannot be described due to damage in the *S. gracilis* sample. In all *Smilodon*, the posterior palatine canal is situated just posterior to the protocone of P4 and midway along the maxilla-palatine suture. There is a single oblong embrasure pit located just medial to the junction of M1 alveolus and P4. These concave pits allow space for the protoconid of m1 during occlusion. The incisive foramina in the *S. gracilis* sample (cast) are damaged and puttied over, but there still remains a narrow trough anterior to the foramina that terminates directly posterior to the first incisor alveolus. In the Camelot sample, the incisive foramina form narrow longitudinal slits that measure about 14 mm medio-laterally and about 16.8 mm anteroposteriorly. The foramina continue anteriorly, forming a narrow channel that tapers to a point just posterior to the first incisors. These foramina increase in both width and length in *S. californicus*, measuring approximately 26-27 mm long and about 10 mm wide. The posterior margin of the incisive foramen lies about in line with the posterior margin of the C alveolus. In the Camelot sample, the incisive foramen lies slightly anterior to the posterior margin of C. The feature was obliterated in the *S. gracilis* sample studied.

Upper Dentition

As the popular name implies, "sabertooth cat" is a term used to characterize (among other features) a pair of hypertrophied upper canine teeth that extend ventrally below the mandible in full occlusion (Figs. 5.2B and 5.3A). Upper canine teeth in the RLB *Smilodon* sample can exceed 138 mm from the distal tip to the alveolar margin; their length is only exceeded by the derived South American *S. populator* and the North American Miocene barbourofelid,

Barbourofelis fricki (Schultz et al., 1970). In the *Megantereon* sample available (N = 3), the mean transverse diameter of the canine measured at the cementum enamel junction (CEJ) is 12.7 mm. Overall length is approximately 83.6 mm measured from the distal tip to the anterodorsal termination of the enamel. The distance from this enamel termination to the alveolar margin is approximately 25.7 mm, with the total length approximately 109.3 mm. Measurements taken using the distal tip directly to the alveolar margin can vary, since the canines may slip out of the canine alveolus postmortem (Tejada-Flores and Shaw, 1984, and chapter 7, this volume). Between the CEJ and the alveolar margin, the cementum area was most likely covered by gingiva (Riviere and Wheeler, 2005). In *S. gracilis*, there is a slight decrease in the transverse diameter at 12.1 mm when compared to the average of 12.7 mm in *Megantereon*, yet the total length has now slightly increased to approximately 84.6 mm, measured at the dorsal enamel termination.

In the Camelot sample the average (N = 3) transverse diameter at the CEJ is 14.8 mm, and the length has increased to approximately 98.4 mm. In the RLB sample the CEJ transverse diameter increases to approximately 15.3 mm, while the enamel length increased to 124.0 mm. Overall, the increase in length is less apparent than canine root development. In *Megantereon*, the root length (including gingiva) measured from the enamel termination to the end of the canine track is 100.8 mm, whereas it is only 87.6 mm in *Smilodon gracilis*. Evidently there was a modification in bite mechanics between *Megantereon* and *S. gracilis*. This same root measurement in the Camelot sample is 112.6 mm and for the RLB sample 148.6 mm. Thus, in the *Smilodon* lineage, there is a 40 mm increase in enamel length versus a 58 mm increase in root length. In all *Smilodon* taxa, incremental medial and lateral growth lines are visible.

Over time, as the morphology of the skull changed, the canine roots also changed. This was likely due to greater stress on the canines, which also affected the bite mechanics. A larger canine would encounter greater mechanical pressures, which, in turn, would require additional anchoring of the tooth in the canine alveolus.

In *Megantereon*, the leading edge of the upper canine has a mild downward curve and then a slight dorsal rise at the distal tip. In *Smilodon gracilis*, in lateral view, the downward bend of the curve increases, but there is no dorsal rise at the tip of the canine. The angle of this ventral curve in *S. gracilis* is almost identical in the Camelot and RLB samples. The anterior keel of the upper canine in *M. cultridens* terminates approximately a centimeter before the cessation of the enamel edge, and it is medially offset along the entire course of the anterior margin of the tooth. In *S. gracilis*, this anterior, slightly serrated, keel is extended, terminating at the proximal edge of the enamel lip anteriorly. The keel does not abruptly curve medially at the anterior gingival sulcus as seen in the Homotherini and the Nimravid *Pogonodon* but follows the length of the anterior dorsal margin, ending somewhere within the last centimeter or two of the enamel crown, which in turn ends within a centimeter or so of the alveolus. The same is true for both the Camelot and RLB sample, scaled up proportionally.

In medial-lateral view, the CEJ in *Megantereon* forms a sinusoidal curve, with the anterior enamel termination about equal to or somewhat ventral of the posterior keel's termination. In the Camelot saber (SCSM 2004.1.2) just dorsal to the CEJ, a lingual groove is present that is nearly parallel to the palate. This feature is also present in *Megantereon* (Wheeler et al., 2004). Identical grooving is frequently seen among the several hundred upper canines of *Smilodon californicus* in the Page Museum collection at RLB (see chapter 7, this volume).

Smilodon upper incisors (Figs. 5.2A, 5.2B, 5.3A, and 5.3C) are unique when compared to those of extant felids, such as *Panthera onca*, in which the upper incisors are reduced to small, transversely positioned, and somewhat spatulate teeth, which are deflected ventrally while in a vertical position. *Megantereon cultridens* exhibits a more rounded incisor arcade compared to *P. onca*, a felid of comparable size. The incisors of *M. cultridens* have an anterior deflection, with the I1 and I2 almost equal in width (3.5 mm to 4.2 mm). The I3 is somewhat wider (7.7 mm). All of the incisors are tightly appressed to each other. Although both CB-20 and BC-6 have extremely worn incisors, Christiansen and Adolfssen

(2007) indicate that all of the incisors of the Senèze *Megantereon cultridens* (NHMB SE-311) have small medial and lateral cusps in addition to the pointed primary cusp, as seen in other derived sabercats. No serrations were in evidence. It is interesting to note that in both CB-20 and BC-106, the two central upper incisors occlude tip to tip with the lower incisor arcade, in a more nipping fashion than an interface occlusion. The I3 does exhibit normal occlusion on its lateral side with the lower canine. In *M. cultridens*, although there is a small (12.3 mm) diastema between I3 and the anterior margin of the upper canine, in anterior view the I3 overlaps the lingual side of the canine by approximately half of its width (3.5 mm), most likely a primitive feature. Unlike in *Megantereon*, the upper incisors of *Smilodon gracilis* (UF 84187 R, UF 87236 L, both specimens missing I1) are more caniniform in shape, pointed, slightly procumbent, with a ventral hook to the tooth arcade in lateral view. The transverse width of I2 and I3, taken at the base of the enamel, measures 5.5 mm and 8.2 mm (UF 87236 L), respectively. Like *M. cultridens*, all of the upper incisors of *S. gracilis* have medial and lateral cusps. The medial cusps rise slightly higher than the lateral cusps, and the ridges coalesce at a central position at the base of each incisor. No serrations were in evidence. Although the I3 canine diastema is slightly smaller (8.9 mm) in *S. gracilis*, there is no overlap of the canine by I3 in anterior view, possibly an advanced feature. In *S. fatalis*, SCSM 2004.1.1, all of the incisors are slightly less appressed to each other than in *S. gracilis* and overlap the lower incisors in occlusion by approximately 7.2 mm. Medial and lateral cusps are present and similar to *S. gracilis* in this respect. No serrations are in evidence. The diastema between I3 and the upper canine is approximately 11.3 mm. There is no overlap of I3 with the canine in anterior view. The incisors of SCSM 2003.75.155 are newly erupted, with the bases of the cusps barely rising above the alveolar margin. These are very pointed, procumbent, and with the hook ventral to the arcade, similar to *S. gracilis*. The left premaxilla of SCSM 2003.75.155 appears to match the left hemimandible SCSM 2003.75.164 in its occlusion, with the I2 and I3 completely interlocked medially and laterally with the lower canine, and i3, forming a tight bond with these teeth. In *S. californicus* the incisor arcade is not as rounded as in *S. fatalis*, and is more similar to *Panthera* in this respect, possibly a derived feature. In anterior view, due to the increase in skull size of *S. californicus* (Table 5.A1), including the overall width of both the upper and lower incisor batteries, the gaps between the upper and lower incisors allowing for occlusion are slightly larger than in *S. fatalis*. Medial and lateral cusps are present. The I3-upper canine diastema in *S. californicus* has increased to approximately 15.7 mm. Despite the larger skull of *S. californicus* in comparison to *S. fatalis*, other than the slightly more transverse tooth battery and the obvious increase overall in incisor length, width, and height, the incisors of *S. californicus* are very similar to those of *S. fatalis*.

In *Smilodon*, the maxillary tooth battery, excluding the canine, comprises three teeth, P3, P4, and M1 (Fig. 5.3A). The third premolar in *Smilodon* is obliquely positioned in the maxilla and is typically composed of three cusps: a small anterior cusplet, a somewhat larger posterior cusplet, and the taller central or principal cusp. There is a slight cingulum along the posterior base (distal side) of the tooth (Berta, 1987). In *Smilodon gracilis*, the AP length of P3 is about 14.3 mm and the tooth tilts posteriorly in the maxilla. It makes close contact with the centerline of the anterior end of P4. In *Megantereon*, P3 is slightly smaller, measuring 12.5–14.5 mm. It also makes contact with the anterior end of P4, slightly overlapping the preparastyle medially. It also has a slight posterior tilt in the crypt and its contact with the anterior part of P4 is closer to the protocone medially in this respect. In the sample of *Smilodon* from Rancho La Brea the P3 is larger, measuring approximately 17–18 mm. In some Rancho La Brea specimens P3 contacts the anterior mid-centerline of P4 while others there is a slight gap between the teeth.

In the Camelot *Smilodon*, the P3 is separated from the canine by approximately 13 mm. Due to the shorter rostrum found in *Megantereon*, this gap or diastema between the canine and P3 is much less, about 7.8 mm. In *S. californicus* this C-P3 diastema averages about 16.2 mm.

The AP length of P4 of SCSM 2003.75.154 is 34.4 mm (R) and 34.2 mm (L). The AP length of P4

in the SCSM sample ranges from 34.4 mm to 37.7mm (Table 5.A2). If the Bee County specimen (USNM 20750) P4 (36.8 mm) is included in the sample set for *S. fatalis* (Slaughter, 1963), then the average AP measurement for all five specimens is 35.96 mm, well below the mean given in Slaughter (1963) of 42.0 mm for *S. californicus*. If one factors in the holotype (AMNH FM 10395) AP of P4 at 33.4 mm, this lowers the average AP length of P4 for *S. fatalis* for the six specimens sampled in this paper to 35.53 mm—again, well below that stated by Slaughter for *S. californicus*. Thus, the P4 AP length of the SCSM sample falls within the lower range expected for *S. fatalis*.

In *Megantereon*, the P4 has a diminutive preparastyle located at the anteriormost end of the tooth. This cusp is no more than a bulge in the enamel at the base of the tooth. In *S. gracilis*, the preparastyle is developed into a small cusp anterior to and appressed against the second slightly taller cusp, the parastyle. The preparastyle forms an isolated anteriorly facing cusp in *S. californicus*. The parastyle is slightly shorter than the paracone. Its lingual surface occludes with the middle to posterior lateral side of the primary cusp of p4. The paracone is the tallest of the P4 cusps and forms a notch where it joins the metacone-metastyle blade (Bryant and Russell, 1995). It occludes with the posterior lateral surface of both the paraconid and protoconid surfaces of m1, forming the carnassial shear. The anterior prominence of the paracone of P4 seems to receive the most force, as many examples of this tooth at RLB are either chipped, broken off, or worn down to the pulp cavity.

The protocone in *Megantereon* appears as a distinct, small, round cusp just lingual and dorsal to the paracone and occludes with the primary cusp of p4. It may have served as a buttress, preventing any overbite in the shearing process. In *S. gracilis*, the protocone is almost nonexistent, other than as a bulge at the base of the root. In UF 206593, the protocone is almost invisible, while in UF 87238 there is a distinct bulge at the base of the paracone but no cusp is evident. In *S. fatalis* the protocone is no longer present. The Camelot sample demonstrates this, although in SCSM 2004.1.1 the lingual side at the base of the paracone has been partially abraded, resulting in a pseudo-protocone effect. As in *S. fatalis*, the protocone is absent in the more derived *S. californicus*.

The fifth cusp of P4, the metacone, receives the most abrasion during the food shearing process. It can (and often does) wear down, exposing the pulp cavity, as illustrated by Miller (1968). It occludes lingually with the anterolateral surface of m1. In all felids, as this wear increases, the P4 can rotate medially, to ensure an active shear surface with the m1 (Bryant and Russell, 1995). The metacone extends into the metastyle. This forms the large posterior shear surface of the tooth. The metastyle occludes medially with the buccal side of the paraconid and protoconid surfaces of m1. The P4 shear action is the same in all species of *Smilodon*. In *Panthera*, the P4

has a very prominent protocone that is maintained throughout the evolution of the clade. The P4 evolved into a longer tooth during the evolution of *Smilodon*, culminating with *Smilodon californicus*, while the P3 was reduced in size.

In the Camelot sample, the M1 is absent, although alveoli for this tooth are present on both sides.

Mandible and Lower Dentition

Five individual hemimandibles (four left and one right) (Figs. 5.5 and 5.6) were recovered in association with the one complete skull and three neurocrania. Although the right hemimandible (SCSM 2003.75.12) and the left hemimandible (SCSM 2003.75.162) are very similar in length (164 mm and 167 mm, respectively)

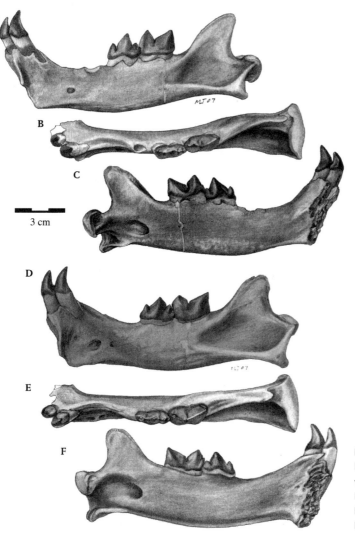

Figure 5.5. Camelot sample hemimandibles: A–C: SCSM 2003.75.163, A = buccal view; B = occlusal view; C = lingual view. D–F: SCSM 2003.75.164, D = buccal view; E = occlusal view; F = lingual view. Drawings by Mary Tanner.

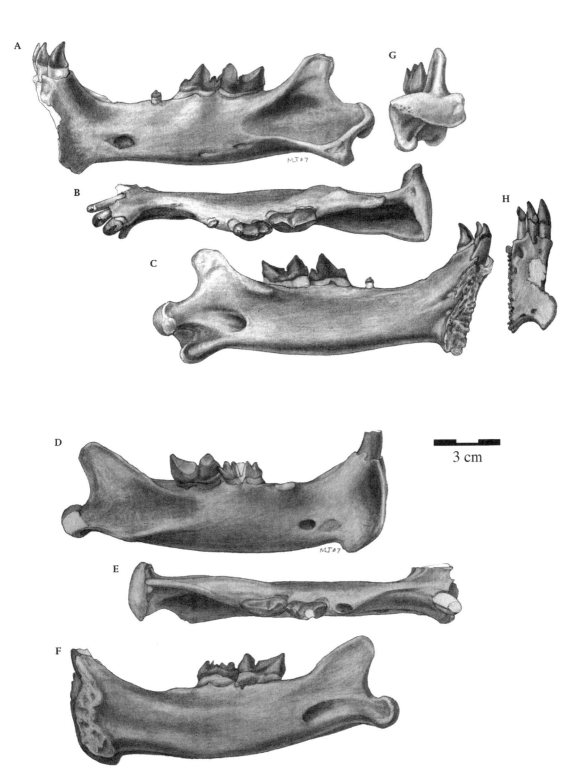

Figure 5.6. Camelot sample hemimandibles. A–C, G–H: SCSM 2004.1.4, A = buccal view; B = occlusal view; C = lingual view; G = posterior view; H = anterior view. D–F: SCSM 2003.75.12, D = buccal view; E = occlusal view; F = lingual view. Drawings by Mary Tanner.

and in carnassial tooth wear, symphyseal articulation is not perfect. Thus it is possible there was a fifth skull that has not been recovered. The Camelot discovery greatly expands the data set for specimens that closely match the holotype of *Smilodon fatalis*. The dental formula for all five Camelot hemimandibles is i3, c, p2, m1, except for SCSM 2003.75.164, in which the p3 alveolus is not present. The i1 is missing in all of the Camelot specimens but the alveolus is present. All hemimandibles have fully erupted teeth with similar labial wear facets on the carnassials, except for SCSM 2003.75.164, in which the wear facets are very small (4 mm in AP length). All five hemimandibles are similar in length, ranging from 165.5 mm to 174.5 mm. Four out of five hemimandibles possessed a single rooted p3, as evidenced by the alveoli. One hemimandible (SCSM 2004.1.4) retains the p3, which has a single, small, and somewhat rounded crown.

Megantereon cultridens has a double-rooted p3, but in most cases the roots are fused just above the alveolar margin (Berta, 1987). The crown is composed of one primary cusp and two smaller anterior and posterior accessory cusps. *Smilodon gracilis* also has a double-rooted alveolus but has a single-rooted p3 above the gum line, with a robust cingulum, reminiscent of the small cusps found in *Megantereon*. The p3 in *S. californicus* at RLB is highly reduced or absent. Of the 1,866 complete and partial mandibles we examined in the Page Museum collections, 60 (3%) have a p3 or its alveolus. A greater percentage (6%) was reported by (Merriam and Stock, 1932) based on an early sample of about 600, but it may have included many juvenile individuals with deciduous alveoli still present.

A small diastema, approximately 4.5 mm long, exists between p3 and p4 in the Camelot sample. This diastema is more variable (1 to 6 mm) in RLB specimens, as is the size of the mandibles. The p3 alveolus in the Camelot specimens ranges from 5.4 to 7.5 mm, but in the few cases where an alveolus for this tooth is present the range is 3 to 7.5 mm at RLB. The Camelot *Smilodon fatalis* alveolus for p3 is fairly uniform, around 5.4 mm long in both anteroposterior and transverse diameters.

The lower canines in *Megantereon*, *Smilodon gracilis*, and both the Camelot and RLB sample form part of the incisor arcade, although in *Megantereon* the lower canines are slightly taller (2 mm) than the incisors, whereas in the RLB and Camelot samples the lower canines barely exceed the height of the incisor arcade. In dorsal view, the lower canine in *Megantereon* has a lateral displacement about even with the lingual side of the carnassial tooth row. In the Camelot sample, due to the increase in lateral deflection of the cheek tooth row, the lower canine is now positioned slightly more medial than the p4 but still in alignment with m1 in that respect. Due to the increase in lateral rotation of both the corpus and cheek tooth row, this lower canine alignment with the cheek tooth row changes over time, with the RLB sample exhibiting the largest medial alignment distance between the lower canine and the carnassial in dorsal view.

In *Megantereon*, the lower incisors are more vertically positioned in the jaw, compared to the moderately procumbent upper incisors. They are somewhat caniniform. All lower incisors are closely appressed, more so than in both the Camelot and RLB sample. Vertical contact with the upper incisors was present in both CB-20 and BC-106. Incisor arcades will be affected by age and attrition, including abrasion, possibly resulting from scraping meat from bone. Although much of the visible wear occurs on the dorsal surface of all the incisors, some minor abrasion does occur on the lingual face of I3 and anterior face of the lower i3 and canine. The incisors form a medial-to-lateral size gradient, with i1 being the smallest. No serrations are present. There is a small single lower lateral cusp on i2, although a bulge in the cingulum in the lower basal area may have represented a second small medial cusp. The i3 in *Megantereon* is about intermediate in size between i2 and the canine. It sits almost vertical in the alveolus and the enameled portion of the tooth has a slight medial tilt. It has one small lateral cusp that rises to about the height of the canine cingulum.

In the Camelot sample, the incisors are less crowded than in *Megantereon* and the roots are slightly recurved posteriorly, compared to the vertical position in *Megantereon*. They are more similar to the RLB sample in this respect. Medial and lateral cusplets are present on both i2 and i3.

The anterior enameled crown of the incisors is pointed, with a slight posterior deflection. The five hemimandibles in the Camelot sample lack the i1, although a narrow alveolus was present in all except for SCSM 2003.75.164. No serrations are present. The i2 has two prominent lower cusps, one lateral and one medial, with a sharp lingual enamel ridge connecting the two cusps at the base of the tooth. Compared to *S. fatalis*, the i2 in the RLB sample is very similar in shape but much larger (Table 5.A1) and its medial tilt toward the i1 appears identical to the Camelot sample. The i3 in the Camelot sample is intermediate in size between i2 and the lower canine. It has both a lateral and a medial cusp, the former being slightly larger. The height of i3 is equal to that of the adjacent teeth including the canine.

In the Camelot sample, the transverse diameter of the lower canine measured at the alveolus has a range of 7.7 to 9.4 mm. It is somewhat larger than the i3 (6.5 to 7.8 mm) and has a pointed posterior curve to the tip, more so than in *Megantereon* and about equal to that seen in the RLB lower canine. The wear facet on the canine is on the lingual surface, so it is similar to *Megantereon* in this respect. In the RLB specimen LACMHC 2002-L&R-250, the lower i3 transverse measurement is 7.9 mm, about twice the size of the i1 (4 mm) and approaching the lower canine in size (9.1 mm).This is only a slight increase in transverse diameter compared to the Camelot sample (7.8 mm) (Table 5.A1). In four of the five Camelot hemimandibles, the lower canine roots have a slightly forward tilt compared to the more vertical position in *Megantereon*. In lateral view, SCSM 2003.75.164 differs in that the lower canine root is more procumbent than the other four, very similar to the RLB sample in this respect. In the RLB sample, the lower canine is part of the incisor arcade similar to the other sabertoothed cats in this study. It is positioned at a level equal to the height of i3, with just a slight increase in size, although now the lower canine has shifted approximately 11 mm medial to the lower carnassial tooth row. This medial shift in lower canine position and lateral carnassial deflection may reflect a change in the bite mechanics of the RLB sample as compared to the Camelot sample.

In many respects the lower cheek teeth of the Camelot *Smilodon fatalis* are very similar to the RLB sample, only smaller. The p4 is composed of a small vertical anterior cusp, a tall central primary cusp, and two somewhat smaller posterior cusps that tilt posteriorly. The most posterior cusp makes contact with the lateral side of the paraconid of m1. The height of the two posterior cusps of the p4 in *Megantereon* reaches the level of the carnassial notch of the m1, so is similar to both the Camelot and RLB samples. In *Megantereon* and the Camelot sample, the primary cusp of p4 has a slight posterior tilt and reaches the level of the paraconid-protoconid complex. This posterior tilt is greatest in the specimens from RLB. As reported earlier, with age and attrition, the p4 will also begin to show signs of wear, beginning with the posterobuccal side of the primary cusp. Due to both the anteroposterior and mediolateral expansion of the p4 in the RLB sample, as compared to the *S. fatalis* sample from the Camelot site (SCSM 2003.75.12 and 162), it appears that this lower premolar makes contact during the shear-slice motion much earlier than in *S. fatalis*, with similar m1 attrition.

Of the Camelot hemimandibles, SCSM 2003.75.12 and SCSM 2003.75.162 exhibit the most extensive wear to the m1. In *Smilodon* the paraconid and protoconid are of equal height, and wear occurs on both shear planes equally, although due to the posterior tilt, in lateral view, the paraconid appears to be slightly taller, especially in the RLB sample. In *Megantereon*, the anterior cusp, at 9.4 mm average anteroposterior length measured at the base of the crown, is smaller than the posterior cusp at 12.0 mm average. In total, the tooth measures 27.4 mm long. In the Camelot sample, the lower m1 measurements for the paraconid range from 9.6 to 10.7 mm, relatively close to *Megantereon*, whereas the protoconid measurements show an overall increase, ranging from 13.4 to 15.4 mm. In the RLB sample the anteroposterior length of the paraconid is approximately 10.0 mm, surprisingly similar to both the Camelot sample and *Megantereon*. The largest increase again occurred in protoconid length, with an average of 15.4 mm for the RLB sample and an approximate overall length of 27 mm. Within the Camelot sample, SCSM 2003.75.164 had the greatest m1 length at

25.6 mm, much smaller than the RLB sample. The m1 length in *S. gracilis* ranged from 20.8 to 25.2 mm, with an average of 23 mm (Berta, 1987), while *Megantereon* had an average m1 length of 21.8 mm, below the average for *S. gracilis*.

In all sabertooth cats, the mandibular symphysis is not fused but connected by bony symphyseal plates. These bony plates differ from species to species. In *Megantereon*, the plates are vertically oriented and the base of the plate lies dorsal to the ventral portion of the genial flange by approximately 18 mm. The vertical length is approximately 44.4 mm and is 18.7 mm wide at its dorsal extent. The plate has an hourglass-shaped taper and is about 12 mm wide at mid-length. It then slightly expands at its base to approximately 13 mm. In *Smilodon gracilis*, the symphyseal plate is slightly smaller dorsally, at 16.7 mm, but taller vertically at 53.5 mm, reflecting the increase in overall size of the hemimandible. In *S. gracilis*, the plate develops a more posterior tilt dorsally than in *Megantereon*. In *S. gracilis*, a slight groove has developed midway down the plate and this groove almost separates the plate into two distinct bony ridges; the lower half is more vertical and similar to *Megantereon* in this respect. The plate terminates about 12 mm dorsal to the ventral extent of the genial flange. In the Camelot sample, due to the overall increase in size of the hemimandible and reduction of the genial flange, the ventral extent of the symphyseal plate is now almost level with the base of the jaw. As in *S. gracilis*, there is a groove midway down and just posterior to the vertical plate that reduces the width from 15 mm to 11 mm. In the RLB sample the symphyseal plates are similar to the Camelot sample in shape, although the vertical extent has increased to approximately 58 mm.

In dorsal view, at the base of the chin between the genial flange and the symphyseal plate there is a distinct fossa in both *Smilodon gracilis* and the Camelot sample. This fossa is absent in *Megantereon* and faint in the RLB sample. In anterior view, on the anterior portion of the hemimandible of *Megantereon*, there are three tiny and somewhat vertical foramina located just dorsal to the symphyseal plate connection. In *S. gracilis* there are three larger, vertically placed foramina in the anterior portion of the jaw. In the Camelot sample the spacing between the foramina increases, and the dorsal-most foramen is located just below the alveolus of i3. In the RLB sample these foramina average 4.5 mm across, and may vary from two to four. They are positioned somewhat similarly to the foramina in the Camelot sample, being staggered from the base of the jaw to a point just ventral to the i2 alveolus.

In all of the *Smilodon* species, the mandibular body curves lingually behind the canines to allow clearance for the upper canines, then with a gradual curve, turns back laterally to the p3. A sharp bony edge runs dorsally along the entire diastema of the jaw, terminating at the alveolus of p3.

All five of the Camelot hemimandibles have a moderate ventrally projecting genial flange, though none have reached a level below the ventral level of the angular process. The genial flange's ventral projection in the RLB sample is about level with the angular process. In *Megantereon* the genial flange projects approximately 8 to 9 mm below the angular process and about 17.5 mm below the symphyseal plate. In *Smilodon gracilis*, the genial flange is shorter, about 9 mm below the plate and slightly lower than ventrally level with the angular process.

In ventral view, just medial to the genial flange, there is a shallow fossa for the insertion of *M. digastricus* in the Camelot sample. This muscle assists in lowering the mandible. In the RLB sample, this area is very broad, yet not as pronounced as in the Camelot sample. In *S. gracilis*, this fossa is present but not as prominent nor as rugose as in the Camelot sample. This fossa is absent in *Megantereon*. There is a small mental foramen located just posterior to the centerline below the c-p3 diastema, measuring approximately 4.5 mm in length, with two or more additional tiny foramina just posterior to the larger foramen. In *S. gracilis* there are two small mental foramina of equal size. The anterior foramen is centrally located in the mandibular corpus below the c-p3 diastema, while the second foramen is at the same level, but posteriorly located just below the anterior root of p3. In four of the five Camelot hemimandibles, a single oblong mental foramen is present, measuring approximately

3 mm in length. All foramina are centrally located below the diastema except in SCSM.75.164. The latter has two mental foramina; the larger, similar in size to those of the first four hemimandibles, lies slightly anterior to the diastema midline and the smaller one posterior to it. In the RLB sample, there is a single, large, oblong mental foramen measuring approximately 4.5 mm long, located below the centerline of the c-p4 diastema.

In *Megantereon*, the masseteric fossa is positioned just ventral to the coronoid process and posteriorly terminates just anterior to the condyle. It does not undercut the condyle nor is it pocketed anteriorly. In lateral view the fossa tapers anteriorly, terminating at a point just below the posterior edge of the m1. The anterior margin of the fossa is formed by the lateral descending anterior ridge of the coronoid process. This ridge terminates laterally at a level about 6 mm below and just anterior to the m1 alveolus. In *Smilodon gracilis* the lateral ridge of the coronoid process terminates about level with the m1 alveolar margin. In *S. gracilis*, the area just ventral to the coronoid forms a deeper pocket than in *Megantereon* and now tapers further anteriorly, terminating about midway below the m1. As in *Megantereon*, it does not undercut the mandibular condyle posteriorly. In the Camelot sample, the masseteric fossa, although similar in shape to that of *S. gracilis*, is much larger, due to the increase in the size of the mandible. It tapers anteriorly, terminating just below the carnassial notch of m1. It is not pocketed anteriorly, although posteriorly the fossa now undercuts the condyle slightly. In *M. cultridens*, *S. gracilis*, and the Camelot sample, the lower lateral margin of the masseteric fossa flares out laterally to a point parallel to the lateral margin of the condyle. In the RLB sample, the masseteric fossa is similar in form to the Camelot sample, although the ventral flare is greater than in the other three species examined.

In all Smilodontini, the coronoid process is very short, barely rising above the protoconid of m1, as compared to the tall, posteriorly deflected process in the conical-toothed cats. This reduction in height reduces overstretching of the masseter muscle (Emerson and Radinsky, 1980). In both *Megantereon* and *Smilodon gracilis*, the coronoid process is more or less parallel to the carnassial tooth row in posterior view. In both the RLB and Camelot samples, the coronoid process is slightly medial to the carnassial tooth row, more so in the RLB sample. The ventral-medial edge of the coronoid process terminates in a rugose bulge, just below the margin of the m1 protoconid. This bulge is somewhat reminiscent of the alveolar torus found in the nimravid species *Nimravus bumpensis* (Matthew, 1910; Toohey, 1959). The Camelot sample also exhibits the same feature, although it is less pronounced. This bulge is reduced in size in *S. fatalis* and is barely discernible in *Megantereon*. In the more basal taxon, *Megantereon*, the coronoid process rises above the protoconid of m1 by approximately 8.7 mm and is about level with the tip of the lower canine. In the Camelot sample the height is approximately 11 mm above the protoconid of m1 but now is slightly lower than the tip of the lower canine. The coronoid is much lower in the RLB sample, close to the level of the top of the carnassial teeth, yet barely rises to the level of the cingulum of the lower canine. Again, with the increase in gape of the RLB sample, this height reduction of the coronoid would help reduce any negative issues with the masseter muscle. In *Megantereon* the posterior deflection of both the anterior and posterior edges of the coronoid is about parallel and the process terminates just anterior to the notch of the mandibular condyle. In *S. gracilis*, the anterior edge of the coronoid still retains its inclined angle, but now the posterior edge of the coronoid is more vertical. In the Camelot sample the posterior edge is also somewhat vertical but less so when compared to the almost vertical posterior edge in the RLB sample.

In ventral view, the angular process of *Megantereon* is a small blunt process situated about 22 mm below the dorsal surface of the mandibular condyle. The ventral extent is at about the same level as the genial flange. Ventrally, in dorsal view, it forms a sharp lateral edge which in turn forms the base of the masseteric fossa. This edge abruptly terminates about 6 mm anterior to the paraconid of m1. It does not deflect medially as in the other Smilodontini. In the Camelot sample the angular process is very stout, has a ventrally flattened medial-lateral expansion

(*M. masseter superficialis* insertion), and inclines ventrally slightly below the lowest level of the genial flange. It has a medial deflection at its distal end. A rugose ridge of bone forms along its distomedial length that serves for the insertion of the *M. pterygoideus*, which assists in elevating the jaw (Naples et al., 2011). In lateral view, the angle terminates about midway below the mandibular condyle. In the RLB sample, the angular process mirrors the Camelot sample, except that it has greater lateral deflection and somewhat less medial deflection. The lateroventral edge forms the ventral margin for the masseteric fossa, as in all Smilodontini. In both the Camelot and RLB samples, the lateroventral margin of the angular process continues anteriorly, culminating at a level about equal to the mandibular condyle just below the carnassial notch of m1, thus forming the anterior portion of the masseteric fossa.

Medially, the mandibular foramen in *Megantereon cultridens* and *Smilodon gracilis* is a small round 3.4 mm aperture, which opens anteriorly and is situated just dorsal to the anterior edge of the angular process and ventral to the mandibular condyle. A small channel forms just posterior to the aperture and deflects slightly ventrally along the dorsomedial margin of the angular process. In the Camelot sample, the dorsoventral measurement has increased to 7.8 mm. The posterior channel now runs only parallel to the jaw and terminates broadly across the corpus at about the centerline of the mandibular condyle. In the RLB sample, the mandibular foramen increases in size to approximately 8.6 mm. The posterior channel is recessed more than in the Camelot sample. Interestingly, in the RLB sample, the mandibular foramen is now much more posteriorly situated nearer the mandibular condyle, or about equal to the centerline of the coronoid process.

In the Camelot sample, the distance between the mandibular foramen, measured from the anterior edge of the aperture, to the posterior end of the condyle, ranges from 45.3 mm in hemimandible SCSM 2003.75.164 to 50.1 mm in SCSM 2004.1.4. This measurement on mandibles from the RLB sample is 46.6 mm (LACMHC 2002.250R&L), similar to the Camelot dimension despite an overall increase in

mandibular length of approximately 36 mm. The mandibular corpus in the RLB sample has increased in length anterior to the mandibular foramen, while posterior to the foramen the length has remained constant.

In dorsal view, the mandibular condyles in all Smilodontini are wide medially and taper laterally, terminating at a point just dorsal to the angular process. There is a condylar notch present, although this notch is not as recessed in *Megantereon* as in the Camelot and RLB samples. The lateral ridge below the condyle continues ventrally onto the angular process. In posterior view, the lateral extension of the condyle is greatest in the RLB sample and least in *Megantereon*, where its lateral excursion only extends to a point parallel to the carnassial tooth row.

Conclusions

Arguably the most significant aspect of the Camelot population is that the holotype of *Smilodon fatalis* is not an outlier or anomaly. It is instead representative of a population of similar specimens, and the population is not that variable. The material includes upper and lower dental, cranial, and postcranial material. Not only is there now a sample of *Smilodon* that closely matches the holotype P4 tooth, there is a population of similar specimens showing the extent of individual variation. Larry Martin's hypothesis (pers. comm., J. P. Babiarz, 2005) was that "the basis for the synonymy of *Smilodon*, over such a tremendous time-span, including the lack of specimens, is not supported." With the Camelot *S. fatalis* discovery, his theory is now supported by the fossil record.

Although we can never be certain that the Smilodontini represents a single anagenetic lineage, beginning with and including its hypothesized earliest ancestors, *Megantereon* (Sardella, 1998; Christiansen and Adolfssen, 2007) and *Rhizosmilodon fiteae* (Wallace and Hulbert, 2013), "nothing exists to prevent an ancestor-descendant relationship between some species of *Megantereon* and *Smilodon*" (Martin et al., 1988:157). *Rhizosmilodon fiteae,* formerly described (in part) as *Megantereon hesperus* (Lund, 1841; Berta and Galiano, 1983) is only known at this time from the Upper Bone Valley Formation in Florida. *Megantereon* was Holarctic and is known from North America, Eurasia, and Africa (Palmqvist et al., 2007). With regard to *Smilodon* itself, due to an early southern dispersal there may actually be two contemporaneous anagenetic lineages, one in South America, deriving from an early dispersal of the genus during the Uquian age of South America (Kurtén and Werdelin, 1990) culminating with the highly derived *S. populator* (Lund, 1842) and another in North America (Merriam and Stock, 1932; Martin, 1967), culminating with the *Smilodon* species found at Rancho La Brea. In either case, the Smilodontini went extinct sometime near the end of the Pleistocene, coincident with the extinction of the megafauna (Martin et al., 1967). The 2003-04 discovery of the Camelot *Smilodon fatalis* sample in Dorchester County, South Carolina, affirms that a late Irvingtonian (Kohn et al., 2005) population belonging to this taxon was present across the North American continent. The initial Hardin County, Texas discovery of a late Irvingtonian maxilla fragment (Slaughter, 1963), including the P4 in its alveolus, and the subsequent naming of it as the holotype of *Trucifelis fatalis* by Leidy (1868), set the stage for the taxonomy and classification of the Smilodontini in North America. The recovery of the contemporaneous South Carolina Camelot *Smilodon* material, which is very similar in morphology to the holotype *S. fatalis*, reinforces the validity of this as an intermediate species. We agree with the hypothesis of L. D. Martin: "Nor are we convinced that the late Irvingtonian material from North America can be easily included with the Rancho La Brea population" (Martin et al., 1988:158). A specimen from Bee County, Texas, a *Smilodon* skull retaining right and left P4, collected in 1949 by George E. Klet, from the Berclair terrace of Medio Creek, Texas (USNM 20750), came from deposits of late Irvingtonian age (Conkin and Conkin, 1962; Slaughter, 1963), as did the holotype. Dr. Lewis Gazin (attributed in Slaughter, 1963:73), found "the P4 similar enough, both in size and form, to leave little doubt that this 'new' skull belongs to Leidy's Texas species." A cranium (SMU-SMP 60006) found in 1920 by Ellis W.

Shuler, from the basal member of the Pemberton Hill-Lewisville Terrace, Texas, was also referred to *S. fatalis* (Slaughter, 1963). Based on the morphology of the Camelot sample, we support the validity of *Smilodon fatalis* as a species intermediate between *Smilodon gracilis* (early Irvingtonian) and *Smilodon californicus* (Webb, 1974; Berta, 1987) from the Rancholabrean NALMA. Based on both the distinct morphology and temporal separation, the Rancholabrean *S. californicus* is distinguished from *S. fatalis* and we argue that it should not be referred to the latter species.

The synonymy of *Smilodon fatalis* (Kurtén and Anderson, 1980) with the more derived *Smilodon californicus* is thus unwarranted, and the taxonomy of Late Pleistocene species of *Smilodon* is in need of revision. Provisional return to the use of *S. californicus* as in this paper may be expedient. Many of the derived characters of *S. californicus* discussed herein, and used to separate it from *S. fatalis* in this work, are absent in specimens assigned to *S. floridanus* (Martin et al., 1988). A decision regarding the status of *S. floridanus* and its relationship to *S. californicus* is a separate issue and beyond the scope of this chapter.

It is of interest to note that Kurtén (1965) suggested that he was in at least partial agreement with Slaughter's (1963) assessment of the assignment of *S. fatalis*, "subject to slight changes" (1965:246). He further stated (246) that "it seems necessary to conclude, therefore, that *Smilodon floridanus* is a synonym of *Smilodon fatalis*" but left open the possible recognition of *S. californicus* as a distinct species, due to its more derived features. Clearly, much more research and new material will be necessary to parse out the taxonomic confusion of this lineage.

Regardless of the conclusion as to whether there are one or two derived species of *Smilodon* (*S. californicus*, *S. floridanus*) in the Middle to Late Pleistocene of North America, we make the argument that *S. fatalis* should be restricted to the late Irvingtonian population. Undoubtedly, members of the Smilodontini will continue to be viewed as one of nature's premiere predators and will forever remain a topic of interesting study.

APPENDIX: MEASUREMENT TABLES

Table 5.A1. Measurement of the skull. *Smilodon fatalis* from the Camelot fauna

(a) = approximate due to missing, damaged, or lateral compression

			Smilodon fatalis					Megantereon cultridens			S. californicus		S. gracilis
No.	Points	Description	SCSM 2003.75.154	SCSM 2003.75.155	SCSM 2003.75.157	SCSM 2004.1.1	Bee-County, TX USNM 20750	BC 106	CB 20	SE 311	LACMHC 2001-249	KUVP 98124	UF 81700
1	1 & 2	Length from anterior end of alveolus of I1 to posterior end of condyles (CBL)	247.0 (a)			253.0	272.9	243.0	239.5		310.0	327.0	244.0
2	1 & 3	Basal length from anterior end of alveolus of I1 to inferior notch between condyles (PR-BA)	238.0 (a)			241.0	261.5	228.0	226.0		297.0	306.0	229.2
3	1 & 4	Length from anterior end of alveolus of I1 to inion (apex of occipital crest)	271.0 (a)			278.0	305.2	250.5	247.0		344.5	355.0	253.8
4	1 & 5	Length from anterior end of alveolus of I1 to anterior end of posterior nasal opening				129.4	138.2	120.0	121.0		165.0	161.0	121.4 (a)
5	1 & 6	Length of palate from anterior end of alveolus of I1 to posterior border of palatine (anterior border of pterygoid fossa) (staphylion) (PL)				169.0	176.8	146.6	147.0		202.0	207.0	
6	7 & 2	Length from posterior end of glenoid fossa to posterior end of condyles	101.0	96.7	104.0	94.8	115.9	82.1	77.3		114.5	117.0	73.9
7	8 & 8.1	Maximum length of nasal bone (anterior posterior distance medial edge)				68.9 (a)	71.9	60.0	59.2 (a)		94.9	97.5	
8	8 & 9	Length of anterior nares (anterior nasal aperture)				55.0	59.2	44.8	48.0		60.8	57.4	29.3 (a)
9	10 & 11	Width of anterior nares (anterior nasal aperture)				43.5	42.2	30.4	27.5		54.3	53.8	33.2 (a)

#		Measurement										
10	12 & 13	Maximum width across the muzzle at canines (CC)		83.1	85.3	97.1	58.4	54.8		104.5	103.3	70.3
11	14 & 15	Maximum width between superior borders of orbits	75.3	60.0 (a)	73.2	89.2	50.3	42.0 (a)		95.3	90.9	52.3
12	16 & 17	Width across postorbital processes (POP)	94.1	80.0 (a)	86.6	106.6	73.0	58.6 (a)		112.9	116.6	64.5 (a)
13	18 & 19	Minimum width of postorbital constriction (POC)	54.8	49.6 (a)	50.4	57.1	41.0	35.8 (a)		58.4	60.5	47.1 (a)
14	20 & 21	Maximum width across zygomatic arches (ZB) (ZW)	166.0 (a)	170.0 (a)	174.0	207.1	120.0 (a)	115.0 (a)	150.8	205.5	225.0	146.4 (a)
15	22 & 23	Minimum anterior palatal width between superior canines	30.8	47.6	40.0 (a)	48.8	48.9	33.0	29.8	61.2	53.8	38.1 (a)
16	24 & 25	Width across palate between posterior ends of alveoli for superior carnassials (PA)			94.5	119.8	71.7	68.8		119.9	124.0	63.8
17	26 & 27	Length of auditory bulla from posterior lacerate foramen to external auditory meatus	42.8	47.3	51.2	39.5	53.5	33.7	37.4	58.2	60.0	36.7
18	28 & 29	Maximum width across mastoid processes (MB)	106.3	119.2	112.8	110.6	134.7	79.2	80.0	139.2	140.5	98.5
19	30 & 31	Maximum width across occipital condyles (CB)	50.1	56.5	57.6	55.5	65.5	42.6	43.7	59.7	72.0	51.5
20	32 & 33	Length from anterior end of alveolus of I3 to posterior end of premaxillary process	76.0 (a)	93.1	90.7	80.3 (a)	105.0			115.6	116.0	
21	6 & 3	Length from posterior border surfaces of palantine (staphylion) to inferior condylar notch	67.7	81.7	75.9	72.1	73.2	108.6		105.0	92.4	100.0
22	34 & 38	Perpendicular height from base of condyles to top of sagittal crest	82.9	85.5	89.7	81.9	106.4	70.6	74.2 (a)	106.2	104.0	70.3

(continued)

Table 5.A1. (continued)

(a) = approximate due to missing, damaged, or lateral compression

No.	Points	Description	*Smilodon fatalis*					*Megantereon cultridens*			*S. californicus*		*S. gracilis*
			SCSM 2003.75.154	SCSM 2003.75.155	SCSM 2003.75.157	SCSM 2004.1.1	Bee-County, TX USNM 20750	BC 106	CB 20	SE 311	LACMHC 2001-249	KUVP 98124	UF 81700
23	34 & 35	Perpendicular height of skull from apex of occipital crest (inion) to dorsal border of foramen magnum	55.2	58.8	62.0	54.6	74.0	46.7 (a)	49.1		73.4	76.3	52.0
24	36 & 37	Maximum width of foramen magnum	27.1	28.2	26.9	24.6	28.1	21.7	22.6		29.0	35.8	
25	35 & 38	Height of foramen magnum	24.3	26.2	24.2	24.7	24.1	23.8	22.6		30.3	34.0	
26	39 & 40	Width of occiput just above level of condyles	75.7	88.0 (a)	72 (a)	70.0	92.1	57.1 (a)	57.9		94.9	106.5	
27	66 & 67	Width between glenoid processes measured from medial edges of glenoid fossae	60.3 (a)	68.2 (a)	69.3	68.5	76.9	44.3	45.2			94.3	
28	68 & 69	Maximum transverse diameter of brain-case measured distal to auditory meatus at base of lambdoidal crest	81.3	83.1	85.9	84.9	99.9	62.8 (a)	59.1 (a)		90.7	95.5	81.8

Table 5.A2. Measurements of the upper dentition. *Smilodon fatalis* from the Camelot fauna

Superior Dentition

(a) = approximate due to missing, damaged, or lateral compression;

N/A = Not Applicable; (x) P3 offset 13mm from C1 centerline

No.	Points	Description	Smilodon fatalis						Megantereon cultridens			Smilodon gracilis			Smilodon californicus		
			SCSM 2003.75.154	SCSM 2003.75.155	SCSM 2003.75.157	SCSM 2004.1.1	Holotype AMNH 10395	Bee-County, TX USNM 20750	BC 106	CB20	SE 311	UF 81700	UF 87238	LACMHC 2001-249	KUVP 98124	LACMHC 2001-24	LACMHC 2001-76
1	41 & 42	Length from anterior end of alveous of C1 to posterior end of alveous P4	97.7	91.6 (a)	87.8	95.5		103.2	81.5	81.3		85.3	85.4 (a)	114	116.8	126.3	102 (a)
2	42 & 43	Length from anterior end of alveous of P3 to posterior end of alveous P4	45.6	49.6	39.6	47.1		53.9	40	41.9		40.6	45.3	53.2	58	63.7	48.8 (a)
3	43 & 44	Length of diastema from posterior end of alveolus of C1 to anterior end of alveolus of P3 (x) = alveoli offset and overlap	17.3	10.1	16	16.7		0 (x)	16.5	15.7	12.2	15	16.7	20.9	20.5	18	15
4	45 & 46	Width of incisor series measured between lateral borders of aveoli of right and left I3	42.6 (a)	50.4		43.5 (a)		47.6	32.6	32.5		43.9 (a)	39	55.9	57.6	63.5	50.5
5	47.1 & 48.1	Maximum anterior posterior diameter of I1 at alveolus (Mesiodistal)	10.1 (a)	10.3		8.8 R		9.5 (a)	6.8	9.5 R	12.2	10.3	7.4	11	13.2		
6	47 & 48	Maximum width of I1 at alveolus (Buccolingual/ Transverse diameter)		5.1		4.7 R		5 (a)	3.8	3.5		5.6	3.8	5.3	6.2	7.6	6.3

(continued)

Table 5.A2. (continued)

No.	Points	Description	Smilodon fatalis						Megantereon cultridens			Smilodon gracilis		Smilodon californicus			
			SCSM 2003.75.154	SCSM 2003.75.155	SCSM 2003.75.157	SCSM 2004.1.1	Holotype AMNH 10395	Bee-County, TX USNM 20750	BC 106	CB20	SE 311	UF 81700	UF 87238	LACMHC 2001-249	KUVP 98124	LACMHC 2001-24	LACMHC 2001-76
7	49.2 & 50.2	Maximum anterior posterior diameter of I2 at alveolus	13 (a)	11		8.9 R		14.9 (a)	7.2	6.9	13	10	7.5	11	14.5		
8	49 & 50	Maximum width of I2 at alveolus	7	6.2		5.9 R		7.5	4.4	4.2		5.4	5	6.4	7.4	9.5	7.9
9	46.3 & 51.3	Maximum anterior posterior diameter of I3 at alveolus	11.9	12.9	10.8 R (a)	12.8 R		16.7	8.7	9.0 R	14.6	12.1	10.1	13.3	16		
10	46 & 51	Maximum width of I3 at alveolus	9.3	10.1	9.3 R	9.7 R		10.5	6.9	7.7		6.5	7.8	9.8	11.1	13	12
11	41 & 44	Anteroposterior length of C1 at alveolus (Mesiodistal diameter)	30 (a)	32.4	N/A	30.8		41.8	23.8	23.8	24.6	23.1	24.2 (a)	44.8	40.4	46.1	39
12	23 & 52	Width of C1 at alveolus	13.8 (a)	14.9	N/A	15.6		19.2	11.7	12.7		13.7	16.5	19.2	18.2	22.9	18
13	43 & 53	Maximum anteroposterior length of P3 at alveolus (base of enamel - b)	13.0 (b)	14.4 (b)	N/A	13.7 (b)		NA	13.8 (b)	12.5 (b)	13.6	13.9	14.9 (b)	15.9 (b)	18.4 (b)	18.5	
14	54 & 55	Width of P3 at alveolus (base of enamel - b)	7.7 (b)	8.1 (b)		7.4 (b)		NA	5.7 (b)	6.1 (b)	7.5	7.6	6.2	8.1	10.3 (b)	10.6	
15	56 & 57	Anteroposterior length of P4 at base of crown	34.4	37.7	36.2	34.7	33.4	36.8	27.1	29.5	31.7	30.4	32.1	38.1	43.8	46	33.4 (a)
16	58 & 59	Maximum width of P4 across protocone/protoradix	13.8	15.8	16.8	15.1	16	19	11	12.2		13.9	14.2	15.6	19.5	19.3	14.2

#		Measurement													
17	60 & 61	Anteroposterior length of P4 paracone at base of crown	12	12.3	12.1	10.5	13.2	9.2	10.9	10.1	11	13.1	13.6	13.7
18	42 & 53	Anteroposterior length of P4 at alveolus	31.3	34.4	33.4	31.7	35	24.9	29.3	28.7	30.3	35.8	39.2	
19	60 & 56	Length of P4 parastyle/ ectoparastyle complex from anterior base of paracone to anterior end of tooth	9.1	9.8	10.1	9.9	11.1	8.4	9.7	8.0	9.2	10.1	11.6	10.1
20	61 & 57	Length of P4 metacone/ metastyle complex blade from base of crown above carnassial notch to most posterior point on base of metacone	13.6	15.3	14	14	12.7	9.8	10.5	12.6	11.9	14.7	18.1	15.2
21	62 & 63	Anteroposterior length of M1 at alveolus	N/A	N/A	9.1 (a)	N/A	NA			7.8	13.6	15.1 R		
22	64 & 65	Width of M1 at alveolus - (a); base of enamel - (b)	N/A	N/A	5.7 (a)	N/A	NA			4.6	5.1	6.9 R		

Table 5.A3. Measurements of the mandible. *Smilodon fatalis* from the Camelot fauna

(a) = approximate due to missing, damaged, or lateral compression; N/A = not applicable

No.	Points	Description	Smilodon fatalis					Megantereon cultridens					S. californicus		S. gracilis (LEISEY)		
			SCSM 2003.75.12-R	SCSM 2003.75.162-L	SCSM 2003.75.163-L	SCSM 2003.75.164-L	SCSM 2004.1.4-L	BC 106		CB 20		SE 311	LACMHC 2002-250		UF 82529/81724		BIOPSI 0153
								R	L	R	L	L	R	L	R	L	L
1	1 & 2	Length of ramus from anterior border of i1/alveolus at symphysis to posterior end of condyle at center	165.0 (a)	165.0 (a)	167.0 (a)	166.0 (a)	175.0	163.0 (a)	158.0 (a)	160.0	162.0	159.2 (a)	207.5	207.5		149.5 (a)	153.2
2	2 & 3	Length of ramus from anterior end of outer flange to posterior end of condyle	161.0	161.0 (a)	162.0	163.0	165.0	151.0	152.0	150.0			195.5	195.0			145.8
3	6 & 7	Distance from alveolus of ci to ventral border of flange	44.6	44.2	43.5	40.2	45.0	59.2	59.4	57.9	57.9		48.3	43.9		55.6 (a)	49.2
4	1 & 26	Length of symphysis measured along anterior border	53.1	49.7	44.8	45.5	51.0	50.1	50.1	49.9	49.9		61.5	61.5			45.6
5	32 & 33	Minimum depth of ramus below diastema	31.1	32.7	27.0	27.6	29.3	36.3	37.8	31.6	32.6		34.6	34.0	29.0	28.2	27.6
6	5 & 9	Depth of ramus below posterior border of m1 alveolus	33.2	34.0	30.4	33.9	33.8	30.5	30.8	34.3	34.5	31.3	39.9	40.4			27.3

No.	Landmarks	Measurement															
7	10 & 11	Depth of ramus below anterior border of p4 alveolus	35.2	34.2	31.1	28.8	32.5	31.4	31.7	29.3	29.9	29.1	39.6	40.5			29.1
8	12 & 13	Thickness of ramus below m1 (Buccolingual diameter)	18.8	18.7	16.9	17.6	19.0	13.0	12.8	12.7	12.8	13.8	21.0	20.7	15.2	15.4	15.6
9	14 & 15	Height from inferior border of angular process to summit of condyle	30.9 (a)	32.4	27.3	32.2	31.6	26.6 (a)	30.0 (a)	28.2 (a)	29.8 (a)		34.7	33.7			
10	14 & 16	Height from inferior border of angular process to summit of coronoid process	61.6 (a)	62.9	57.2	62.2	60.3	56.3	57.3	64.0	63.2 (a)		69.6	69.0	54.1		54.0 (a)
11	17 & 18	Transverse width of condyle	39.7 (a)	40.4	38.1	37.8	39.4	36.6	34.8	31.6 (a)	33.6		48.4	48.4			31.6
12	15 & 19	Maximum depth of condyle	15.9	16.6	15.3	15.9	16.3	16.8	16.3	12.5	12.6		13.3	18.8			13.9
13		Length of Diastema between p3 and p4	4.5	3.9	3.9	N/A	4.5	2.7	1.5	3.8	4.3	N/A	N/A	4.2			0.5
14		Depth of ramus below anterior border of p3 alveolus	32.1	32.0	28.2	N/A	30.8	31.8	32.5	30.5	31.9	N/A	N/A	30.7			27.5

Table 5.A4. Measurements of the lower dentition. *Smilodon fatalis* from the Camelot fauna

(a) = approximate due to missing, damaged, or lateral compression; (N/A) = not applicable; (*) = p3 present; (±) = Endentulous

| No. | Points | Description | *Smilodon fatalis* | | | | | *Megantereon cultridens* | | | | | *S. californicus* | | *S. gracilis* (LEISEY) | | |
			SCSM 2003.75.12- R	SCSM 2003.75.162- L	SCSM 2003.75.163- L	SCSM 2003.75.164- L	SCSM 2004.1.4- L	BC 106 R	BC 106 L	CB 20 R	CB 20 L	SE 311 L	LACMHC 2002-250 R	LACMHC 2002-250 L	UF 82529 / 81724 R	UF 82529 / 81724 L	BIOPSI 0153 ± L
1	21 & 5	Length from anterior border of c1 alveolus to posterior border of m1 alveolus	106.1	110.2	110.0	105.8	111.7	100.4	100.6	98.3	98.3		137.2	137.3			95.3
2	4 & 8	Length of diastema from posterior border of c alveolus to anterior border of p3 alveolus	43.3	42.1	38.9	N/A	43.0	43.3	43.7	41.9	42.5	32.4	N/A	N/A			32.3
3	4 & 10	Length of diastema from posterior border of ci alveolus to anterior border of p4 alveolus (p3*)	55.7*	54.2*	51.8*	46.9	52.7*	52.8*	55.3*	53.2*	54.3*		70.9	70.9			43.7*
4	8.1 & 10	Length of diastema from posterior border of p3 alveolus to anterior border of p4 alveolus (p3*)	4.5	3.9	3.9	N/A	4.5	2.7	1.5	3.8	4.3		N/A	N/A			0.45
5	8 & 5	Length from anterior border of p3 alveolus to posterior border of m1 alveolus	54.3	54.5	57.9	N/A	59.3	50.0	50.0	48.3	48.9		N/A	N/A	54.3	54.0	54.9
6	10 & 5	Length from anterior border of p4 alveolus to posterior border of m1 alveolus	41.8	42.7	45.7	47.0	47.4	38.5	37.9	37.5	37.9		52.3	52.3			43.9
7	1 & 20	i1 maximum anteroposterior diameter at alveolus	5.5 (a)	5.3 (a)	8.5 (a)	7.7 (a)	7.4 (a)	4.9	4.7	5.8	5.0		7.3	7.4			8.5 (a)
8	1.1 & 27	i1 maximum transverse diameter at alveolus	1.9 (a)	1.9 (a)	1.9 (a)	2.6	2.0 (a)	3.4	3.4	2.2	2.2		3.5	3.4			3.2 (a)
9	3.2 & 20.2	i2 maximum anteroposterior diameter at alveolus	6.6 (a)	7.5 (a)	9.0 (a)	9.0 (a)	8.5	7.2	6.3	7.1	6.3		10.0	9.1			9.0 (a)
10	27.2 & 28	i2 maximum transverse diameter at alveolus	4.5 (a)	4.3 (a)	4.2 (a)	5.4 a	4.4	2.9	3.3	3.1	2.8		9.1	9.8			9.2 (a)

No.	Specimen	Measurement														
11	3.3 & 20.3	i3 maximum anteroposterior diameter at alveolus	8.7 (a)	9.5 (a)	8.9	9.2	9.5	7.1	7.7	7.1	7.0			11.0	10.0	11.5 (a)
12	28.3 & 29	i3 maximum transverse diameter at alveolus	6.5 (a)	7.8 (a)	6.5	7.8	6.5	5.8	5.5	5.1	5.3			8.0	8.1	6.2 (a)
13	30 & 31	c1 maximum anteroposterior diameter at base of enamel	10.8		11.9	11.7	11.4	9.6	8.8	10.1	9.3			13.2	14.0	10.2
14	30.1 & 31.1	c1 maximum transverse diameter at base of enamel	8.5		8.5	9.4	8.7	7.0	7.5	7.8	7.5			9.1	9.1	
15	34 & 35	c1 maximum transverse diameter at alveolus	9.4	9.9	7.8	7.7	7.9	7.8	7.6	7.3	7.7			13.5	14.5	6.0 (a)
16	8 & 8.1	p3 maximum anteroposterior diameter at alveolus	7.4	6.8	7.5	N/A	5.4*	8.8	9.0	7.7	7.9			N/A	N/A	9.3
17	8 & 8.1	p3 maximum anteroposterior diameter at base of enamel	N/A	N/A	N/A	N/A	5.5*	9.0	9.8	8.0	8.0			11.9	9.5	
18	36 & 37	p3 maximum transverse diameter at alveolus	5.2	5.5	5.8	N/A	5.3*							N/A	5.0	5.0
19	36 & 37	p3 maximum transverse diameter at base of enamel	N/A	N/A	N/A	N/A	5.4*	4.5	4.8	4.7	4.6			5.5	5.0	
20	10 & 23	p4 maximum anteroposterior diameter at base of enamel	22.1	20.3	22.1	22.5	22.3	17.6	17.6	17.3 (a)	17.8	20.9	26.0	26.9	18.7	18.6 (a)
21	38 & 39	p4 maximum transverse diameter at base of enamel	9.8	8.6	8.5	10.0	9.1	7.5	7.5	7.2 (a)	7.4	8.6	11.4	11.7	9.0 (a)	10.0 (a)
22	25 & 23.1	m1 maximum anteroposterior diameter at base of enamel	23.4	23.2	23.9	25.6	25.9	21.0	20.1	19.9	20.1	21.8	27.4	27.2	23.7	24.0
23	40 & 41	m1 maximum transverse diameter at base of enamel	11.0	10.5	10.8	11.3	11.9	8.9	8.5	9.7	10.1	9.7	13.2	13.2	13.2	9.9
24	23.1 & 24	m1 length of paraconid blade at base of crown	9.6	9.9	9.8	9.9	10.7	9.5	9.1	9.1	9.9			11.3	11.8	
25	24 & 25	m1 length of protoconid blade at base of crown	13.8	14.1	13.5	15.1	15.4	11.6	13.6	11.0	10.2			15.4	15.8	

REFERENCES

Antón, M., and A. Galobart. 1999. Neck function and predatory behavior in the scimitar toothed cat *Homotherium latidens* (Owen). Journal of Vertebrate Paleontology 19:771-784.

Antón, M., and L. Werdelin. 1998. Too well restored? The case of the *Megantereon* skull from Senèze. Lethaia 31:158-160.

Baskin, J. A., and R. G. Thomas. 2007. South Texas and the great American interchange. Transactions of the Gulf Coast Association of Geological Societies 57:37-45.

Berta, A. 1987. The sabercat *Smilodon gracilis* from Florida and a discussion of its relationships (Mammalia, Felidae, Smilodontini). Bulletin of the Florida State Museum, Biological Science 31:1-63.

Berta, A. 1995. Fossil carnivores from the Leisey Shell Pits, Hillsborough County, Florida. Bulletin of the Florida Museum of Natural History 37:463-500.

Berta, A., and H. Galiano. 1983. *Megantereon hesperus* from the Late Hemphillian of Florida with remarks on the phylogenetic relationships of Machairodonts (Mammalia, Felidae, Machairodontinae). Journal of Paleontology 57:892-899.

Bryant, H. N., and A. P. Russell. 1995. Carnassial functioning in nimravid and felid sabretooths: theoretical basis and robustness of inferences; pp. 116-135 in J. J. Thomason (ed.), Functional Morphology in Vertebrate Paleontology. Cambridge University Press, Cambridge, UK.

Christiansen, P., and J. S. Adolfssen. 2007. Osteology and ecology of *Megantereon cultridens* SE311 (Mammalia, Felidae, Machairodontinae), a sabrecat from the Late Pliocene—Early Pleistocene of Senèze, France. Zoological Journal of the Linnean Society 151:833-884.

Christiansen, P., and J. M. Harris. 2012. Variation in craniomandibular morphology and sexual dimorphism in pantherines and the sabercat *Smilodon fatalis*. PLoS One 7(10):e48352.

Conkin, J. E., and B. M. Conkin. 1962. Pleistocene Berclair terrace of Medio Creek, Bee County, Texas. American Association of Petroleum Geologists Bulletin 46:344-353.

Duckler, G. L. 1997. Parietal depressions in skulls of the extinct saber-toothed felid *Smilodon fatalis*: evidence of mechanical strain. Journal of Vertebrate Paleontology 17:600-609.

Emerson, S. B., and L. Radinsky. 1980. Functional analysis of sabertooth cranial morphology. Paleobiology 6:295-312.

García-Perea, R. 2002. Andean mountain cat, *Oreailurus jacobita*: morphological description and comparison with other felines from the Altiplano. Journal of Mammalogy 83:110-124.

Hillson, S. 1986. Teeth; in D. Brothwell, B. Cunliffe, S. Fleming, and P. Fowler (eds.), Cambridge Manuals in Archaeology. Cambridge University Press, New York, 376 pp.

Hulbert, R. C. Jr. 2001. Florida fossil vertebrates; pp. 25-33 in R. C. Hulbert Jr. (ed.), The Fossil Vertebrates of Florida. University Press of Florida, Gainesville, Florida.

Knight, J. L., and L. D. Martin. 2005. Preliminary Report on the Late Irvingtonian Camelot Local Fauna, Dorchester County, South Carolina. Abstracts, Tate Conference, Casper College, Casper, Wyoming, p. 12.

Knight, J. L., and L. D. Martin. 2007. Comments on the Predator-Prey Population Structure, Paleoecology, and Taphonomy of the Camelot Local Fauna (Late Irvingtonian), Berkeley County, South Carolina. Abstracts, Tate Conference, Casper College, Casper, Wyoming, p. 63.

Kohn, M. J., M. P. McKay, and J. L. Knight. 2005. Dining in the Pleistocene—who's on the menu? Geology 33:649-652.

Kurtén, B. 1965. The Pleistocene Felidae of Florida. Bulletin of the Florida State Museum, Biological Sciences 9:215-273.

Kurtén, B., and E. Anderson. 1980. Pleistocene Mammals of North America. Columbia University Press, New York, 422 pp.

Kurtén, B., and L. Werdelin. 1990. Relationships between North and South American *Smilodon*. Journal of Vertebrate Paleontology 10:158-169.

Leidy, J. 1868. Notice of some vertebrate remains from Harden Co., Texas. Proceedings of the Academy of Natural Sciences of Philadelphia 20:174-176.

Lund, P. W. 1841. Tillæg til de to sidste afhandlinger over Brasiliens dyreverden för sidste jordomvæltning. Det Kongelige Danske videnskabernes naturvidenskabelige og mathematiske afhandlinger 8:273-296.

Lund, P. W. 1842. Blik paa Brasiliens Dyreverden för sidste jordomvæltning. Fjerde afhandling: Fortsættelse af pattedyrene. Det Kongelige Danske videnskabernes selskabs naturvidenskabelige og mathematiske afhandlinger 9:137-208.

Martin, L. D., and S. A. Czerkas. 2000. The fossil record of feather evolution in the Mesozoic. American Zoologist 40:687-694.

Martin, L. D., J. P. Babiarz, and V. L. Naples. 2011. The osteology of a cookie-cutter cat, *Xenosmilus hodsonae*; pp. 43-97 in V. L. Naples, L. D. Martin, and J. P. Babiarz (eds.), The Other Saber-tooths: Scimitar-tooth Cats of the

Western Hemisphere. Johns Hopkins University Press, Baltimore, Maryland.

Martin, L. D., C. B. Schultz, and M. R. Schultz. 1988. Saber-Toothed Cats from the Plio-Pleistocene of Nebraska. Transactions of the Nebraska Academy of Sciences 16:153-163.

Martin, L. D., J. P. Babiarz, V. L. Naples, and J. Hearst. 2000. Three ways to be a saber-toothed cat. Naturwissenschaften 87:41-44.

Martin, P. S. 1967. Prehistoric overkill; pp. 75-120 in P. S. Martin and H. E. Wright Jr. (eds.), Pleistocene Extinctions, the Search for a Cause. Yale University Press, New Haven, Connecticut.

Martin, P. S., and H. E. Wright Jr. (eds.). 1967. Pleistocene Extinctions: The Search for a Cause. Yale University Press, New Haven, Connecticut, 453 pp.

Matthew, W. D. 1910. The phylogeny of the Felidae. Bulletin of the American Museum of Natural History 28:289-316.

Merriam, J. C., and C. Stock. 1932. The Felidae of Rancho La Brea. Carnegie Institution of Washington, Publication No. 422:1-231.

Miller, G. J. 1968. On the age distribution of *Smilodon californicus* Bovard from Rancho La Brea. Los Angeles County Museum of Natural History, Contributions in Science 131:1-17.

Morgan, G. S., and R. C. J. Hulbert Jr. 1995. Overview of the geology and vertebrate biochronology of the Leisey Shell Pit local fauna, Hillsborough County, Florida. Bulletin of the Florida Museum of Natural History 37:1-92.

Naples, V. L., and L. D. Martin. 2000. Evolution of hystricomorphy in the Nimravidae (Carnivora, Barbourofelinae): evidence for complex character convergence with rodents. Historical Biology 14:169-188.

Naples, V. L., L. D. Martin, and J. P. Babiarz (eds.). 2011. The Other Saber-tooths: Scimitar-tooth Cats of the Western Hemisphere. Johns Hopkins University Press, Baltimore, Maryland, 236 pp.

Palmqvist, P., V. Torregrosa, J. A. Pérez-Claros, B. Martínez-Navarro, and A. Turner. 2007. A re-evaluation of the diversity of *Megantereon* (Mammalia, Carnivora, Machairodontinae) and the problem of species identification in extinct carnivores. Journal of Vertebrate Paleontology 27:160-175.

Riviere, H. L., and H. T. Wheeler. 2005. Cementum on *Smilodon* sabers. Anatomical Record 285:634-642.

Sardella, R. 1998. The Plio-Pleistocene Old World dirk-toothed cat *Megantereon* ex gr. *cultridens* (Mammalia, Felidae, Machairodontinae), with comments on taxonomy, origin and evolution. Neues Jahrbuch für Geologie und Paläontologie, Abhandlungen 207:1-36.

Schultz, C. B., M. R. Schultz, and L. D. Martin. 1970. A new tribe of saber-toothed cats (Barbourofelini) from the Pliocene of North America. Bulletin of the University of Nebraska State Museum 9:1-31.

Slater, G. J., and B. Van Valkenburgh. 2008. Long in the tooth: evolution of sabertooth cat cranial shape. Paleobiology 34:403-419.

Slaughter, B. H. 1963. Some observations concerning the genus *Smilodon*, with special reference to *Smilodon fatalis*. Texas Journal of Science 15:68-81.

Tejada-Flores, A. E., and C. A. Shaw. 1984. Tooth replacement and skull growth in *Smilodon* from Rancho La Brea. Journal of Vertebrate Paleontology 4:114-121.

Toohey, L. 1959. The species of *Nimravus* (Carnivora, Felidae). Bulletin of the American Museum of Natural History 118:71-112.

Van Valkenburgh, B., and C. B. Ruff. 1987. Canine tooth strength and killing behaviour in large carnivores. Journal of Zoology 212:379-397.

Wallace, S. C., and R. C. Hulbert Jr. 2013. A new machairodont from the Palmetto fauna (Early Pliocene) of Florida, with comments on the origin of the Smilodontini (Mammalia, Carnivora, Felidae). PLoS One 8:e56173.

Webb, S. D. 1974. The status of *Smilodon* in the Florida Pleistocene; pp. 149-157 in S. D. Webb (ed.), Pleistocene Mammals of Florida. University Presses of Florida, Gainesville, Florida.

Wheeler, H. T., H. L. Riviere, T. J. Fremd, and J. P. Babiarz. 2004. Convergent evolution in the enamel and gingiva of the nimravid *Pogonodon* and the felid *Smilodon* revealed in new material from John Day Fossil Beds National Monument. Geological Society of America Abstracts with Programs 36:53.

Woodard, G. D., and L. F. Marcus. 1973. Rancho La Brea fossil deposits: a re-evaluation from stratigraphic and geological evidence. Journal of Paleontology 47:54-69.

6 Understanding Killing Behavior in *Smilodon fatalis*: The Role of Computational Biomechanics

STEPHEN WROE AND WILLIAM C. H. PARR

Introduction

There are few, if any, mammalian fossil species that have excited more scientific or popular interest than the great sabertooth cat, *Smilodon fatalis*. For over 160 years, researchers have debated almost every aspect of its ecology and behavior. Was it a social predator? What was its primary prey? How big was it? Did it eat bones? Why did it go extinct (Christiansen and Harris, 2005; Van Valkenburgh, 2007; Meachen-Samuels and Van Valkenburgh, 2010; Andersson et al., 2011; Christiansen, 2011, 2012; DeSantis et al., 2012; Meachen-Samuels, 2012; Meachen et al., 2014; Van Valkenburgh et al., 2016)? However, perhaps the single most asked question has been, how did it kill (Matthew, 1910; Emerson and Radinsky, 1980; Akersten, 1985; Bryant, 1996; Antón et al., 2004; McHenry et al., 2007; Wroe et al., 2008)? Some have even suggested that it was not an active predator at all (Marinelli, 1939). Only one theme appears to have remained constant: that the cranio-dental and postcranial morphology of the sabertooth cat was so distinct from that of any living predator that it must have been doing something different.

Some of the earliest interpretations of killing behavior in *Smilodon fatalis* presented a portrait of a stabbing, slashing predator, wildly wielding its huge saberteeth as a set of knives (Warren, 1853; Matthew, 1910). Debate has also centered on which part of the prey's anatomy the killing bite was applied to, the throat or the abdomen (Akersten, 1985). Central to many interpretations has been the conclusion that the jaw adducting muscles of *S. fatalis* were relatively weak, and that jaw muscle-delivered bite reaction forces must have been weak too, especially at initial wide gapes, where mechanical advantage is minimized. However, the claim that the sabertooth cat's bite was weak has also been contested (Therrien, 2005).

Over the last decade or so, however, a somewhat limited consensus appears to have been reached. Studies of cranio-dental mechanics combined with evidence that both the forelimb and the head-depressing musculature of *Smilodon fatalis* were particularly well developed have led most researchers to conclude that it was an active predator that killed by restraining typically large prey and applying a bite to the throat (Christiansen, 2006; McHenry et al., 2007). In these studies, the head depressing neck muscles are thought to play a significant role in the deployment of a 'canine-shear' bite (Akersten, 1985). In this scenario, the head is rotated about the cranio-cervical joint and opposition from the mandible and lower canines allows the upper canines to puncture prey tissues. The prey animal is dispatched quickly through rapid blood loss as major blood vessels and perhaps the windpipe are severed.

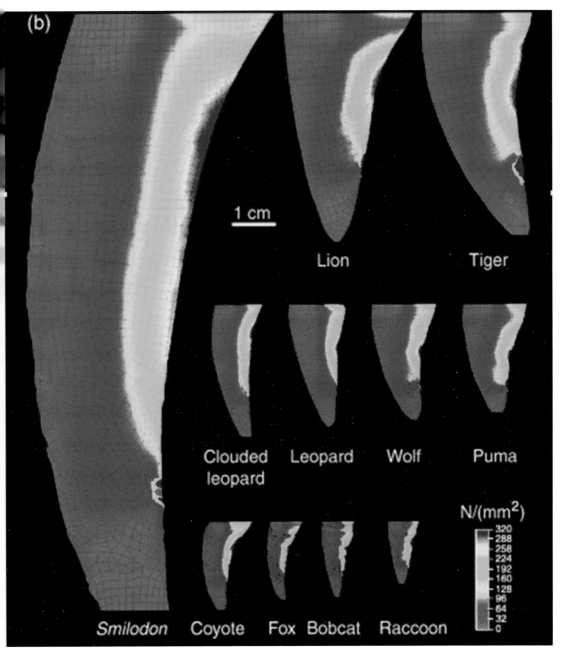

(b)

1 cm

Lion

Tiger

Clouded
leopard

Leopard

Wolf

Puma

N/(mm²)

320
288
258
224
192
160
128
96
64
32
0

Smilodon

Coyote

Fox Bobcat

Raccoon

Plate 6.1. Finite element analysis mesh models taken from whole canine teeth showing the distribution of stresses. Teeth to scale. From Freeman and Lemen (2007).

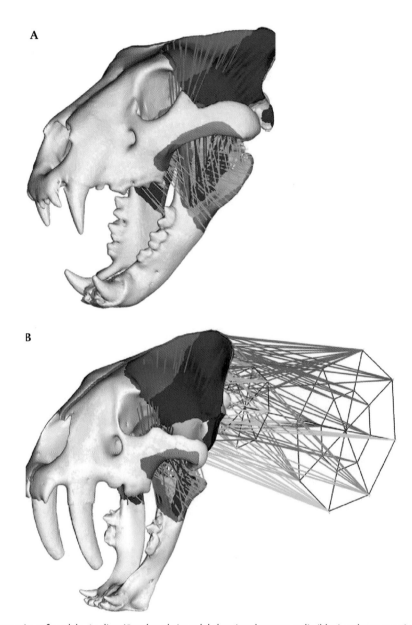

Plate 6.2. Construction of models. A = lion (*Panthera leo*) model showing the temporalis (blue) and masseter (red) systems muscle 'beams' and attachment areas. B = *Smilodon fatalis* model showing the neck assembly. The different colored beams on the neck correspond to neck muscle groups in felids. From McHenry et al. (2007). Note that although the crania of A and B are approximately the same size, the mandible of *S. fatalis* is smaller and more gracile than that of the lion.

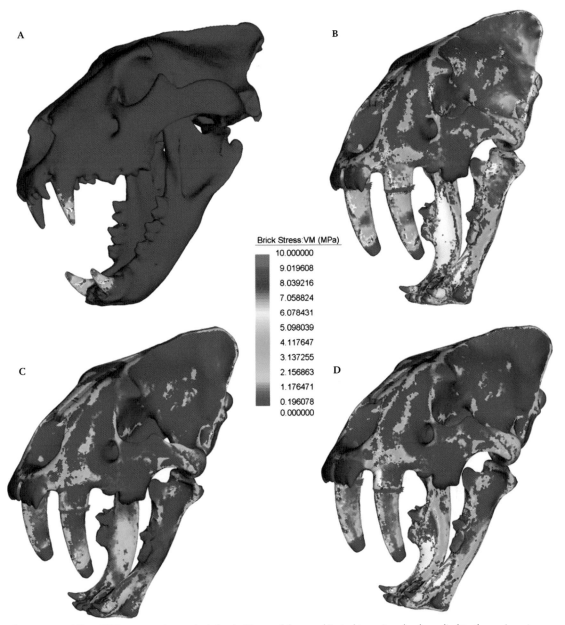

Plate 6.3. Von Mises (VM) stress under extrinsic loads. The models are subjected to various loads applied to the canines. Jaw and neck muscles are used to brace the skull but do not apply forces. A = lion (*Panthera leo*) with 2,000 N lateral force (extrinsic load case 1: lateral shake). B = *Smilodon fatalis* with 2,000 N lateral force (extrinsic load case 1). C = *S. fatalis* with 100 Nm axial moment (extrinsic load case 2: twist). D = *S. fatalis* with 2,000 N anterior force (extrinsic load case 3: pull-back). From McHenry et al. (2007).

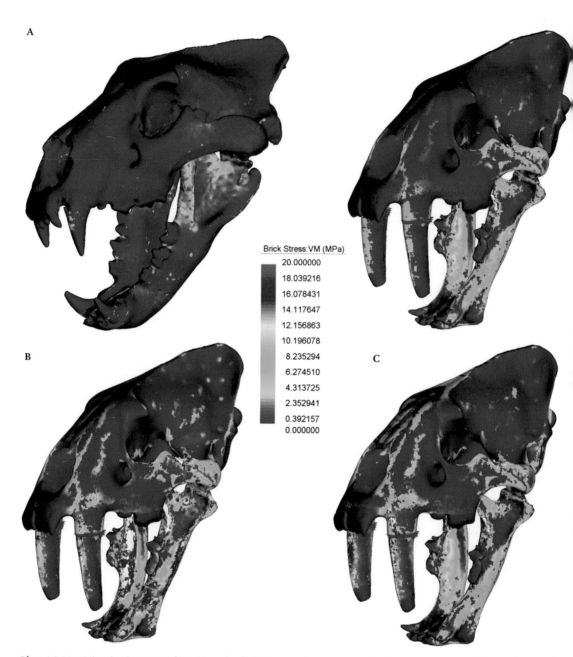

Plate 6.4. Von Mises (VM) stress under intrinsic loads (bilateral canine bites). A = bite force predicted by 3D dry skull method, adjusted to account for pennation; shown are lion (*Panthera leo*) biting at 3,388 N (left) and *Smilodon fatalis* biting at 1,104 N (right). B, C = *S. fatalis* biting at the forces calculated from ref. 13 for the regression of bite force on body mass for a 229 kg felid (2,110 N), powered by jaw adductors only (B) and by neck and jaw muscles (C). From McHenry et al. (2007).

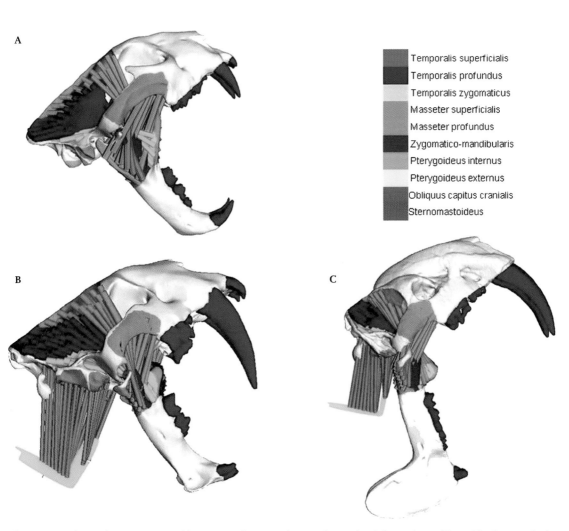

A

Temporalis superficialis
Temporalis profundus
Temporalis zygomaticus
Masseter superficialis
Masseter profundus
Zygomatico-mandibularis
Pterygoideus internus
Pterygoideus externus
Obliquus capitus cranialis
Sternomastoideus

B

C

Plate 6.5. Muscle simulations. A = jaw-adducting muscles in *Panthera pardus*; B = head-depressing and jaw-adducting muscles in *Smilodon fatalis*; C = head-depressing and jaw-adducting muscles in *Thylacosmilus atrox*. From Wroe et al. (2013).

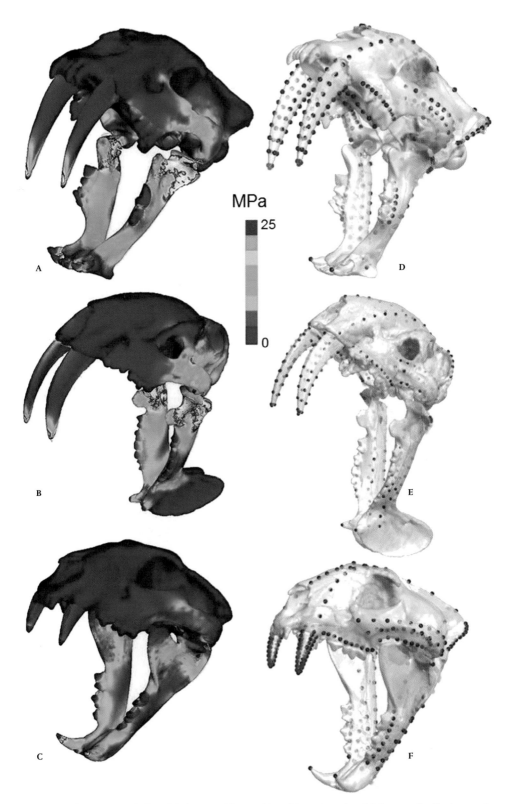

Plate 6.6. Stress distributions in scaled models for jaw-adductor-driven bites. Von Mises (VM) stress distributions and mean landmark point VM stresses given respectively for: A, D = *Smilodon fatalis*; B, E = *Thylacosmilus atrox*; C, F = *Panthera pardus*. MPa = Megapascals. Muscle forces scaled to bite reaction forces predicted on the basis of body mass. From Wroe et al. (2013).

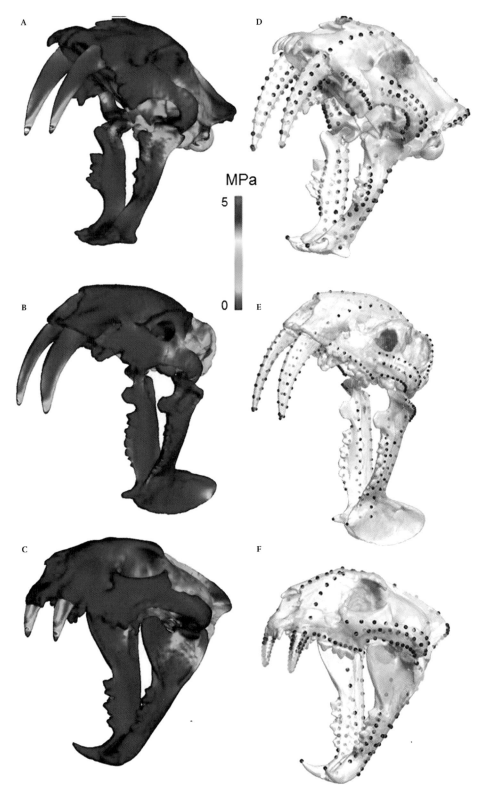

MPa

5

0

Plate 6.7. Stress distributions in scaled models for neck-muscle-driven bites. Von Mises (VM) stress distributions and mean landmark point VM stresses given respectively for: A, D = *Smilodon fatalis*; B, E = *Thylacosmilus atrox*; C, F = *Panthera pardus*. MPa = Megapascals. Muscle forces scaled to bite reaction forces predicted on the basis of body mass. From Wroe et al. (2013).

This interpretation has appeal. There are obvious advantages to the predator. It is more energetically efficient for mammalian carnivores to take relatively large prey, perhaps critically so for big predators (Carbone et al., 1999), but there are caveats, the most obvious of these being that large prey are dangerous and hard to kill. It may take a large conical-toothed cat, such as a lion, up to 13 minutes to kill a large herbivore through a clamp-and-hold bite to the throat or muzzle, using canines that may not even fully pierce the skin (Schaller, 1972). Until it is dead, the prey poses a potentially lethal threat to the predator (Schaller, 1972). Moreover, the commotion associated with killing may well attract aggressive herd members or other predators, and formidable as it undoubtedly was, the Late Pleistocene *Smilodon fatalis* lived in habitats that may have overlapped with those of some of the largest mammalian carnivores ever to have existed, including the short-faced bear and American lion, not to mention competition with conspecifics.

Thus analyses of craniodental mechanics have played a pivotal role in our arrival at this point. Traditional beam theory has long provided important contributions and continues to do so. The majority have been consistent with the canine-shear bite hypothesis, concluding that jaw adductor-driven bite forces were relatively low. However, others have maintained that bite force in *Smilodon fatalis* was not weak because the mandible is robustly shaped and comprises a high proportion of stiff cortical bone (Akersten, 1985). More recently, it has been argued that the canine shear-bite hypothesis is flawed and that rotation about the cranio-cervical joint is not feasible (Brown, 2014).

The Contribution of Computational Biomechanics

Over the last decade or so a relatively new computational approach, finite element analysis (FEA), has been applied to questions surrounding killing behavior in *Smilodon fatalis*. Originally developed by the aerospace industry over 60 years ago, FEA can now be applied to the more complex shapes and volumes that face students of biology and biomedicine thanks to major advances in computing power (Rayfield et al., 2001).

In FEA, a discrete number of geometrically simple elements to which material properties can be attributed approximates complex structures. Loaded and restrained in such a way as to simulate actual or hypothesized behaviors, patterns, and magnitudes of stress, strain and displacement can be calculated and displayed in 2 or 3 dimensions (Rayfield, 2007).

The ability to easily predict stress and strain throughout the structure is of particular importance because it allows the researcher to assess whether the structure under investigation is well adapted to sustaining proposed loads in either an absolute or comparative sense. Validation studies based on experimental data are critical if we are to determine absolute results, such as under which loadings a structure would fail. This is less crucial in comparative FEA, where the objective is to determine whether one structure is relatively better adapted than another to a particular behavior or loading and restraint scenario.

With models typically based on computed tomography, FEA is of obvious value in biology, and paleontology in particular, because it is a nondestructive process. Over the last decade it has been widely applied to the study of vertebrate mechanics (Rayfield, 2005; Wroe et al., 2007; Wroe, 2008; Wroe et al., 2008b; Slater and Van Valkenburgh, 2009; Tseng, 2009; Wroe et al., 2010; Tseng and Flynn, 2015; Attard et al., 2016; Tseng et al., 2016).

So what has FEA contributed to the debate over killing behavior in *Smilodon fatalis*, and perhaps just as importantly, how might it assist in the future? To date, four relevant studies have been conducted.

The first of these was the analysis presented by Freeman and Lemen (2007). These authors incorporated both beam theory and FEA to analyze and predict mechanical performance in canine teeth for a sample of 11 carnivoran species, including *Smilodon fatalis*. Models were painstakingly created by sectioning teeth at 0.635 mm intervals with a low-speed saw. Cross sections were hand drawn under a camera lucida, then scanned and digitized into a computer.

Results for some of the more common species were validated by taking the canines from wild-caught specimens to the point of failure using a materials testing system. Combining experimental values with values calculated from the FEA, the

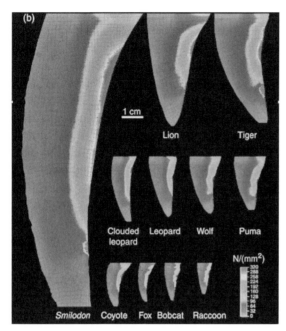

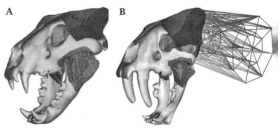

Figure 6.2. Construction of models. *See color plates following page 108.* A = lion (*Panthera leo*) model showing the temporalis and masseter systems muscle 'beams' and attachment areas. B = *Smilodon fatalis* model showing the neck assembly. The beams on the neck correspond to neck muscle groups in felids. From McHenry et al. (2007). Note that although the crania of A and B are approximately the same size, the mandible of *S. fatalis* is smaller and more gracile than that of the lion.

Figure 6.1. Finite element analysis mesh models taken from whole canine teeth showing the distribution of stresses. Teeth to scale. *See color plates following page 108.* From Freeman and Lemen (2007).

authors then modeled the log—log relationship between body mass and tooth strength with standard linear regression.

They concluded that the canine of a 'typical' mammalian predator could support around four times its body mass (Fig. 6.1). The canines of *Smilodon fatalis* on the other hand would have failed at around 2.2 times its body mass. More recently, analysis of tooth fracture mechanics has broadly supported this conclusion (Lawn et al., 2013).

Importantly, Freeman and Lemen (2007) determined that FEA was a more accurate predictor than beam theory (BTA) based on comparison with their experimentally derived results, stating, "We conclude that if BTA is to be used, the results must be verified using FEA. Given the greater power and the growing availability of FEA software, this method is preferable over BTA" (Freeman and Lemen, 2007:186).

In the same year, the senior author and others published a study in which we used FEA to compare mechanical performance in an African lion (*Panthera leo*) and *Smilodon fatalis* (McHenry et al., 2007). This was the first use of FEA in a study incorporating a fossil mammalian carnivore.

These were 3D models incorporating both crania and mandibles, wherein we accounted for multiple (8) properties of bone (from stiff cortical to less stiff cancellous), and approximated both jaw adducting and head depressing musculature using pre-tensioned trusses (Fig 6.2). We applied both intrinsic and extrinsic loads to simulate 'biting' driven by both jaw and cervical muscles, and further to simulate stress distributions and magnitudes generated by struggling prey. These loadings were proportionate to estimates of body mass for the two cats modeled. Although we simulated the cervical musculature, its inclusion was to provide a more realistic restraint for the skull, and we did not attempt to load the neck muscles.

Different analyses were conducted with jaw-adducting loads based first on specimen-specific estimates of muscle cross-sectional area from the skulls themselves, and second, scaled to body mass on the basis of regression data for felids from Wroe et al. (2005). Scaling is important and is a subject we return to later.

From these analyses we found that under loadings simulating lateral or torsional forces, as might be expected in encounters with unrestrained, struggling prey, the sabertooth cat's skull showed much higher stresses than the lion's (Fig. 6.3). These findings were certainly consistent with the canine shear-bite hypothesis, implying that the prey of *Smilodon fatalis* had to be secured before the killing bite was delivered.

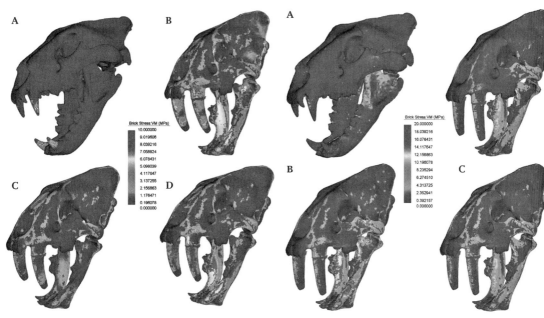

Figure 6.3. Von Mises (VM) stress under extrinsic loads. The models are subjected to various loads applied to the canines. Jaw and neck muscles are used to brace the skull but do not apply forces. *See color plates following page 108.* A = lion (*Panthera leo*) with 2,000 N lateral force (extrinsic load case 1: lateral shake).B = *Smilodon fatalis* with 2,000 N lateral force (extrinsic load case 1). C = *S. fatalis* with 100 Nm axial moment (extrinsic load case 2: twist). D = *S. fatalis* with 2,000 N anterior force (extrinsic load case 3: pull-back). From McHenry et al. (2007).

Figure 6.4. Von Mises (VM) stress under intrinsic loads (bilateral canine bites). *See color plates following page 108.* A = bite force predicted by 3D dry skull method, adjusted to account for pennation; shown are lion (*Panthera leo*) biting at 3,388 N (left) and *Smilodon fatalis* biting at 1,104 N (right). B, C = *S. fatalis* biting at the forces calculated from ref. 13 for the regression of bite force on body mass for a 229 kg felid (2,110 N), powered by jaw adductors only (B) and by neck and jaw muscles (C). From McHenry et al. (2007).

Bite reaction forces derived from specimen-specific unscaled modeling of biting at the canines showed that the sabertooth cat's bite was less than one-third as powerful as that of the lion (~1100 N vs. 3400 N, respectively). In another analysis, applying jaw-muscle adducting forces that would be expected for a cat the size of our *Smilodon fatalis* specimen, we found very high stresses in both its cranium and mandible relative to the lion (Fig. 6.4).

Overall, we interpreted these results as indicative of a predator that needed to apply its killing bite with some precision, if not finesse. We concluded that the prey animal needed to be well restrained first. This is consistent with what we know of the sabertooth cat's massive and robust postcranial anatomy, which at least in some respects is superficially more bearlike than catlike (Wroe et al., 2008; see also chapter 11, this volume). Strength in the forelimbs of *Smilodon fatalis* appears to be particularly extreme (Gonyea, 1978; Meachen-Samuels and Van Valken-

burgh, 2010; Martín-Serra et al., 2014). Moreover, it has since been demonstrated, in a comparison of sabertooth carnivores from three carnivoran families, that taxa with longer, more slender canines tend to have more robust forelimbs, with this hyper-robusticity especially marked in *S. fatalis* (Meachen-Samuels, 2012).

Some years later, we and other colleagues published a further FEA-based analysis incorporating a skull model of *Smilodon fatalis* (Wroe et al., 2013). In this study, we compared the sabertooth cat to another sabertooth mammalian carnivore, the South American metatherian *Thylacosmilus atrox*, as well as another extant conical-toothed cat, the leopard (*Panthera pardus*).

Although it is a considerably smaller animal (for our specimens ~80 kg vs. ~260 kg, respectively), based on morphology alone *Thylacosmilus atrox* shows even more extreme adaptations than *Smilodon fatalis*. Relative to the size of the animal the laterally

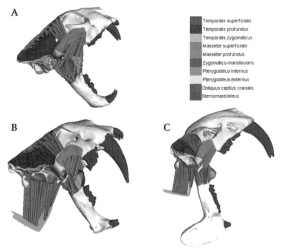

Temporalis superficialis
Temporalis profundus
Temporalis zygomaticus
Masseter superficialis
Masseter profundus
Zygomatico-mandibularis
Pterygoideus internus
Pterygoideus externus
Obliquus capitus cranialis
Sternomastoideus

Figure 6.5. Muscle simulations. *See color plates following page 108.* A = jaw-adducting muscles in *Panthera pardus*; B = head-depressing and jaw-adducting muscles in *Smilodon fatalis*; C = head-depressing and jaw-adducting muscles in *Thylacosmilus atrox*. From Wroe et al. (2013).

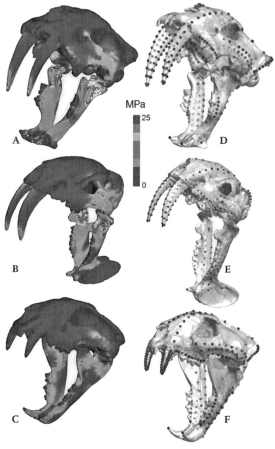

Figure 6.6. Stress distributions in scaled models for jaw-adductor-driven bites. *See color plates following page 108.* Von Mises (VM) stress distributions and mean landmark point VM stresses given respectively for: A, D = *Smilodon fatalis*; B, E = *Thylacosmilus atrox*; C, F = *Panthera pardus*. MPa = Megapascal. Muscle forces scaled to bite reaction forces predicted on the basis of body mass. From Wroe et al. (2013).

compressed upper canines are longer, the gape wider, and the skull generally more gracile. Furthermore, the area available for the attachment of jaw muscles in *T. atrox* is relatively tiny (Turnbull, 1976). Previous estimates of jaw-driven bite force were extremely low given the size of the animal (Wroe et al., 2005).

This study differed from the McHenry et al. (2007) analysis in that the action of head-depressing musculature was simulated as well as the previously applied intrinsic and extrinsic forces (Fig. 6.5). Where comparable, with respect to *Smilodon fatalis* the results were consistent with those from McHenry et al. (2007). Applying jaw-adductor-driven bite forces scaled to body mass, the saber-tooth cat showed considerably more stress than did the conical-toothed leopard (Fig. 6.6). Similarly, *Panthera pardus* outperformed *S. fatalis* in extrinsic loadings that simulated lateral shaking or torsion (Fig. 6.7).

Of further interest, in the simulation of head depression using cervical musculature, which is also proportional to body mass, we found that *Smilodon fatalis* was not clearly more stressed than *Panthera pardus*. From this, we concluded that although the sabertooth cat was not as well adapted to deliver high jaw-muscle-driven bites or to resist lateral or torsional forces, it was well adapted to resist the

forces that would be generated where the neck musculature was deployed to drive the upper canines ventrally with the cranium in rotation about the atlanto-occipital joint.

Lastly, we concluded that relative mechanical performance indicated even more extreme adaptation in the metatherian sabertooth carnivore than in the *Smilodon fatalis* (Figs. 6.6 and 6.7). Thus, we found that the bite reaction forces for *S. fatalis* (519 Newtons [N]), were not much greater than for the leopard (484 N), an animal around one-third its body mass. However, jaw-muscle-driven bite force at the canines for *Thylacosmilus atrox* was a tiny 38 N, and its performance under extrinsic loads was

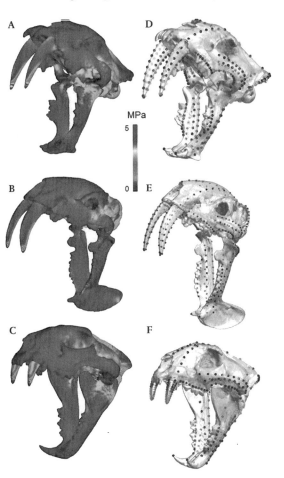

Figure 6.7. Stress distributions in scaled models for neck-muscle-driven bites. *See color plates following page 108.* Von Mises (VM) stress distributions and mean landmark point VM stresses given respectively for: A, D = *Smilodon fatalis*; B, E = *Thylacosmilus atrox*; C, F = *Panthera pardus*. MPa = Megapascal. Muscle forces scaled to bite reaction forces predicted on the basis of body mass. From Wroe et al. (2013).

poorer than that of the sabertooth cat. In only one simulation did the metatherian outperform *S. fatalis*, and this was in its ability to sustain forces generated by head depressing cervical musculature (Fig. 6.7).

These results further supported those of McHenry et al. (2007) regarding *Smilodon fatalis* in that they showed that the sabertooth cat was relatively well adapted to drive its upper canines into prey using its neck muscles. More broadly, we concluded that *Thylacosmilus atrox* represented a still more extreme sabertooth morphotype than *S. fatalis*.

In the same year, a further study applied FEA in an analysis including *Smilodon fatalis* (Piras et al., 2013). These authors integrated FEA with geometric morphometric (GMM) analyses, a growing trend in form-function studies (Parr et al., 2012). This was a wide-ranging study that included mandibles from 35 extant (all felid) and 59 extinct (felid, nimravid, and barbourofelid) species. The objective of the study was to identify and quantify any relationships between modularity (e.g., integration) and mechanical performance in the mandible of cats and cat-like mammalian carnivores. The mandibles were divided into two modules (with functional and developmental a priori reasoning): the tooth-bearing corpus and the ascending ramus. Ten landmarks and 27 semi-landmarks were applied to the outline of the mandible as a whole.

The very large sample size clearly sets this study apart from the previous FEA-based works by Freeman and Lemen (2007), McHenry et al. (2007), and Wroe et al. (2013) because different questions were asked. It was further distinguished from these studies in that the FEA models were two-dimensional as opposed to three-dimensional. Moreover, in this study the FE models were created by interpolating the landmarks and semilandmarks around the outlines of the mandibles, once the landmarks had been scaled to unit centroid size (the Procrustes coordinates). In other words, the FE models, from their creation, were scaled to the same size. The authors then rescaled all of these models to the same unit size using centroid size (as calculated from the Procrustes analysis of the original landmark points). This effectively geometrically scales all of the models to the same nominal size. The authors then applied a scaled bite force to each of the models using centroid size. The results of these FE analyses agree with McHenry et al. (2007) and Wroe et al. (2013) in that the sabertooth mandibles, including *Smilodon fatalis*, exhibit higher von Mises (VM) stress than is observed in the mandibles of other clades.

The results were of considerable interest to students of form and function in mammalian carnivores, and in sabertooth carnivores in particular. Piras et al. (2013) found that mechanical performance and morphological integration were linked in the mandible: specifically the per-clade average measure of integration and the per-clade average mandibular VM stress and coronoid 'surface traction' results were significantly related. The integration levels were higher in the sabertooth clades than in the conical-toothed clades. This means that there was a higher degree of integration between corpus and ascending ramus modules of the mandible in the sabertooth clades compared to the conical toothed clades. The authors then found a relationship between this high integration of the two mandible modules and the mandibular VM stress in the sabertooth clades. Although a significant link does not necessarily mean that there is a direct evolutionary cause-and-effect relationship between the two measures of mandibular integration and stress, it certainly adds weight to the hypothesis that macroscopic bone shape factors (e.g., interspecific overall bone shape, bone modular shape, and bone modular covariation [integration]) are influenced by functionally generated stress (e.g., during feeding). The authors did not draw any direct inferences on killing behavior in *Smilodon fatalis* based on their results; however, the study demonstrates again that computational biomechanics and the application of FEA currently provide some of the best tools for testing such wider-scope evolutionary form-function hypotheses.

Concluding Remarks

To date much of the contribution of computational biomechanics to the determination of killing behavior in *Smilodon fatalis* has largely been to further refine and delimit the range of potential behaviors that the sabertooth cat was best adapted to perform. These studies have all been relative to the extant conical-toothed cats (Freeman and Lemen, 2007; McHenry et al., 2007), and the metatherian sabertooth, *Thylacosmilus atrox* (Wroe et al., 2013).

Considered together, these analyses have shown that the sabertooth cat's huge upper canine teeth were not adapted to resist high bending loads relative to its body size, and that neither its cranium nor its mandible were well adapted to resist the high loadings that might be expected in encounters with relatively large, unrestrained prey. Similarly, FEA-based study strongly suggests that although *Smilodon fatalis* compares favorably with conical-toothed cats in the ability of its skull to resist forces generated by head depressing cervical musculature, its skull is not well adapted to resist jaw-muscle-driven forces that would be expected for a conical-toothed cat of its great size. The conclusion: *S. fatalis* killed using a 'bite' in which the neck muscles played a much more important role than in extant cats and that large prey would have been restrained before the killing bite was delivered.

One point here is deserving of further consideration and explanation, the contention that jaw-adductor-driven bite force in *Smilodon fatalis* was relatively weak. This proposition was previously put forward on the basis of comparative analysis using beam theory adjusted for differences in body mass (Wroe et al., 2005). Finite element analyses have been consistent with findings by Wroe et al. (2005), suggesting a bite reaction force comparable to that of a large leopard in a 200 kg-plus *S. fatalis*. Moreover, the FEA provides further support for this argument because it shows that the sabertooth cat's jaws could not sustain the forces that would be generated by a conical-toothed cat of comparable body mass.

It may be argued that a bite on par with that of a 70 kg leopard cannot be called weak (Christiansen, 2006). We would disagree, but at the very least it can most certainly be called relatively weak for an animal of its size, and this has decided functional implications for the animal, which is the subject of interest here.

To the best of our knowledge, arguments to the effect that the *Smilodon fatalis* had a strong bite do not properly account for size. For example, based on beam theory Therrien (2005) argued that the bites of *S. fatalis* and other carnivoran sabertooths were comparable to, or higher than those of similar-sized extant felids. The inherent assumption here is that mandibular length is a reasonable proxy for the size of the animal. However, certainly for *S. fatalis*, the cranium and mandible are relatively short for a felid of its body mass (McHenry et al., 2007). A *S. fatalis* with a mandible the same length as that of a *Panthera leo* would be far heavier than the lion.

In short, it may be that a mandible of *Smilodon fatalis* scaled to the same length as that of a *Panthera leo* might be able to resist comparable forces to that of the lion (this is certainly testable with FEA). But then we would be comparing two very different sized cats, a relatively small *S. fatalis* and a relatively large *P. leo*. In terms of identifying and understanding differences in predatory behavior we do not see the point, and the fact remains that relative to its body mass, most studies, including FEA-based ones, have concluded that the bite of the sabertooth cat is relatively weak—and from this it can be reasonably inferred that its killing behavior was different.

Lastly, while we would argue that while computational biomechanics has made a significant contribution to our understanding of killing behavior in *Smilodon fatalis*, clearly many questions remain unanswered and much more can be done. Perhaps most obviously, modeling could be extended to include a far greater range of extinct species, especially other dirk-toothed and scimitar-toothed species, to test more comprehensively for differences in mechanical performance that could shed light on differences in killing behavior. The incorporation of other computational approaches, such as multibody dynamics, could offer a means of obtaining a more accurate approximation of loading conditions (Moazen et al., 2009). Such approaches could comprehensively test the canine-shear bite hypothesis.

REFERENCES

Akersten, W. A. 1985. Canine function in *Smilodon* (Mammalia, Felidae, Machairodontinae). Natural History Museum of Los Angeles County, Contributions in Science 356:1-22.

Andersson, K., D. B. Norman, and L. Werdelin. 2011. Sabretoothed carnivores and the killing of large prey. PLoS One 6:e24971.

Antón, M., M. J. Salesa, J. F. Pastor, I. M. Sánchez, S. Fraile, and J. Morales. 2004. Implications of the mastoid anatomy of larger extant felids for the evolution and predatory behaviour of sabretoothed cats (Mammalia, Carnivora, Felidae). Zoological Journal of the Linnean Society 140:207-221.

Attard, M. R. G., L. A. B. Wilson, T. H. Worthy, P. Scofield, P. Johnston, W. C. H. Parr, and S. Wroe. 2016.

Moa diet fits the bill: virtual reconstruction incorporating mummified remains and prediction of biomechanical performance in avian giants. Proceedings of the Royal Society of London B: Biological Sciences 283:1822.

Brown, J. G. 2014. Jaw function in *Smilodon fatalis*: a reevaluation of the canine shear-bite and a proposal for a new forelimb-powered Class 1 lever model. PLoS One 9:e107456.

Bryant, H. N. 1996. Force generation by the jaw adducture musculature at different gapes in the Pleistocene Sabretoothed felid *Smilodon*; pp. 283-299 in K. M. Stewart and K. L. Seymour (eds.), Palaeoecology and Palaeoenvironments of Late Cenozoic Mammals: Tributes to the Career of C. S. (Rufus) Churcher. University of Toronto Press, Toronto, Canada.

Carbone, C., G. M. Mace, S. C. Roberts, and D. W. Macdonald. 1999. Energetic constraints on the diet of terrestrial carnivores. Nature 402:286-288.

Christiansen, P. 2006. Sabertooth characters in the clouded leopard (*Neofelis nebulosa* Griffiths 1821). Journal of Morphology 267:1186-1198.

Christiansen, P. 2011. A dynamic model for the evolution of sabrecat predatory bite mechanics. Zoological Journal of the Linnean Society 162:220-242.

Christiansen, P. 2012. The making of a monster: postnatal ontogenetic changes in craniomandibular shape in the great sabercat *Smilodon*. PLoS One 7:e29699.

Christiansen, P., and J. M. Harris. 2005. Body size of *Smilodon* (Mammalia, Felidae). Journal of Morphology 266:369-384.

DeSantis, L. R. G., B. W. Schubert, J. R. Scott, and P. S. Ungar. 2012. Implications of diet for the extinction of saber-toothed cats and American lions. PLoS One 7:e52453.

Emerson, S. B., and L. Radinsky. 1980. Functional analysis of sabertooth cranial morphology. Paleobiology 6:295-312.

Freeman, P. W., and C. Lemen. 2007. An experimental approach to modeling the strength of canine teeth. Journal of Zoology 271:162-169.

Gonyea, W. J. 1978. Functional implications of felid forelimb anatomy. Acta Anatomica 102:111-121.

Lawn, B. R., M. B. Bush, A. Barani, P. J. Constantino, and S. Wroe. 2013. Inferring biological evolution from fracture patterns in teeth. Journal of Theoretical Biology 338:59-65.

Marinelli, W. 1939. Der Schädel von *Smilodon*, nach der Funktion das Kieferapparates analysiert. Paleobiologica 6:246-272.

Martín-Serra, A., B. Figueirido, and P. Palmqvist. 2014. A three-dimensional analysis of the morphological evolution and locomotor behaviour of the carnivoran hind limb. BMC Evolutionary Biology 14:129.

Matthew, W. D. 1910. The phylogeny of the Felidae. Bulletin of the American Museum of Natural History 28:289-316.

McHenry, C. R., S. Wroe, P. D. Clausen, K. Moreno, and E. Cunningham. 2007. Supermodeled sabercat, predatory behavior in *Smilodon fatalis* revealed by high-resolution 3D computer simulation. Proceedings of the National Academy of Sciences of the United States of America 104:16010-16015.

Meachen-Samuels, J. A. 2012. Morphological convergence of the prey-killing arsenal of sabertooth predators. Paleobiology 38:1-14.

Meachen-Samuels, J. A., and B. Van Valkenburgh. 2010. Radiographs reveal exceptional forelimb strength in the sabertooth cat, *Smilodon fatalis*. PLoS One 5:e11412.

Meachen, J. A., F. R. O'Keefe, and R. W. Sadleir. 2014. Evolution in the sabre-tooth cat, *Smilodon fatalis*, in response to Pleistocene climate change. Journal of Evolutionary Biology 27:714-723.

Moazen, M., N. Curtis, P. O'Higgins, M. E. H. Jones, S. E. Evans, and M. J. Fagan. 2009. Assessment of the role of sutures in a lizard skull: a computer modelling study. Proceedings of the Royal Society of London B: Biological Sciences 276:39-46.

Parr, W. C. H., S. Wroe, U. Chamoli, H. S. Richards, M. R. McCurry, P. D. Clausen, and C. R. McHenry. 2012. Toward integration of geometric morphometrics and computational biomechanics: new methods for 3D virtual reconstruction and quantitative analysis of Finite Element Models. Journal of Theoretical Biology 301:1-14.

Piras, P., L. Maiorino, L. Teresi, C. Meloro, F. Lucci, T. Kotsakis, and P. Raia. 2013. Bite of the cats: relationships between functional integration and mechanical performance as revealed by mandible geometry. Systematic Biology 62:878-900.

Rayfield, E. J. 2005. Aspects of comparative cranial mechanics in the theropod dinosaurs *Coelophysis*, *Allosaurus*, and *Tyrannosaurus*. Zoological Journal of the Linnean Society 144:309-316.

Rayfield, E. J. 2007. Finite element analysis and understanding the biomechanics and evolution of living and fossil organisms. Annual Review of Earth and Planetary Sciences 35:541-576.

Rayfield, E. J., D. B. Norman, C. C. Horner, J. R. Horner, P. M. Smith, J. J. Thomason, and P. Upchurch. 2001. Cranial design and function in a large theropod dinosaur. Nature 409:1033-1037.

Schaller, G. B. 1972. The Serengeti Lion: A Study of Predator-Prey Relations. University of Chicago Press, Chicago, Illinois, 504 pp.

Slater, G. J., and B. Van Valkenburgh. 2009. Allometry and performance: the evolution of skull form and function in felids. Journal of Evolutionary Biology 22:2278-2287.

Therrien, F. 2005. Feeding behaviour and bite force in sabretoothed predators. Zoological Journal of the Linnean Society 145:393-426.

Tseng, Z. J. 2009. Cranial function in a late Miocene *Dinocrocuta gigantea* (Mammalia, Carnivora) revealed by comparative finite element analysis. Biological Journal of the Linnean Society 96:51-67.

Tseng, Z. J., and J. J. Flynn. 2015. Are cranial biomechanical simulation data linked to known diets in extant taxa? a method for applying diet-biomechanics linkage models to infer feeding capability of extinct species. PLoS One 10:e0124020.

Tseng, Z. J., C. Grohé, and J. J. Flynn. 2016. A unique feeding strategy of the extinct marine mammal *Kolponomos*: convergence on sabretooths and sea otters. Proceedings of the Royal Society of London B: Biological Sciences 283: 20160044.

Turnbull, W. D. 1976. Restoration of masticatory musculature of *Thylacosmylus*; pp. 169-185 in Athlon: Essays on Palaeontology in honour of Loris Shano Russell. Royal Ontario Museum, Ontario, Canada.

Van Valkenburgh, B. 2007. Deja vu: the evolution of feeding morphologies in the Carnivora. Integrative and Comparative Biology 47:147-163.

Van Valkenburgh, B., M. W. Hayward, W. J. Ripple, C. Meloro, and V. L. Roth. 2016. The impact of large terrestrial carnivores on Pleistocene ecosystems. Proceedings of the National Academy of Sciences of the United States of America 113:862-867.

Warren, J. C. 1853. Remarks on *Felis smylodon*. Proceedings of the Boston Society of Natural History 4:256-258.

Wroe, S. 2008. Cranial mechanics compared in extinct marsupial and extant African lions using a finite-element approach. Journal of Zoology 274:332-339.

Wroe, S., M. B. Lowry, and M. Antón. 2008. How to build a mammalian super-predator. Zoology 111:196-203.

Wroe, S., C. McHenry, and J. Thomason. 2005. Bite club: comparative bite force in big biting mammals and the prediction of predatory behaviour in fossil taxa. Proceedings of the Royal Society of London B: Biological Sciences 272:619-625.

Wroe, S., P. Clausen, C. McHenry, K. Moreno, and E. Cunningham. 2007. Computer simulation of feeding behaviour in the thylacine and dingo as a novel test for convergence and niche overlap. Proceedings of the Royal Society of London B: Biological Sciences 274:2819-2828.

Wroe, S., T. L. Ferrara, C. R. McHenry, D. Curnoe, and U. Chamoli. 2010. The craniomandibular mechanics of being human. Proceedings of the Royal Society of London B: Biological Sciences 277:3579-3586.

Wroe, S., U. Chamoli, W. C. Parr, P. Clausen, R. Ridgely, and L. Witmer. 2013. Comparative biomechanical modeling of metatherian and placental saber-tooths: a different kind of bite for an extreme pouched predator. PLoS One 8:e66888.

Wroe, S., D. R. Huber, M. Lowry, C. McHenry, K. Moreno, P. Clausen, T. L. Ferrara, E. Cunningham, M. N. Dean, and A. P. Summers. 2008b. Three-dimensional computer analysis of white shark jaw mechanics: how hard can a great white bite? Journal of Zoology 276:336-342.

7 An Engineering Experiment Testing the Canine Shear-Bite Model for *Smilodon*

H. TODD WHEELER

Introduction

Among the various theories presented for the killing
bite technique of the iconic hypercarnivore sabertooth cat, *Smilodon*, one is so comprehensive as to
constitute a testable hypothesis, the canine shear-bite
(Akersten, 1985). This chapter takes a different
approach to this question, as an engineering
experiment based on the morphology of the skull,
upper canine, and mandible. I use a fabricated
mechanical attachment based on casts of the upper
and lower dentition of *Smilodon* from Rancho La Brea
to measure the force required to bite a suitable prey
proxy, using any desired sequence conforming to a
chosen bite model. Discretion is required because,
limited only by the original geometry, the device can
easily exceed the force of the original and do the
impossible. In this experiment *Bison bison bison* is
used as a proxy for *B. bison antiquus*. *Bison bison
antiquus* is the most abundant large herbivore at
Rancho La Brea, and Coltrain et al. (2004) demonstrated that bison was one of the primary prey
species for both *Smilodon* and *Panthera atrox* at
Rancho La Brea. Since both cats shared the same
prey base the smaller size of *Smilodon* might suggest
that it was a more efficient predator.

The Rancho La Brea collection facilitated the
experiment by providing an abundant, accessible

sample of *Smilodon* from which to select representative
individuals. Smilodon (referred to variously as
S. fatalis, S. floridanus, and S. californicus) has been
the primary taxon used to study canine function in
sabertooth cats, and the leading work on the subject
has been described with sufficient detail to present a
testable hypothesis. The model tested has been
labeled the canine shear-bite (Akersten, 1985). Using
that work for the given conditions, a machine was
constructed to the dimensions and parameters of the
skull, mandible, and canines of Smilodon, capable of
replicating any proposed biting action, using sabers
of the same size and shape as those in Smilodon.
Using the bite model theories proposed to reproduce
saber movement through cadavers permitted the
forces required to pierce the skin and determine the
resultant injury by necropsy. This permitted the
required force to be determined experimentally. Bites
targeting both throat and abdomen were evaluated.

Freeman and Lemen, (2007) also determined
canine strength experimentally. The approach used
by others, such as McHenry et al. (2007), has used
sophisticated FEA analysis to determine the strength
of the skull, but relies on the assumption that the bite
theory is correct, and the available force is adequate.
We found this was not the case.

Our objective to verify this theory experimentally
under real world conditions, would require that the

necessary force required be a comfortable margin of safety less than that of the tooth strength.

This chapter summarizes my observations based on the experimental mechanical device constructed. In addition, a summary of my observations on wear found on the upper canine is presented. My conclusion is that overall the canine shear-bite model proposed by Akersten (1985) is unworkable. It still remains the best study of the subject, however, and many insights presented in that paper are remarkably accurate. It was immediately obvious that we know a lot more about the fossil (and extant) cats than we do about the properties, elasticity, and strength of the prey and their soft tissues. The basic unanticipated experimental observation entails the extreme distortion of prey tissues as bite force is applied and the independent movement of hide and subcutaneous structures. In order to be successful, the sabertooth cat killing bite has to fatally injure an animal that is (figuratively) within an armor-like leather bag.

Materials and Methods

A mechanism (called a 'Robocat') was developed by the author to simulate movements of the skull, lower jaw, and neck during an act of predation by *Smilodon* based on the conditions laid out by the canine shear-bite model proposed by Akersten (1985) and alternative theories. The canine shear-bite model is the only detailed model presented in sufficient detail to be tested, but I have tried a few variations as experience suggested.

Robocat's construction and use has been previously described in detail (Wheeler, 2011). This machine was attached to a Bobcat X-331 hydraulic trackhoe. The applied force was determined by reading the hydraulic circuits that were moving the boom; an additional cylinder from which to measure applied force was added to move the mandible. Roller chain linkage was employed for the sprocket mounted on the temporo-mandibular joint (TMJ) axis of the mandible, which negated the use of lever arms and allowed the applied force to be independent of gape angle. Hydraulic pressure in the boom and mandible was measured with four Omega DPi8-DC digital units supplemented by a research

grade 8-inch mechanical pressure gage with a 20 psi resolution.

Canine function in *Smilodon* (Akersten, 1985) was determined based on precise geometry using an accurate scale drawing of an individual specimen of *Smilodon fatalis* from Rancho La Brea. While scaling for size probably does not affect the results, it should be noted that there is a significant size difference, sufficient to affect force measurements, between the smaller pre-Wisconsinan *Smilodon* and the larger Late Pleistocene *S. fatalis* from Rancho La Brea. A high level of precision has not yet been achieved in this testing, but the device is constructed to that more precise standard. Fabrication was held to a ± 3 mm tolerance. Test parameters are scaled to the post-late glacial maximum (LGM) full-sized cat, and the fossil specimens from Rancho La Brea were chosen accordingly. For his layout drawing, Akersten (1985) utilized skull LACMHC 2001-2. Because of readily available high quality casts, we used skull LACMHC 2001-249 and mandible 2002-L and R-250 in the design and construction of Robocat. The cast skull and mandible is about 3% smaller than LACMHC 2001-2, so the drawings were scaled down accordingly. To build a suitably proportioned device to test the canine shear-bite model of Akersten (1985) I examined specimens in the Rancho La Brea collection, which includes over 2,000 individuals of *Smilodon fatalis*.

In addition, the cast skull did not have associated canines, so reproductions of canines were cast separately from unrelated individuals and additional steel and steel composite were added for use with the skull weldment. For the fabricated canine's dimensions and margin, the sharpness (radius) was carefully copied from eight sabers borrowed from the La Brea Tar Pits and Museum at Rancho La Brea. Provision was made to bolt in the canines in the field.

The correct position of a *Smilodon* canine as it sits in the maxillary alveolus can be difficult to determine. Postmortem tooth loss (PMTL) is common enough to have its own acronym in forensics, particularly with large, single-rooted teeth (Oliveira et al., 2000). During decomposition, as the periodontal ligament decays, the canine becomes mobile in the alveolus and can either shift in position or become separated from the skull. Saber slip would seem to be a subset of PMTL. The correct placement of the upper canine in this context is a critical (but largely unrecognized) problem, especially when working with the Rancho La Brea collection. Most *Smilodon* skulls from Rancho La Brea were found without maxillary canines (postmortem loss), and the recovery of isolated canines is common in all Rancho La Brea sites. Of those skulls excavated with intact canines, many have partially shifted their position ventrally in the alveolus prior to preservation (Tejada-Flores and Shaw, 1984). The correct canine location appears be with the dorsal tip of the posterior enamel margin where it tapers up to a point about 5-15 mm from the alveolus. Canines that have the enamel margin closer to or within the root socket are probably not fully erupted and those with the enamel more ventrally positioned have likely shifted position within the alveolus. The skull LACMHC 2001-2, previously chosen by Akersten to establish geometry, was used as the design reference. Because of the slightly smaller size of the cast we used (LACMHC 2001-249) reference points were scaled down 3% from the canines of LACMHC 2001-2 to set up initially, and to field-check tip location using a simple jig. Tip location ventral to the alveolus and anterior to the TMJ also scale with relative skull size.

The initial objective was to refine the cutting margin incidence angle and the path it follows. The only successful experimental cuts observed have utilized a draw cut, with a downward component (along the tooth axis) to achieve an incision (perpendicular to the tooth axis). The plan was to carefully examine canine wear orientation using fossil specimens to determine these angles. As is not uncommon in experimentation, the results were quite different than what was expected.

Discussion

I did not employ high-resolution dental microwear texture analysis (DMTA) on SEM microwear data, as previous authors have done (Goillot et al., 2009; DeSantis et al., 2012); rather than quantitative relative wear, I focused on qualitative wear orientation present on upper canines from the Rancho La Brea collection. Conventional light microscopy at low to moderate magnification was employed to look for visible, large-scale wear. One can observe maxillary canine wear because the fine serrations on the anterior and posterior edges appear sharp on newly erupted canines and dull on more mature teeth; the tips are worn dull as well, as observed in the Rancho La Brea collection (Akersten, 1985). In addition, a number of upper canines have grooves worn into the dentine (Fig. 7.1) exposed on the medial embayment not covered by enamel between the cemento-enamel junction (CEJ) and the alveolus (the Bell Curve region) (Riviere and Wheeler, 2005). The grooves start in the center of the tooth and progress toward the anterior and posterior edges as the groove lengthens with depth. The bottom tends to be straight rather than following the curve of the tooth. As the groove widens with depth the apical side blends with the dentin surface approaching the alveolus, and the crown side also blends with the dentin surface until it meets the CEJ with a sharp polished corner. The groove orientation is parallel in general to the occlusal plane but is presumably following the direction of movement of the tongue. The groove can be more easily felt than seen at first, and can reach a depth exceeding 8 mm. The wear pattern as it progresses follows a set pattern and is proportional to the overall size of the tooth. Apparently the grooves were created by the tongue wearing away soft dentine, ordinarily protected by gingiva. This phenomenon is observed on only a few upper canines, usually in old individuals, and frequently in individuals with other oral conditions associated with advanced periodontal disease and loss of the gingiva.

This grooving is seen in all Smilodontini as well as in nimravids (*Pogonodon, Nimravus, Dinictis,* and *Hoplophoneus*) that have the Bell-Curved CEJ diagnostic of the dirk-tooth morphology (Wheeler et al., 2004). It has not been found in machairodonts with a stair-stepped CEJ, a feature diagnostic of Homotheriini (Werdelin and Sardella, 2006). Very fine scratches, oriented parallel to the occlusal plane, are found at 20X to 100X. In uncommon instances small tongue grooves have also been observed in the extinct *Dinofelis* and the conical-toothed felids *Puma concolor* (JODA 8022) (Fig. 7.2) and *Panthera atrox* (LACMP23 8734) from Rancho La Brea.

Figure 7.1. Typical tongue-grooved upper canines of *Smilodon fatalis*. A = LACM 50603; B = LACM 50601; C = LACMHC 50627; D = LACMHC 2000-L-21.

The purpose of the exposed cementum overlaying the dentine in the enamel-free embayment on the lateral side of the canine remains the subject of speculation. Riviere and Wheeler (2005) established that the feature exists and infer that it would have been covered with gingiva, concluding that it provided tactile feedback to aid in use of the tooth. While it is difficult to prove, anecdotal evidence was obtained during biting experiments at the 2008 Sabertooth Symposium in Pocatello, Idaho, when the author tested a "Russ Kommer" model CLK 2500 knife (Fig. 7.3) on a large mammal carcass. The knife in question has a thumb-sized hole in the grip and the blade is located in the same relationship as the enamel-free embayment to the margins of dirk-tooth sabers. With the thumb in the hole touching the hide and the blade hooked under the hide, the blade can

easily be guided with your eyes closed and will cut like undoing a zipper. With regard to *Smilodon*, it seems that the enamel-free embayment would be an ideal location for tactile input to guide a saber into a prey target that is outside the field of vision. Because the Kommer knife blade is much sharper than the canine of *Smilodon*, it is not helpful as an analog to the canine of *Smilodon*; in fact it is deceptive because with a sharp enough knife just about any cut can be performed.

The common perception is that a blade can be moved perpendicular to its edge. For an accurate replica of a *Smilodon* upper canine to pierce and cut into bison hide, the reality is different, and this is readily apparent only when attempting to force a blunt, saber-like blade through hide and tissue. It is clear that some other process is involved when one

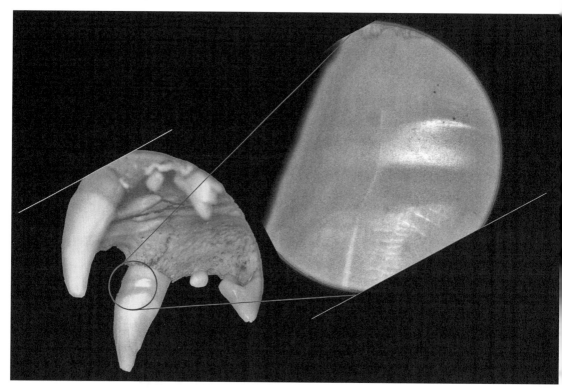

Figure 7.2. Slightly tongue-grooved upper canines of *Puma concolor* (JODA 8022).

Figure 7.3. "Russ Kommer" CLK 2500 knife with thumb window.

observes that the amount of force required for piercing or cutting the prey's hide approaches or exceeds the breaking strength of the tooth. The canine of *Smilodon* will withstand 7,000 newtons (N)

of force (Freeman and Lemen, 2007). We used a maxillary upper canine replica with a 0.1–0.2 mm margin sharpness (typical of a moderately worn tooth) which requires over 11,000 N. McHenry et al. (2007), in their finite element analysis (FEA) used 2,000 N. A perpendicular cut with a stalled blade and force that low is possible but requires a sharp knife edge, something that is difficult to achieve in nature even with an attrition-sharpened tooth. Experience with knives creates unrealistic expectations of cutting by teeth, so the purpose of this experimental work is to avoid such errors. With blade experiments producing actual cuts, force is a very poor substitute for sharpness (McCarthy et al., 2007; Gilchrist et al., 2008; McCarthy et al., 2010).

There is another reason for the CEJ to have the prominent Bell-Curved feature (which while large relative to the overall size of the tooth is not that much more curved than many other 'normal' canines and other similar teeth). There is a process called abfraction that has recently come under study in human dentistry (Bartlett and Shah, 2006; Romeed et al., 2012) in which a notch is formed at the CEJ that is caused by bending forces applied to the tooth. Abfraction in teeth forms in a similar manner to stress cracks and probably arises from differing elasticity and strength properties of enamel versus cementum. There is also a stress increase at the edge of the root socket where the emergent tooth abruptly loses the support of the alveolar bone. Where these features coincide, there is an increased probability of damage to the tooth.

Results in the search for directional scratches were surprising. No directional wear scratches in the enamel on the cutting margins of the maxillary canine crowns were found, and there were only slight directional marks near the CEJ of LACMHC 50627 (Fig. 7.4). The movement of the soft tissue of the prey while being consumed can also produce a polish with few directional marks on the enamel apart from the rare mark resulting from grit on the prey. However, directional scratches were observed, with varying clarity, in the exposed dentine. In *Megantereon* [BIOPSI 2001.1.0154 (Fig. 7.5)] distinct directional scratches (inset) are seen oriented on the tooth. The *Smilodon* specimen with faint marks in the enamel seen in Figure 7.4 shows similarly oriented scratches more distinctly in dentin [LACMHC 50627 (Fig.7.6)]. Tongue grooves are fairly clear, and visible scratches are oriented with the long axis of the tongue. All scratches were either oriented on the tooth axis and parallel to the tangent of the arc of the mandible or were parallel to and within the tongue groove in the tooth. No evidence of cutting (move-ment through the hide perpendicular to the tooth axis) was found at all. Tooth wear markings with an orientation parallel to the tooth axis were observed, which is consistent with evidence for biting. The test observation was that biting occurred easily without the geometric problems anticipated by Akersten (1985) in his presentation of the canine shear-bite model. The resulting wound inflicted on the cadavers appeared severe despite the stall resulting in failed cutting bite attempts.

Biting (i.e., closing the mouth from full gape using mandibular opposition), supplemented by the force from the head depressing musculature, appears viable and there is some fossil evidence that it did occur. Figure 7.6 illustrates a typical tongue groove, with Figure 7.6A the full frame, Figure 7.6B an inset at higher magnification, and Figure 7.6C an additional enlargement showing faint scratches oriented with the tongue groove and additional wear scratches parallel to the tooth axis. There is no evidence based on maxillary canine wear and scratch orientation that shear cutting by the canines occurred. It has been shown experimentally that it is possible to cut bison and cervid hide with a repli-cated saber, but only with great difficulty, approach-ing or going beyond the breaking strength of real teeth as established in laboratory tests (Freeman and Lemen, 2007).

What occurs when the Robocat machine starts to close the jaws on part of a test cadaver is that the sabers bite into the prey, forming a deep dimple at

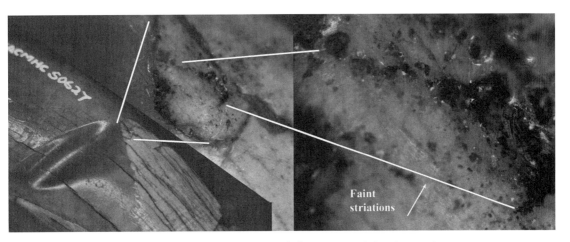

Figure 7.4. Upper canine of *Smilodon fatalis* (LACMHC 50627) with faint tongue striations in enamel.

2 cm

Figure 7.5. BIOPSI 2001.1.0154. Right-grooved *Megantereon* upper canine; rotated study.

the crown tip that is over 30 mm deep before any piercing or cutting begins (and demonstrates the impracticality of the stab concept). A closed-mouth stab attempt uses up the distance from crown tip of the saber to the mandible before any piercing can occur and so these kinds of bite models (Wilson et al., 2013) are unworkable against bison, large cervids, and similar sized prey. Unless the hide is supported by bone (e.g., a skull, scapula, or pelvis), the hide and tissue deflect away from the tip of the saber until the closing mandible comes up against the body of the prey. Although skulls of predators exist in the fossil record with saber shaped holes that could have been inflicted with a closed mouth, they are rare indicators of social behavior and have little to do with the typical prey killing mechanics in sabertooth carnivores.

As the bite progresses, piercing occurs rapidly once penetration starts and as the hide rebounds. The force required is not enormous but too brief for us obtain a measurement. (Future testing will use better equipment to correct this measuring deficiency.) *Smilodon* and sabertooths in general have a relatively weak bite with limited use of the coronoid process, but at the high gape involved, hystricomorphy (Naples and Martin, 2000) and head depression musculature are probably more important anyway. The coronoid process functions as a lever arm and lever arms don't work well through a 90° arc. McHenry et al. (2007) present a clear depiction of the known felid musculature, but at full gape there is a very poor mechanical advantage to the conventional insertion points.

The hystricomorph concept infers muscle locations lacking analogs in extant low-gape cats (Naples and Martin, 2000). It postulates muscle paths similar to rodents and similar high-gape morphologies attaching anteriorly on the mandible. The exact details of this soft tissue conformation may never be known and was probably supplemented by the head-depressing musculature. The downward external force on the skull cannot add to the opposition force from the mandible to increase the bite force, but it can act directly against the prey to augment the total resultant force acting to drive the upper canines through the hide.

As observed from the experiments, the saber margins wedge the wound open to allow a path for the tooth going in, without trapping it within the tissues of the intended prey. The presence of cutting margins on the canines does not mean they were necessarily used to cut through the hide perpendicular to the tooth axis. Long, slender piercing teeth are not always round (Freeman and Lemen, 2007); teeth also occur with edged margins that allow them to create the wound. A concern when biting with a long, slender tooth is extracting it from the wound afterward. The withdrawal of such a tooth may be the greatest shortfall of the bite technique in sabertooth carnivores.

The steps described so far are common to all sabertooth cat bites; what happens next is subject for debate. I suggest that nothing (that is, no further wounding) is a viable option. The Akersten (1985) model proposes that the hide was 'gathered up'

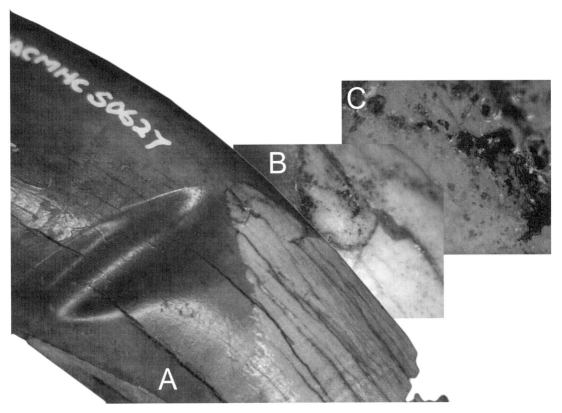

Figure 7.6. Tongue-grooved upper canine of *Smilodon fatalis* (LACMHC 50627), three magnification view. A = tooth and groove orientation; B = at 60× locates region with striations; C = at 210× shows faint tooth axis oriented striations in enamel. Some striations oriented with tongue groove are visible, but most are out of field to the bottom.

inside of the mouth and ripped from the prey animal in order to leave a gaping wound from which the animal would bleed to death. Multiple attempts have demonstrated that this doesn't work. Experiments have shown that getting the incisors hooked into the hide stalls the entire process. Attempts to cut through bison or cervid hide indicate a different process is followed. After piercing the hide, it is necessary to pull the saber against it until all available slack is taken up. As additional force is applied, the hide becomes so stretched at the tooth margin that it cuts (or it may tear if the tooth is too blunt or the hide too tough). If the posterior margin of the upper canine (or 'blade' of the canine) is moved (apically or distally) a cut may develop at the tooth, and be sustained as long as motion continues. If at the start of the process multiple teeth (incisors) penetrate the hide some of the force is diverted from the anterior margin of the canine and the attempt

becomes unsuccessful. In a sense, the incisors drag a little slack behind the canine. Particularly, if the canines exit the other side of the prey, the edge may turn so that the blade is pulling against a fold in the hide. At that point, I suspect even a knife may have trouble cutting. The observation that the hide must be held tight before any cutting can occur is one of the greatest limiting factors in theoretical killing bite model development.

Subsequent tests attempting any type of bite cutting perpendicular to the tooth axis using any technique were unsuccessful. Anecdotal indications are that a cutting bite is possible with an extremely sharp knife. My observation is that the force required as the sharpness approaches reality escalates so rapidly that such a bite is unfeasible (McCarthy et al., 2007; Gilchrist et al., 2008; McCarthy et al., 2010). A force measurement of 11,000 newtons was under ideal conditions and obtained on a large cervid. A successful cut with a stalled tooth on a on bison was never obtained, so how much greater the force needed for this animal has not yet been determined.

Conclusion

Development of bite models for dirk-tooth saber-tooth cats taking full advantage of the derived morphology of machairodonts is seductive. It is reasonable that a dramatic specialization such as this warrants an equally exotic use. Following my mechanical experiments, the evidence suggests this might not be the case. What an extinct sabertooth cat or virtual or real simulation is capable of accomplishing under ideal conditions is not what matters. A predator needs to kill on a reasonably regular basis without being injured in the process. A reality-based killing technique has to incorporate a reasonable safety factor with margin for error. A simple bite technique is possible, arguably has supporting fossil evidence, and has to be included as a viable killing solution for sabertooth cats. The parameters used by Akersten (1985) for the canine shear-bite model are a good framework to start with, but need to be refined based on a better understanding of the forces required. The mechanical properties of hide, soft tissue, and other anatomical parts of the prey involved remain virtually unknown. We have barely scratched the surface in obtaining that understanding.

ACKNOWLEDGEMENTS

The author would like to thank Aisling B. Farrell and Shelley M. Cox of the Department of Rancho La Brea, La Brea Tar Pits and Museum, Natural History Museum of Los Angeles County (formerly the George C. Page Museum of La Brea Discoveries) for facilitating access to and help with the collections at the museum.

REFERENCES

Akersten, W. A. 1985. Canine function in *Smilodon* (Mammalia, Felidae, Machairodontinae). Natural History Museum of Los Angeles County, Contributions in Science 356:1-22.

Bartlett, D. W., and P. Shah. 2006. A critical review of non-carious cervical (wear) lesions and the role of abfraction, erosion, and abrasion. Journal of Dental Research 85:306-312.

Coltrain, J. B., J. M. Harris, T. E. Cerling, J. R. Ehleringer, M.-D. Dearing, J. Ward, and J. Allen. 2004. Rancho La Brea stable isotope biogeochemistry and its implications for the paleoecology of the Late Pleistocene, coastal southern California. Palaeogeography, Palaeoclimatology, Palaeoecology 205:199-219.

DeSantis, L. R. G., B. W. Schubert, J. R. Scott, and P. S. Ungar. 2012. Implications of diet for the extinction of saber-toothed cats and American lions. PLoS One 7:e52453.

Freeman, P. W., and C. A. Lemen. 2007. An experimental approach to modeling the strength of canine teeth. Journal of Zoology 271:162-169.

Gilchrist, M. D., S. Keenan, M. Curtis, M. T. Cassidy, G. Byrne, and M. Destrade. 2008. Measuring knife stab penetration into skin simulant using a novel biaxial tension device. Forensic Science International 177:52-65.

Goillot, C., C. Blondel, and S. Peigné. 2009. Relationships between dental microwear and diet in Carnivora (Mammalia)—implications for the reconstruction of the diet of extinct taxa. Palaeogeography, Palaeoclimatology, Palaeoecology 271:13-23.

McCarthy, C. T, A. N. Annaidh, and M.D. Gilchrist. 2010. On the sharpness of straight edge blades in cutting soft solids: Part II—analysis of blade geometry. Engineering Fracture Mechanics 77:437-451.

McCarthy, C. T., M. Hussey, and M. D. Gilchrist. 2007. On the sharpness of straight edge blades in cutting soft solids: Part I—indentation experiments. Engineering Fracture Mechanics 74:2205-2224.

McHenry, C. R., S. Wroe, P. D. Clausen, K. Moreno, and E. Cunningham. 2007. Supermodeled sabercat, predatory behavior in *Smilodon fatalis* revealed by high-resolution 3D computer simulation. Proceedings of the National Academy of Science of the United States 104:16010-16015.

Naples, V. L., and L. D. Martin. 2000. Evolution of hystricomorphy in the Nimravidae (Carnivora, Barbourofelinae): evidence for complex character convergence with rodents. Historical Biology 14:169-188.

Oliveira, R. N., R. F. Melani, J. L. Antunes, E. R. Freitas, and L. C. Galvão. 2000. Postmortem tooth loss in human identification processes. Journal of Forensic Odonto-Stomatology 18:32-36.

Riviere, H. L., and H. T. Wheeler. 2005. Cementum on *Smilodon* sabers. Anatomical Record 285:634-642.

Romeed, S. A., R. Malik, and S. M. Dunne. 2012. Stress analysis of occlusal forces in canine teeth and their role in the development of non-carious cervical lesions: abfraction. International Journal of Dentistry 234845:1-7

Tejada-Flores, A. E., and C. A. Shaw. 1984. Tooth replacement and skull growth in *Smilodon* from Rancho La Brea. Journal of Vertebrate Paleontology, 4:114-121.

Werdelin, L., and R. Sardella. 2006. The *"Homotherium"* from Langebaanweg, South Africa and the origin of *Homotherium*. Palaeontographia Abteilung A 277:123-130.

Wheeler, H. T. 2011. Experimental paleontology of the scimitar and dirk-toothed killing bites; pp. 19-34 in V. L. Naples, L. D. Martin, and J. P. Babiarz (eds.), The Other Sabertooths: Scimitar-Tooth Cats of the Western Hemisphere. Johns Hopkins University Press, Baltimore, Maryland.

Wheeler, H. T., H. L. Riviere, T. J. Fremd, and J. P. Babiarz. 2004. Convergent evolution in the enamel and gingiva of the nimravid *Pogonodon* and the felid *Smilodon* revealed in new material from John Day Fossil Beds National Monument. Geological Society of America Abstracts with Programs 36:53.

Wilson, T., D. E. Wilson, and J. M. Zimanske. 2013. Pneumothorax as a predatory goal for the sabertooth cat (*Smilodon fatalis*). Open Journal of Animal Sciences 3:42-45.

8

The Evolution of the Skull, Mandible, and Teeth of Rancho La Brea *Smilodon fatalis* as They Relate to Feeding Adaptations

JULIE A. MEACHEN AND WENDY J. BINDER

Introduction

Smilodon fatalis is one of the most charismatic of the megafaunal species that went extinct at the end of the Ice Age. Although the most intriguing and puzzling part of its anatomy may be the large saber canines, these large saber teeth did not work in isolation to kill prey. *Smilodon fatalis* depended on a suite of structures, including the skull, jaws, cheek teeth, and limbs, used precisely in concert to dispatch prey (Akersten, 1985; Anyonge, 1996; Antón et al., 1998; McHenry et al., 2007; Meachen-Samuels and Van Valkenburgh, 2010). This chapter focuses on the skull, mandibles, and teeth of *S. fatalis* and how those structures evolved over time in North America.

There is no better record for a 30,000-year window of *Smilodon* evolution than at the Rancho La Brea tar pits. Rancho La Brea (RLB) is a Late Pleistocene (approx. 40,000-10,000 years before present) fossil site in Los Angeles, California that offers a view of the Pleistocene fauna and flora through ephemeral, deadly asphalt seeps that lasted for a few hundred to a few thousand years at a time and trapped large numbers of big carnivores. *Smilodon fatalis* is the second most common fossil vertebrate trapped in these sites, closely behind the dire wolf (*Canis dirus*) in number of individuals recovered (Stock, 1992). This site provides both an extended window of time

and the best sample size of *S. fatalis* in the world and is therefore optimal for a synthetic study of this species' feeding adaptations and evolution.

Rancho La Brea Is the Pits

Rancho La Brea is not a single fossil accumulation; rather, the fauna is preserved in several Late Pleistocene asphalt deposit sites (also called 'pits') which were active traps at different times. By comparing each of these sites, we can compare slices in time to assess what changes had taken place in a given species. In our previous studies, we concentrated on relatively few sites at RLB, some more important and significant than others. These include the following, in chronological order: Pit 77 (40-32 kybp), Pit 2051 (30-20 kybp), Pit 91 (29-25 kybp), Pit 3 (20-15 kybp), Pit 13 (18-17 kybp), and Pit 61/67 (14-13 kybp) (Friscia et al., 2008; O'Keefe et al., 2009). The age ranges of these sites are not all encompassing; rather they represent the ranges when the majority of the dated fossils were deposited. Additionally, some of these sites are only known from five to six dates and need extensive further dating to robustly determine their durations. Until that is possible, we use these dates as guides to when these sites were deposited.

Pit 77 is one of the oldest sites at RLB. It only has six radiocarbon dates, ranging from 32,000 to 40,000 years before present (O'Keefe et al., 2009).

With additional dating, the majority of fossils may be attributed to a narrower age range or possibly even two separate depositional events. As an older site, this deposit is a good baseline for *S. fatalis* at RLB. There are no published floral data to corroborate the climate during the deposition of Pit 77 at RLB; however, the Greenland ice core data show a cyclical climate with both high and low temperatures in this age range, making it difficult to pinpoint the climate during the time this site was active (Jouzel et al., 2007).

All of the faunal remains from Pit 2051 at Rancho La Brea are housed at the University of California Museum of Paleontology in Berkeley. It is unclear whether Pit 2051 is from one or multiple deposits, as the site includes a wide range of dates, but the majority of specimens fall between 20,000 and 30,000 years before present (O'Keefe et al., 2009). This site has a very large sample size of *Smilodon fatalis*. Pit 2051 also lacks floral data to determine the climate; however, the Greenland ice core data show a trend of decreasing temperatures in this time window, with very few high temperature spikes (there are a few toward the end of the age range), suggesting that the predominant climate during the deposition of this site was cooler and wetter. The date ranges of Pit 2051 also include the lowest temperatures on record in the Northern Hemisphere in the last 40,000 years (at approximately 22,500 years) (Jouzel et al., 2007).

Pit 91 was one of the last sites to be excavated; therefore it was done with more precise and meticulous methodology than many of the previous sites, which were excavated in the early twentieth century. This site also has the best radiocarbon record, with more than 45 dates. The age range is large, from 14,000 to 45,000 years before present, and it is likely that there are at least two, or possibly three, separate depositional episodes in Pit 91. However, the majority of radiocarbon dates from this site belong to a depositional event from 25,000 to 29,000 years before present (78% of dated fossils), with a small depositional event (20% of fossils dated) between 35,000 to 45,000 years before present and another small depositional event around 14,000 years before present (2% of dated fossils) (Friscia et al., 2008). The flora of Pit 91 was dominated by coastal

sage scrub with conifers at higher elevations. The temperature at the time of deposition of Pit 91 was in a transitional state, with a cooler and more mesic environment than we see today in southern California, but not as cold as earlier or later climates, such as the last glacial maximum (Coltrain et al., 2004).

Pit 3 has a relatively large depositional window (approximately 15,000 to 20,000 years before present) and currently encompasses the total time span of Pit 13 within it (O'Keefe et al., 2009). One interesting aspect of Pit 3 is the tooth breakage and wear data from *Smilodon fatalis*. This site appears to have more evidence of tooth breakage and wear in *S. fatalis* than any other site examined (Binder and Van Valkenburgh, 2010), as is discussed later in this chapter. Pit 3 spans the last glacial maximum with a cool and mesic environment dominated by juniper woodlands (Coltrain et al., 2004).

Pit 13 is unique for two reasons. Preliminary dates suggest that it was deposited during the last glacial maximum (17,000–18,000 ybp) (O'Keefe et al., 2009), and specimens show extreme pit wear, which is a taphonomic process of bones rubbing together such that large scores and divots are scraped into the bones, giving them a whittled appearance (Friscia et al., 2008). Additionally, dire wolves (*Canis dirus*) as well as *Smilodon fatalis* have a very high incidence of tooth breakage and wear in this site (Binder et al., 2002; Binder and Van Valkenburgh, 2010). Although Pits 3 and 13 overlap in time, they do show some important distinctions, such as differential tooth breakage and wear values in dire wolves and *Smilodon*. Pit 13 accumulated during a climate regime similar to that of Pit 3 (Coltrain et al., 2004).

Pit 61/67 is the youngest Pleistocene site and closest to the extinction event at RLB. The five radiocarbon dates available place this site between 13,000 and 14,000 years before present (O'Keefe et al., 2009), so animals from this site accumulated directly before or during the end-Pleistocene extinction events. This site is one of the largest deposits. It was originally thought to be two separate sites (Pits 61 and 67), but it later was found to actually be one very large site (Stock, 1992). This deposit has an enormous sample size of both *Canis dirus* and *Smilodon fatalis*. Pit 61/67 is hypothesized to have been deposited during the Bølling-Allerød warming interval and therefore during a warmer, drier climate regime than the other pits (O'Keefe et al., 2009). Conifers became scarce and the modern chaparral environment dominated the landscape (Coltrain et al., 2004).

The remainder of this chapter presents an integrated overview of the feeding adaptations of *Smilodon fatalis* over time at RLB and how this enigmatic species adjusted to a changing environment until it no longer was able to, resulting in its extinction.

Tooth Breakage and Wear over Time in *Smilodon fatalis*

Van Valkenburgh and Hertel (1993) published a study on tooth breakage and wear in the carnivores at RLB and found that these carnivores broke their teeth a lot more than expected. Not only did they break their teeth frequently, but they also showed greater tooth wear in comparison to living carnivore species. This interesting observation led Van Valkenburgh and Hertel to believe that carcasses were picked clean, resulting in more tooth contact with harder material like bone, suggesting that there must have been more competition for prey in the Late Pleistocene. While there were more large herbivores available at this time, there was also a greater diversity and abundance of carnivores competing for resources, implying that survival was more difficult for large carnivores during the Late Pleistocene. *Smilodon fatalis* with its super-specialized dentition had a breakage rate of approximately 20% overall, which is on par with dire wolves and much higher than that of extant large canids and felids (Van Valkenburgh and Hertel, 1993). This was the first study to alert the paleontological community to the notion that the Pleistocene was special in this regard. However, this study did not examine the data within the context of trends through time.

A study by Binder et al. (2002) found that dire wolves—based on tooth wear and breakage evidence from Pit 13—experienced nutritional stress, although evidence from Pit 61/67 showed that they seemed to be under less stress immediately prior to their extinction. This was counterintuitive, demonstrating that nutritional stress was not the proximate cause of their extinction; whether this was true for *Smilodon* was unknown until a study comparing similarly

temporal sites was completed (Binder and Van Valkenburgh, 2010). The results of the 2010 study showed that *S. fatalis* had higher incidences of tooth breakage in the earlier sites at Rancho La Brea, and that it was reduced in the later sites. However, the rates of breakage and wear were higher in *Smilodon fatalis* than in dire wolves in every case, suggesting that *S. fatalis* consumed more bone in the Pleistocene than any extant cat species.

Mandibular Shape Change in *Smilodon fatalis*

Another way to determine how this sabertooth predator changed through the Pleistocene is to examine the evolution of a key component of the prey-killing apparatus, the mandible. The mandible can provide interesting and useful information on prey killing and feeding in carnivores. Many studies have examined the mandible as a prey-killing tool in carnivores in general (Biknevicius and Ruff, 1992a, 1992b; Biknevicius and Van Valkenburgh, 1996; Biknevicius and Leigh, 1997; La Croix et al., 2011) and in sabertooth predators in particular (Therrien, 2005). The shape of the mandible can provide information on lever arms of the jaw-closing masseter and temporalis muscles (Van Valkenburgh and Koepfli, 1993; Meachen-Samuels and Van Valkenburgh, 2009), dietary evolution via proportion of grinding to shearing teeth (Van Valkenburgh and Ruff, 1987; Van Valkenburgh, 1988; Van Valkenburgh and Koepfli, 1993; Biknevicius and Van Valkenburgh, 1996; Friscia et al., 2007), and resistance to bending and torsion during killing and feeding (Biknevicius and Ruff, 1992a, 1992b; Biknevicius and Leigh, 1997).

While it is always best to use the mandible in conjunction with the skull to obtain a complete picture of killing and feeding, sometimes that is not possible. For example, in fossil assemblages, complete mandibles are fossilized far more frequently than complete skulls; therefore, many fossil sites have a better-preserved record of mandibles than skulls, making them a useful tool for dietary reconstruction. Additionally, biomechanically the mandible is an overall simpler structure than the skull, with fewer additional functions other than feeding, and much of the important information about killing and feeding can be gained from only two dimensions in mandibles, making photography and 2D geometric morphometrics a viable option to obtain this information. It can also be modeled as a hollow beam to determine resistance to bending and torsional stresses (Biknevicius and Ruff, 1992a, 1992b).

Smilodon fatalis individuals did not use their mandibles in an identical fashion to extant large felids. Large, living, 'conical tooth' felids, attack prey, assisted by their forearms, and position themselves so that they can use their skull and jaws to strangle prey with a chokehold or a muzzle bite (Schaller, 1967, 1972; Ewer, 1973). This process may take tens of minutes and during this time the prey is struggling. The round cross-section of the (conical) canines is built to withstand these forces (Van Valkenburgh and Ruff, 1987). Sabertooth cats with their elongated, labiolingually flattened saber canines would break their teeth with alarming frequency if they killed prey in the same manner. We do not see this in the fossil record; therefore, *Smilodon* would have killed prey differently. A few studies indicate how *Smilodon* may have killed its prey, including the increased use of its neck musculature (Antón and Galobart, 1999), increased strength, and use of the forelimbs in prey capture and subduing (Meachen-Samuels and Van Valkenburgh, 2010; Meachen-Samuels, 2012). One paper even proposes a new bite model for *Smilodon* incorporating the large upper canines, the increased neck musculature, and the incredibly strong forelimbs (Brown, 2014). Since the 1990s most hypotheses on *Smilodon* prey killing agree that *S. fatalis* was an ambush predator that surprised prey and held it down with the forelimbs while depressing the canines into the throat of prey using neck musculature and forelimbs (Anyonge, 1996; Antón and Galobart, 1999; Martin et al., 2000; Meachen-Samuels and Van Valkenburgh, 2010; Wroe et al., 2013; Brown, 2014).

The shape of *Smilodon* mandibles is very distinct from that of conical toothed cats and reflects their unique killing style (Therrien, 2005). *Smilodon fatalis* (and many other sabertooth) mandibles have a mandibular flange, a ventrally oriented projection from the anterior portion of the mandible that buttresses the upper canine when the mouth is closed. This flange is thought to protect the canine, and its shape varies from species to species in

sabertooth predators. In *S. fatalis* it is present but not very elongated or pronounced (Therrien, 2005).

Another important difference between *S. fatalis* and living cats is the height of the coronoid process; this portion of the mandible is the insertion point for the temporalis muscle at the posterior end of the jaw. Because of the size of the upper canines, jaw gape in *Smilodon* must be unusually large in order to fit prey between the tips of the saber-like canines and the lower canines—approximately 105.8 degrees (by comparison living cats have a gape of around 65 to 70 degrees) (Wroe et al., 2013). This gape necessitates a very long temporalis muscle combined with an insertion point close to the joint to avoid over-stretching this long muscle, and consequently the coronoid process is short in *Smilodon* (Emerson and Radinsky, 1980). Coronoid height may negatively correlate with upper canine length in saber tooth predators, as the two are biomechanically linked.

Many paleontologists have assessed that *Smilodon* may have used a killing strategy very different from that of the larger modern felids, but in all likelihood its feeding adaptations would be very similar since both groups of cats are adapted to a hypercarnivorous diet, composed of over 90% large vertebrate prey (Van Valkenburgh, 1996). The material properties of skin, muscle, and bones have not changed in the last 50,000 years, so both types of cats would be adapted to cutting skin and slicing flesh with their sharp carnassial teeth. We know that neither *Smilodon* nor modern cats ate as much bone as other carnivores (such as dogs) because none of them possess the post-carnassial molars they would need to crack hard objects (Van Valkenburgh, 1996; DeSantis et al., 2012).

Although *Smilodon fatalis* is unique with respect to living cat species, this does not mean that this extinct species always remained the same with regard to mandibular shape or prey-killing techniques. Both of these may have changed in response to changing biotic or abiotic environments throughout *S. fatalis*' evolutionary history and changes in the types of prey species available.

Meachen et al. (2014) looked at how *Smilodon fatalis* mandibles changed over the course of the Pleistocene at RLB. They examined *S. fatalis* mandibles from Pits 77, 91, 2051, 13, and 61/67 using two-dimensional geometric morphometrics, a technique that relies more heavily on shape than size to determine the differences between the sites. They found pronounced differences between sites, showing variation over time (Fig. 8.1). In the earliest site (Pit 77), *S. fatalis* showed a generalized, more plesiomorphic morphology. The mandibles found in this early site had a shape profile with more rounded anterior edges leading up to the canine, as in modern cats (Therrien, 2005). They also had dorsoventrally shallower mandibles, indicating less reinforcement of the mandible in this plane. One other interesting feature of the mandibles from Pit 77 is the lack of a pronounced mandibular flange. Pit 77 also had the smallest individuals measured, with a very large standard deviation in mean size, suggesting that selection for size was not strong at this time. Pit 2051 showed the same trend as Pit 77, with the same pattern of less derived morphological traits. The individuals from Pit 2051 are still small, but the standard deviation is now lower, and individuals comparable in size to the smallest ones from Pit 77 are absent, indicating selection against small size (Meachen et al., 2014).

Although Pit 91 and Pit 2051 overlap in time, the trends in saber length and gape are the opposite in these two sites. *Smilodon fatalis* from Pit 91 shows more derived sabertooth traits with shorter coronoid processes, larger mandibular flanges, deeper jaws, and larger overall mandible size, whereas those from Pit 2051 have longer coronoid processes and less pronounced mandibular flanges.

In Pit 61/67 this derived trend is most pronounced, with the shortest coronoid processes, the largest mandibular flanges, the deepest mandibles (dorsoventrally), and the largest overall size. Pit 13 is between the two extremes both morphologically and temporally (Meachen et al., 2014).

Although the available radiocarbon dates make determining an exact time period difficult, these morphological differences seem to correlate well to climate changes in the Late Pleistocene. The less derived morphologies correlate with cooler, wetter periods and the more derived morphologies are correlated with warmer, drier periods. These morphological changes would have likely affected hunting style in some way and may have been in response to changes in vegetation and herbivore species present or herbivore behavior patterns. For example, based on *Smilodon fatalis* morphology (extremely robust limbs, short tail) it was not likely to chase its prey for long distances. Rather, the evidence indicates it was an ambush predator that surprised its prey from close range. Vegetation changes would have played a role in where *Smilodon* hid while waiting for prey and would have had a

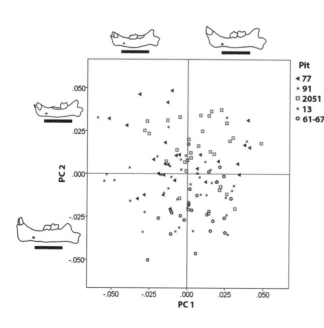

Figure 8.1. PC1 vs. PC2 for 2D geometric morphometric data of *Smilodon fatalis* mandibles. Small mandible diagrams on the sides of the plot show the relative shape of mandibles found in that quadrant. Reproduced with permission from Meachen et al. (2014).

large effect on which herbivores were present in the region at the time.

During cooler, wetter periods, pine trees dominated the southern California landscape, whereas in warmer, drier periods chaparral and coastal sage scrub dominated (Coltrain et al., 2004). The body size of *Smilodon fatalis* may have fluctuated through time based on which prey species were most abundant and the nature of the habitat (open or closed). In a drier, warmer environment, open habitat species dominated the landscape. Species that frequent this environment are generally larger-bodied grazers, such as mammoths, camels, horses, and bison. In a wetter, cooler environment, smaller, closed-habitat ungulates would include tapirs, deer, and peccaries. This may explain the size difference between the cooler, wetter sites (Pits 2051 and 77) and the warmer, drier sites (Pits 91 and 61/67). Although Pits 2051 and 91 overlap in time, macrofaunal samples from Pit 91 indicate that the environment at that time was a combination of coastal sage scrub (open habitat) and pine forest (closed habitat) (Coltrain et al., 2004).

A study by Van Valkenburgh et al. (2004), showed a link between large body size and dietary specialization in carnivores; their term for this phenomenon is the "macroevolutionary ratchet." This means that as large carnivores increased in size, their diet needed to be larger to accommodate their energy needs, and this resulted in a one-way ratcheting of diet specialization toward extremely large prey. This suggests that as *Smilodon* became larger, it required more derived sabertooth-like morphologies to kill the large prey in these warmer, drier sites (Van Valkenburgh et al., 2004).

The deposition of Pit 61/67 likely partially coincided with the beginning of the Pleistocene extinction event in North America. The aforementioned ratchet could have resulted in the relatively extreme morphology of *Smilodon fatalis* seen at that site, which may have led to its demise. If *S. fatalis* was becoming more specialized just as its prey was becoming scarcer, this is a clear example of the macroevolutionary ratchet, whereby they evolved to a point of specialization that could not be reversed in time to make them viable in a world with increasingly fewer, smaller-bodied prey.

Although it is very difficult to pinpoint one event as the cause of the megafaunal extinctions, the warming climate and subsequent morphological specialization would likely have had a cumulative effect on *S. fatalis*, possibly contributing to its extinction.

Changes in Mandibular Cortical Thickness in *Smilodon fatalis*

Previous work on the mandible of *Smilodon fatalis* has led to questions about the internal structure of this critical element of the prey-killing apparatus. Mandibular cortical thickness can give information that cannot be obtained from external measurements. For example, the depth of the mandible can be measured externally, but how much of this depth is dense cortical bone versus hollow space can only be ascertained through internal imaging (e.g., radiographs or CT scans).

Using x-rays we can image the internal structure of the mandible and model it as an asymmetrical hollow beam. This technique has been used successfully to examine the internal structure in the mandibular corpus in numerous extant carnivorans (Biknevicius and Ruff, 1992a, 1992b; Biknevicius and Leigh, 1997). An asymmetrical model of the mandible was found to be most accurate, and it can be measured by using two radiographic views of the mandible, one in the transverse plane and one in the sagittal plane. The cortical bone measurements can be used to ascertain the cortical area (CA), second moments of area—bending rigidity in the transverse (I_y) and sagittal plane (I_x) (including greatest and least bending rigidity), and also mean moments of area—and torsional rigidity (rigidity in nonaxial loadings) ($J/2$) in the mandible (Biknevicius and Ruff, 1992a, 1992b; Heinrich and Biknevicius, 1998). While the asymmetrical model is closest to the actual bone distribution, a hollow symmetrical model, which can be determined with only a lateral radiograph, permits an estimation of the extent of the medullary cavity and cortical bone cortex equivalently. This approach only deviates 7% to 9% from the true values of the second moments of area, which can be accurately measured in fossil mandibles using the ventral cortical area or width of the mandible (Fig. 8.2).

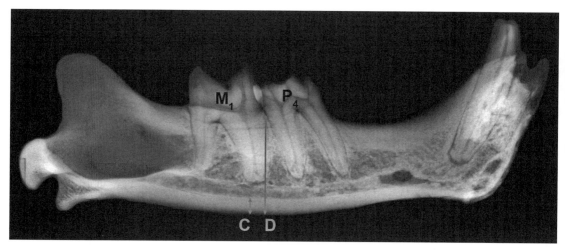

Figure 8.2. Radiograph of mandible of *Smilodon fatalis*, with relative cortical width measurements. C = cortical bone width; D = mandible width. Reproduced with permission from Binder et al. (2016).

Resistance to bending and torsional stress in the mandible can tell us what types of foods carnivores are eating. For example, Biknevicius and Ruff (1992a) found that precarnassial bending strength relative to jaw length was greater in felids than it was in canids, reflecting their modes of prey killing. Felids bite with their anterior jaws and hold on until the animal dies, whereas canids chase their prey and nip with their anterior teeth with no prolonged killing bite. However, canids had thicker postcarnassial bone than felids, reflecting their consumption of hard foodstuffs (i.e., bones) with their molars, whereas felids do not as they lack functional postcarnassial molars. Reinforcement of cortical strength can also extend to changes in diet within an animal's lifetime. For instance, if one dog eats more hard foodstuffs than another of the same species, the durophagous individual should have a more reinforced mandible due to increased cortical bone deposition in life than its soft food-eating counterpart. Repeated stress and strain on cortical bone should cause more bone to be laid down over the course of an animal's lifetime (Hylander, 1979; Chou et al., 2015).

If bone remodels over an animal's lifetime we may see differences in cortical thickness in the mandibles of *Smilodon fatalis* between the different sites at RLB. Binder et al. (2016) looked at cortical thickness in *Smilodon* using lateral radiographs to determine relative bone strength (using the equivalent of a hollow symmetrical model). Cortical thickness near

the carnassial tooth was found to be lowest in Pit 13 and highest in Pit 61/67, a trend similar to earlier results, suggesting that carnivores preserved in Pit 13 were dealing with unusual and difficult circumstances. Given that breakage and wear was highest in Pit 13, this is counterintuitive to what we would expect. In view of studies that examined nutritional stress in neonatal rats (Romano et al., 2010; Hattori et al., 2013), this suggests that Pit 13 *Smilodon* individuals may have been under nutritional stress early in life and were thus unable to deposit the thicker, stronger cortical bone throughout their lifetimes. It is curious that Pit 61/67 has the thickest cortical bone, especially since this pit was deposited during the Pleistocene extinction event. We also see an intermediate toothrow length in Pit 13, and a much longer toothrow in Pit 61/67, again suggesting that after Pit 13, nutritional stress declined, and jaws, like other morphological features, could grow larger when nutritional stress was gone.

Skull Phenotypic Integration and Evolution in *Smilodon fatalis*

Phenotypic integration relates to how well correlated structures are within a set of morphological characters or a whole organism. Two structures in the skull may be highly correlated with one another while a third may not with regard to growth or development. These interrelated traits can be referred to as modular units, with strong correlations of traits

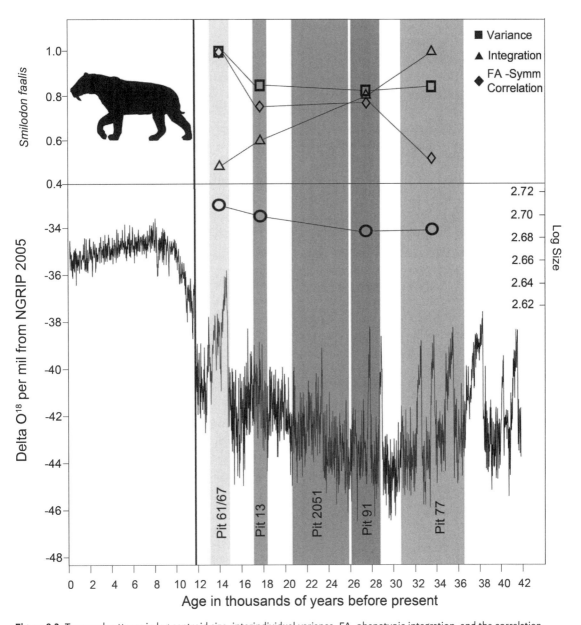

Figure 8.3. Temporal patterns in log centroid size, interindividual variance, FA, phenotypic integration, and the correlation between fluctuating asymmetry and overall integration for *Smilodon fatalis* from RLB. Modified with permission from Goswami et al. (2015).

within a module, but weak correlations between modules (Goswami et al., 2015). For example, monotremes show high correlation between the structure of the femur and the humerus, which could be considered a module in this group; but marsupials show little correlation between their hind limbs and forelimbs, so this would not be considered a module in marsupials (Bennett and Goswami, 2011; Kelly and Sears, 2011). Osteological phenotypic integration is driven by heterochrony shifts and gene-expression patterns (Goswami et al., 2009; Sears et al., 2012; Goswami et al., 2015).

Fluctuating asymmetry is a quantitative approach that can evaluate the contribution of development to phenotypic integration. Developmental errors can lead to random asymmetries that can be measured with this technique, so the degree of asymmetry relates to the relative developmental health. Skull asymmetry can be compared across time using different sites at RLB to see if there are trends of asymmetry and whether they correspond to other morphological variation.

Goswami et al. (2015) found *Smilodon* crania to be significantly different in shape across time, with a size increase over time at RLB from the oldest to the more recent pits (Fig.8.3). In addition, variance between individuals also increased over time, as did fluctuating asymmetry (with Pit 13 having the highest amount). Finally, phenotypic integration, which is significantly correlated with fluctuating asymmetry, decreased over this same time period in the younger sites. These results support the conclusions of other studies showing higher levels of nutritional stress in the Late Pleistocene, and that *Smilodon* may have been responding to increases in extrinsic forces such as temperature change during this time.

Conclusions

Smilodon, along with its prey, habitat, environment, and climate, was actively evolving in the Late Pleistocene. Extrinsic factors, such as climate, cause changes in vegetation. These vegetation changes directly affect herbivores, thus creating a chain reaction that affects the hunting and killing strategies of the large carnivores dependent on large herbivore populations to survive.

These environmental relationships are not simple, however, and different sites at Rancho La Brea do not show identical trends. *Smilodon* appears to have always had 'tough times' in the Pleistocene, with considerable amounts of tooth breakage and wear, far more than is seen in modern large carnivores. At the same time, there is evidence that in the most recent site at RLB, Pit 61/67, *Smilodon* was recovering from nutritional deficiencies, with longer jaws, less tooth breakage, and more robust mandibles with reinforced cortical bone for killing and chewing. Their crania appear to support this, as they changed too, demonstrating an increase in size, decrease in fluctuating asymmetry, and more specialization over time. Evidence supports the hypothesis of a macroevolutionary ratchet: that *Smilodon* were getting

larger, more specialized, and less nutritionally stressed in the latest site at RLB prior to their extinction. They may also have been forced into the untenable situation of depending completely on the availability of very large prey, which may have contributed to their demise.

This chapter does not examine *Smilodon* sociality. There has been some contention about this issue in the past (McCall et al., 2003; Carbone et al., 2009; Meachen-Samuels and Binder, 2010). The *S. fatalis* data from Meachen et al. (2014) and floral data from Coltrain et al. (2004) suggested that *Smilodon* may have been modifying its hunting strategy based on habitat structure and available prey species. One interesting avenue for further investigation is the hypothesis that *Smilodon* may have changed its social structure depending on environmental conditions, similar to modern coyotes (Gese et al., 1988; Arjo and Pletscher, 1999).

Other future studies may include examining differences in saber length between the sites at Rancho La Brea—an aspect that, surprisingly, has not yet been explored. Other morphological differences in postcranial measurements between the sites at Rancho La Brea also have yet to be investigated. Although our focus in this work is on one of the better-studied fossil species, there are still many questions remaining to be answered.

REFERENCES

Akersten, W. A. 1985. Canine function in *Smilodon* (Mammalia, Felidae, Machairodontidae). Natural History Museum of Los Angeles County, Contributions in Science 356:1-22.

Antón, M., and A. Galobart. 1999. Neck function and predatory behavior in the scimitar toothed cat *Homotherium latidens* (Owen). Journal of Vertebrate Paleontology 19:771-784.

Antón, M., R. García-Perea, and A. Turner. 1998. Reconstructed facial appearance of the sabretoothed felid *Smilodon*. Zoological Journal of the Linnean Society 124:369-386.

Anyonge, W. 1996. Microwear on canines and killing behavior in large carnivores: saber function in *Smilodon fatalis*. Journal of Mammalogy 77:1059-1067.

Arjo, W. M., and D. H. Pletscher. 1999. Behavioral responses of coyotes to wolf recolonization in northwestern Montana. Canadian Journal of Zoology 77:1919-1927.

Bennett, C. V., and A. Goswami. 2011. Does developmental strategy drive limb integration in marsupials and monotremes? Mammalian Biology 76:79–83.

Biknevicius, A., and S. Leigh. 1997. Patterns of growth of the mandibular corpus in spotted hyenas (*Crocuta crocuta*) and cougars (*Puma concolor*). Zoological Journal of the Linnean Society 120:139–161.

Biknevicius, A., and C. Ruff. 1992a. The structure of the mandibular corpus and its relationship to feeding behaviors in extant carnivorans. Journal of Zoology, London 228:479–507.

Biknevicius, A., and C. Ruff. 1992b. Use of biplanar radiographs for estimating cross-sectional geometric properties of mandibles. Anatomical Record 232:157–163.

Biknevicius, A. R., and B. Van Valkenburgh. 1996. Design for killing: craniodental adaptations of predators; pp. 393–428 in J. L. Gittleman (ed.), Carnivore Behavior, Ecology, and Evolution. Cornell University Press, Ithaca, New York.

Binder, W. J., and B. Van Valkenburgh. 2010. A comparison of tooth wear and breakage in Rancho La Brea sabertooth cats and dire wolves across time. Journal of Vertebrate Paleontology 30:255–261.

Binder, W. J., K. S. Cervantes, and J. A. Meachen. 2016 Measures of relative dentary strength in Rancho La Brea *Smilodon fatalis* over time. PLoS One 11(9): e0162270.

Binder, W. J., E. N. Thompson, and B. Van Valkenburgh. 2002. Temporal variation in tooth fracture among Rancho La Brea dire wolves. Journal of Vertebrate Paleontology 22:423–428.

Brown, J. G. 2014. Jaw function in *Smilodon fatalis*: a reevaluation of the canine shear-bite and a proposal for a new forelimb-powered Class 1 lever model. PLoS One 9:e107456.

Carbone, C., T. Maddox, P. J. Funston, M. G. L. Mills, G. F. Grether, and B. Van Valkenburgh. 2009. Parallels between playbacks and Pleistocene tar seeps suggest sociality in an extinct sabretooth cat, *Smilodon*. Biology Letters 5:81–85.

Chou, H. Y., D. Satpute, A. Müftü, S. Mukundan, and S. Müftü. 2015. Influence of mastication and edentulism on mandibular bone density. Computer Methods in Biomechanics and Biomedical Engineering 18:269–281.

Coltrain, J. B., J. M. Harris, T. E. Cerling, J. R. Ehleringer, M.-D. Dearing, J. Ward, and J. Allen. 2004. Rancho La Brea stable isotope biogeochemistry and its implications for the palaeoecology of Late Pleistocene, coastal southern California. Palaeogeography Palaeoclimatology Palaeoecology 205:199–219.

DeSantis, L. R. G., B. W. Schubert, J. R. Scott, and P. S. Ungar. 2012. Implications of diet for the extinction of saber-toothed cats and American lions. PLoS One 7:e52453.

Emerson, S. B., and L. Radinsky. 1980. Functional analysis of sabertooth cranial morphology. Paleobiology 6:295–312.

Ewer, R. F. 1973. The Carnivores. Cornell University Press, Ithaca, New York.

Friscia, A. R., B. Van Valkenburgh, and A. R. Biknevicius. 2007. An ecomorphological analysis of extant small carnivorans. Journal of Zoology 272:82–100.

Friscia, A. R., B. Van Valkenburgh, L. Spencer, and J. M. Harris. 2008. Chronology and spatial distribution of large mammal bones in pit 91, Rancho La Brea. Palaios 23:35–42.

Gese, E. M., O. J. Rongstad, and W. R. Mytton. 1988. Relationship between coyote group size and diet in southeastern Colorado. Journal of Wildlife Management 52:647–653.

Goswami, A., V. Weisbecker, and M. R. Sánchez-Villagra. 2009. Developmental modularity and the marsupial-placental dichotomy. Journal of Experimental Zoology Part B: Molecular and Developmental Evolution 312B:186–195.

Goswami, A., W. J. Binder, J. A. Meachen, and F. R. O'Keefe. 2015. From shrews to sabre-toothed cats: a deep time perspective on phenotypic integration and modularity. Proceedings of the National Academy of Sciences of the United States of America 112:4891–4896.

Hattori, S., J. H. Park, U. Agata, T. Akimoto, M. Oda, M. Higano, Y. Aikawa, Y. Nabekura, H. Yamato, I. Ezawa, and N. Omi. 2013. Influence of food restriction combined with voluntary running on bone morphology and strength in male rats. Calcified Tissue International 93:540–548.

Heinrich, R. E., and A. Biknevicius. 1998. Skeletal allometry and interlimb scaling patterns in mustelid carnivorans. Journal of Morphology 235:121–134.

Hylander, W. L. 1979. Mandibular function in *Galago crassicaudatus* and *Macacu fascicularis*: an *in vivo* approach to stress analysis of the mandible. Journal of Morphology 159:253–296.

Jouzel, J., V. Masson-Delmotte, O. Cattani, G. Dreyfus, S. Falourd, G. Hoffmann, B. Minster, J. Nouet, J. M. Barnola, J. Chappellaz, H. Fischer, J. C. Gallet, S. Johnsen, M. Leuenberger, L. Loulergue, D. Luethi, H. Oerter, F. Parrenin, G. Raisbeck, D. Raynaud, A. Schilt, J. Schwander, E. Selmo, R. Souchez, R. Spahni, B. Stauffer, J. P. Steffensen, B. Stenni, T. F. Stocker, J. L. Tison, M. Werner, and E. W. Wolff. 2007. Orbital and millennial Antarctic climate variability over the past 800,000 years. Science 317:793–796.

Kelly, E. M., and K. E. Sears. 2011. Reduced phenotypic covariation in marsupial limbs and the implications for mammalian evolution. Biological Journal of the Linnean Society 102:22-36.

La Croix, S., K. E. Holekamp, J. A. Shivik, B. L. Lundrigan, and M. L. Zelditch. 2011. Ontogenetic relationships between cranium and mandible in coyotes and hyenas. Journal of Morphology 272:662-674.

Martin, L. D., J. P. Babiarz, V. L. Naples, and J. Hearst. 2000. Three ways to be a saber-toothed cat. Naturwissenschaften 87:41-44.

McCall, S., V. Naples, and L. Martin. 2003. Assessing behavior in extinct animals: was Smilodon social? Brain Behavior and Evolution 61:159-164.

McHenry, C. R., S. Wroe, P. D. Clausen, K. Moreno, and E. Cunningham. 2007. Supermodeled sabercat, predatory behavior in Smilodon fatalis revealed by high-resolution 3D computer simulation. Proceedings of the National Academy of Sciences of the United States of America 104:16010-16015.

Meachen, J. A., F. R. O'Keefe, and R. W. Sadleir. 2014. Evolution in the sabertooth cat Smilodon fatalis in response to Pleistocene climate change. Journal of Evolutionary Biology 27:714-723.

Meachen-Samuels, J. A. 2012. Morphological convergence of the prey-killing arsenal of sabertooth predators. Paleobiology 38:1-14.

Meachen-Samuels, J. A., and W. J. Binder. 2010. Sexual dimorphism and ontogenetic growth in the American lion and sabertoothed cat from Rancho La Brea. Journal of Zoology 280:271-279.

Meachen-Samuels, J., and B. Van Valkenburgh. 2009. Craniodental indicators of prey size preference in the Felidae. Biological Journal of the Linnean Society 96:784-799.

Meachen-Samuels, J., and B. Van Valkenburgh. 2010. Radiographs reveal exceptional forelimb strength in the sabertooth cat, Smilodon fatalis. PLoS One 5:e11412.

O'Keefe, F. R., E. V. Fet, and J. M. Harris. 2009. Compilation, calibration, and synthesis of faunal and floral radiocarbon dates, Rancho La Brea, California. Natural History Museum of Los Angeles County, Contributions in Science 518:1-16.

Romano, T., J. D. Wark, and M. E. Wlodek. 2010. Calcium supplementation does not rescue the programmed adult bone deficits associated with perinatal growth restriction. Bone 47:1054-1063.

Schaller, G. B. 1967. The Deer and the Tiger: A Study of Wildlife in India. University of Chicago Press, Chicago, Illinois.

Schaller, G. B. 1972. The Serengeti Lion: A Study of Predator Prey Relations. University of Chicago Press, Chicago, Illinois.

Sears, K. E., C. K. Doroba, X. Cao, D. Xie, and S. Zhong. 2012. Molecular determinants of marsupial integration and constraint; pp. 257-278 in R. J. Asher and J. Müller (eds.), From Clone to Bone: The Synergy of Morphological and Molecular Tools in Palaeobiology. Cambridge University Press, Cambridge, UK.

Stock, C. 1992. Rancho La Brea: A Record of Pleistocene Life in California. Revised by J. M. Harris, Natural History Museum of Los Angeles County, Science Series 37:1-113.

Therrien, F. 2005. Feeding behaviour and bite force of sabretoothed predators. Zoological Journal of the Linnean Society 145:393-426.

Van Valkenburgh, B. 1988. Trophic diversity in past and present guilds of large predatory mammals. Paleobiology 14:155-173.

Van Valkenburgh, B. 1996. Feeding behavior in free-ranging, large African carnivores. Journal of Mammalogy 77:240-254.

Van Valkenburgh, B., and F. Hertel. 1993. Tough times at La Brea: tooth breakage in large carnivores of the Late Pleistocene. Science 261:456-459.

Van Valkenburgh, B., and K. Koepfli. 1993. Cranial and dental adaptations to predation in canids. Symposium of the Zoological Society of London 65:15-37.

Van Valkenburgh, B., and C. B. Ruff. 1987. Canine tooth strength and killing behavior in large carnivores. Journal of Zoology 212:379-397.

Van Valkenburgh, B., X. Wang, and J. Damuth. 2004. Cope's rule, hypercarnivory, and extinction in North American canids. Science 306:101-104.

Wroe, S., U. Chamoli, W. C. H. Parr, P. Clausen, R. Ridgely, and L. Witmer. 2013. Comparative biomechanical modeling of metatherian and placental saber-tooths: a different kind of bite for an extreme pouched predator. PLoS One 8:e66888.

9 Analyzing the Tooth Development of Sabertooth Carnivores: Implications Regarding the Ecology and Evolution of *Smilodon fatalis*

M. ALEKSANDER WYSOCKI AND ROBERT S. FERANEC

Introduction

Smilodon fatalis, the most renowned sabertooth carnivore, has been the subject of numerous investigations that have revealed many insights about this species and sabertooth carnivores in general (Gonyea, 1976; Emerson and Radinsky, 1980; Tejada-Flores and Shaw, 1984; Akersten, 1985; McHenry et al., 2007; Christiansen, 2012; Wroe et al., 2013). The sabertooth condition is a unique set of features not present in any modern carnivorous species, but it has evolved at least six times in the past, occurring within the Felidae, Nimravidae, Barbourofelidae, Thylacosmilidae, Gorgonopsia, and creodonts (Antón, 2013). The term 'creodont' is used here in an informal sense because the order Creodonta is currently considered polyphyletic and the positions of the sabertooth taxa are equivocal.

Sabertooth carnivores are a form of terrestrial predator primarily characterized by the presence of elongated and laterally compressed upper canine teeth (Cope, 1880; Gonyea, 1976; Emerson and Radinsky, 1980). Sabertooths have evolved multiple times and exhibit various forms of sabertooth canine morphology, such as the long, narrow, and finely serrated upper canines of dirk-toothed felids, as in *Smilodon fatalis*, and the relatively short, broad, and coarsely serrated canines of the scimitar-toothed felids, as in *Homotherium serum* (Martin, 1980;

Meachen-Samuels, 2012). In addition to these morphological differences, there are differences in the rates of tooth growth among the various sabertooth species (Feranec, 2008; Forasiepi and Sánchez-Villagra, 2014). Developmental timing is critical for the formation of morphological structures, and the ontogenetic ages at which those structures become functional is crucial for individuals in that it influences ecology and life history. In the case of *S. fatalis* and other sabertooth carnivores, the upper canines were one of those essential morphological features (Akersten, 1985; McHenry et al., 2007).

Dental epithelial stem cell niche regulation may have the evolutionary flexibility that would allow for changes to tooth morphologies and tooth growth rate, as well as to the timing of root development. This suggests that it may be closely linked to the evolution of the different forms of saberteeth and the various patterns and rates of tooth development observed in the different sabertooth species (Tummers and Thesleff, 2003; Wang et al., 2007). Moreover, if the timing of the transition from tooth crown formation to root formation were eliminated altogether, this would generate a hypselodont (open-rooted or ever-growing) tooth, such as that seen in *Thylacosmilus atrox* (Riggs, 1934; Tummers and Thesleff, 2003; Yokohama-Tamaki et al., 2006).

The absence of a modern sabertooth carnivore has made it challenging to understand the sabertooth condition and many questions remain about sabertooth carnivores. Questions include how sabertooth carnivores developed their dentitions, how sabertooths functioned as parts of their respective ecosystems, and how the sabertooth condition evolved. Although many studies have focused on understanding the function of the upper canines, this investigation examines the developmental patterns of teeth in *S. fatalis* and other sabertooth carnivores. This study explores what aspects of *S. fatalis* tooth development are merely typical of felids (i.e., the result of phylogeny) and which are related to ecology (i.e., the result of its sabertooth condition), and begins to address questions such as how *S. fatalis* cubs matured into proficient adult predators. By analyzing the differences in tooth developmental patterns in sabertooth carnivores, the current investigation seeks to establish a better understanding of the ecology and evolution of *S. fatalis*.

Methods

For this study we examine all of the published tooth eruption sequences of sabertooth carnivores and construct additional parts of eruption sequences from specimen descriptions. We also review many of the tooth eruption sequences of conical-toothed felid species to further understand tooth development in *Smilodon fatalis*. Given that some of the analyzed tooth eruption sequences are based on the observation of live animals and others are based on fossils and/or modern skulls, this study compares tooth eruption sequences that have different degrees of precision. This study also includes a survey of the specimens of immature sabertooth carnivores within the collections of the American Museum of Natural History (AMNH). Notably, the analysis of FM104728 and FM104729 from Edson Quarry, Sherman County, Kansas (Miocene) made it possible to construct a portion of the *Machairodus* tooth eruption sequence. This investigation focuses on understanding *S. fatalis* tooth development in relation to the tooth development of other sabertooth carnivore species in order to better understand its ecology and evolution. It is

not intended to be a comprehensive review of tooth development and replacement patterns in all taxa that contain sabertooth-bearing species; in particular, the analysis of tooth eruption in herbivorous sabertooth species (e.g., musk deer) is beyond the scope of this investigation.

Results

Felidae

The Felidae is a carnivoran family that first evolved in the Early Oligocene in Europe and is still extant (Martin, 1998). Currently, the family consists of 38 species. It is found on all continents except Antarctica and ranges in body mass from a few kilograms, as in the rusty-spotted cat, *Prionailurus rubiginosus*, to a few hundred kilograms, as in the tiger, *Panthera tigris* (Sunquist and Sunquist, 2002). Although the sabertooth morphology is no longer present within this taxon, felid species are still particularly useful for studying this condition. The Felidae display multiple tooth forms, including conical-toothed species such as modern lions and tigers (*P. leo* and *P. tigris*), scimitar-toothed species such as *Homotherium serum*, and dirk-toothed species such as *Smilodon fatalis*. For *S. fatalis*, Tejada-Flores and Shaw (1984) use the fabulous record of specimens at Rancho La Brea in southern California to identify the eruption sequence of teeth (Fig. 9.1) relative to permanent upper canine (C) length. In a complementary study, Wysocki et al. (2015) use this sequence along with data from stable oxygen isotopes and CT-scanning to identify the absolute timing of development for the teeth and a few other skeletal features (e.g., the fusion of the parietals) in *S. fatalis* (Fig. 9.2). Surprisingly, few comprehensive modern studies examine the pattern and timing of tooth development. However, one such study by Smuts et al. (1978), which examines an ontogenetic series of modern lions (*P. leo*), can serve as a basis for a more thorough comparison with *S. fatalis*. The following highlights major developmental events within large conical-toothed felids compared to the sabertooth species, *S. fatalis*.

Limitations of the respective fossil records make it difficult to ascertain the precise timing of initial dC eruption relative to the eruption of the rest of the deciduous dentition, but it is apparent that dC

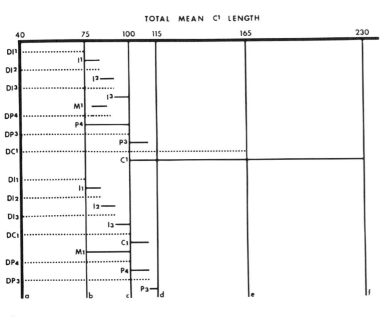

Figure 9.1. Tooth replacement sequence compared with total mean length in mm of the permanent upper canine in *Smilodon fatalis* from Rancho La Brea. Vertical lines mark major events in the growth of the dentition. a = presence of the entire deciduous dentition; b = beginning of permanent tooth eruption; c = beginning of C eruption; d = permanent dentition fully erupted except for the C; e = loss of the dC; F = the C fully erupted. From Tejada-Flores and Shaw. 1984. Tooth replacement and skull growth in *Smilodon* from Rancho La Brea. Journal of Vertebrate Paleontology 4:114–121. © 2015 The Society of Vertebrate Paleontology. Reprinted and distributed with permission of the Society of Vertebrate Paleontology.

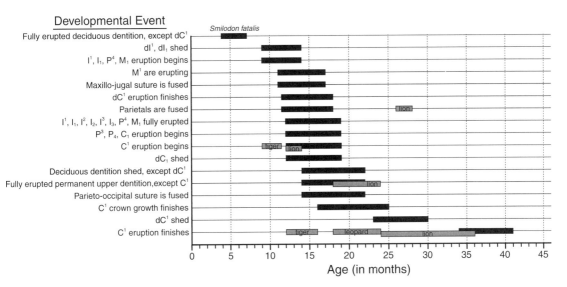

Figure 9.2. Absolute ontogenetic age ranges for developmental events in *Smilodon fatalis*. Figure from Wysocki et al. (2015). Using a novel absolute ontogenetic age determination technique to calculate the timing of tooth eruption in the saber-toothed cat, *Smilodon fatalis*.PLoS One 10:e0129847. doi:10.1371/journal.pone.0129847. © 2015 Wysocki et al. This is an open access article distributed under the terms of the Creative Commons Attribution License, which permits unrestricted use, distribution, and reproduction in any medium. http://creativecommons.org/licenses/by/4.0/

eruption began with the rest of the deciduous dentition and prior to permanent tooth eruption in *Smilodon fatalis, S. populator,* and *Homotherium serum* (Rawn-Schatzinger, 1983; Tejada-Flores and Shaw, 1984; Christiansen, 2012). It is evident that the dC of

S. fatalis did not finish erupting and was not shed until later in life than it is in many extant conical-toothed felids (Schaller, 1972; Crowe, 1975; Smuts et al., 1978; Currier, 1983; Tejada-Flores and Shaw, 1984; Smith, 1993; Stander, 1997). This is likely due to the greater crown height of the *S. fatalis* dC (Smuts et al., 1978; Tejada-Flores and Shaw, 1984). In particular, the *S. fatalis* dC was still erupting as many permanent teeth (i.e., I1, i1, I2, i2, P4, m1, and M1) began to erupt (Tejada-Flores and Shaw, 1984). Initial C eruption coincided with initial c eruption in *S. fatalis*, just as it does in several other felid species (*H. serum, Panthera tigris, P. uncia,* and *Lynx rufus*) (Pocock, 1916; Mazák, 1981; Rawn-Schatzinger, 1983; Tejada-Flores and Shaw, 1984; Jackson et al., 1988). Although the established eruption sequence for *P. leo* indicates that c eruption precedes C eruption, it should be noted that this study also describes the c and C as starting to erupt within the same window of 12 to 15 months of age (Smuts et al., 1978). The apparent discrepancy, consisting of *P. leo* c eruption preceding C eruption, may reflect the superb precision of the Smuts et al. (1978) study rather than a major anomaly among felid tooth eruption sequences. Overall, the available tooth eruption sequences suggest that initial eruptions of the c and C are nearly simultaneous in felid species, whether conical-toothed or sabertooth (Pocock, 1916; Smuts et al., 1978; Mazák, 1981; Rawn-Schatzinger, 1983; Tejada-Flores and Shaw, 1984; Jackson, 1988).

Similar to the pattern of initial permanent tooth eruption exhibited by many extant conical-toothed felids, the first permanent teeth to erupt in the dirk-toothed felid *Smilodon fatalis*, as well as the scimitar-toothed *Homotherium serum*, were the permanent first incisors (I1 and i1) (Schneider, 1959; Mazák, 1981; Smuts et al., 1978; Currier, 1983; Rawn-Schatzinger, 1983; Tejada-Flores and Shaw, 1984; Jackson et al., 1988). Permanent first incisor (I1, i1) eruption coincided with the initial eruption of the permanent carnassials (P4 and m1) in *S. fatalis* (Tejada-Flores and Shaw, 1984). As Tejada-Flores and Shaw (1984) note, this is a prominent difference between the eruption sequences of *S. fatalis* and *Panthera leo* because initial permanent carnassial (P4 and m1) eruption occurs after all of the permanent incisors (I1, i1, I2, i2, I3, and i3) have started to erupt in *P. leo* (Smuts et al., 1978). Comparing the relative timing of permanent carnassial eruption of *S. fatalis* to those of additional felid species reveals that the initial eruptions of the P4 and m1 also occur relatively earlier in the *S. fatalis* eruption sequence than

in *P. tigris, P. uncia, Lynx rufus,* and *H. serum*—species in which all of the permanent incisors (I1, i1, I2, i2, I3 and i3) begin eruption prior to initial P4 and m1 eruption (Pocock, 1916; Mazák, 1981; Rawn-Schatzinger, 1983; Jackson et al., 1988). The discovery of other sabertooth felid tooth eruption sequences might explain whether this discrepancy in the timing of permanent carnassial eruption between *S. fatalis* and *H. serum* is related to their different morphologies (i.e., dirk-toothed felid versus scimitar-tooth felid) or a feature of tooth development that was produced by the disparate selective forces of two distinctive ecological niches. The stages of tooth eruption evident in *Machairodus* specimens FM104728 and FM104729 at the American Museum of Natural History, New York suggest that the initial eruptions of the i1, i2, i3, and m1 occurred before the initial eruptions of the c and p4 in *Machairodus*. From these specimens it cannot be determined exactly when the m1 began to erupt in relation to the permanent lower incisors (i1, i2, i3), but it is possible that permanent incisor eruption overlapped to some extent with permanent carnassial eruption in *Machairodus*.

Nimravidae

Although cat-like in overall form, the Nimravidae is a carnivoran clade that evolved separately from and earlier than the true cats of the Felidae (Martin, 1998; Antón, 2013). The earliest Nimravid taxa derive from the Late Eocene of Asia and North America (McKenna and Bell, 1997). Similar to the Barbourofelidae, the specimen FSP-ITD 342 of *Eusmilus* cf. *E. bidentatus* (paleontology collections at Faculté des Sciences Fondamentales et Appliquées, Université de Poitiers, Poitiers, France) shows that the dC of this nimravid began erupting after all of the other deciduous teeth began erupting (i.e., dI1, di1, dI2, di2, dI3, di3, dP3, dp3, dP4, dp4, and dc) (Bryant, 1988; Peigné and De Bonis, 2003). Additionally, Bryant (1988) indicates that dC eruption was delayed in *E. cerebralis*, possibly to the extent that initial dC eruption overlapped with the eruption of the permanent dentition. For *Dinictis,* the first permanent teeth to erupt were the upper and lower permanent first incisors (I1 and i1), which were followed by the eruption of the permanent second and third incisors (I2, i2, I3, and i3) (Bryant,

1988). This is a pattern of tooth development that is generally similar to the tooth development of the Felidae, in which eruption of the permanent teeth begins with the first incisors (I1 and i1) and is followed by the eruption of the second incisors (I2 and i2) and third incisors (I3 and i3) (Pocock, 1916; Schneider, 1959; Smuts et al., 1978; Mazák, 1981; Currier, 1983; Rawn-Schatzinger, 1983; Tejada-Flores and Shaw, 1984; Jackson, 1988). However, the P2, p2, M1, and m1 of *Dinictis* also began to erupt with the I2, i2, I3, and i3, which is not characteristic of felid tooth eruption (Pocock, 1916; Smuts et al., 1978; Mazák, 1981; Rawn-Schatzinger, 1983; Bryant, 1988; Jackson, 1988). Bryant (1988) also notes that the tooth eruption sequences of *Dinictis* and felids differ in that *Dinictis* had initial C eruption that occurred after the initial eruption of all other permanent teeth, including the c. Initial eruptions of the C and c occur at approximately the same time in most felids, including *S. fatalis* (Pocock, 1916; Smuts et al., 1978; Mazák, 1981; Rawn-Schatzinger, 1983; Tejada-Flores and Shaw, 1984; Bryant, 1988; Jackson, 1988). Not only is initial C eruption delayed, initial c eruption is relatively delayed in the eruption sequence of *Dinictis* compared to tooth eruption of several felids (Pocock, 1916; Smuts et al., 1978; Mazák, 1981; Rawn-Schatzinger, 1983; Tejada-Flores and Shaw, 1984; Bryant, 1988; Jackson, 1988). Thus, the tooth eruption sequences of *S. fatalis* and the other felids share some general similarities with the tooth eruption sequences of the nimravids, but they differ dramatically when it comes to the relative timing of deciduous and permanent canine eruptions (Schneider, 1959; Smuts et al., 1978; Mazák, 1981; Currier, 1983; Rawn-Schatzinger, 1983; Tejada-Flores and Shaw, 1984; Bryant, 1988; Jackson, 1988).

Barbourofelidae

The Barbourofelidae is a sabertooth taxon that first evolved in the Early Miocene of Africa (Morlo et al., 2004; Antón, 2013). This taxon includes species possessing some of the most derived morphological characteristics of sabertooth carnivores (Schultz et al., 1970; Bryant, 1988; Slater and Van Valkenburgh, 2008; Meachen-Samuels, 2012) and it displays noteworthy patterns of tooth development. One of the most striking aspects of tooth development in

barbourofelids when compared to that of felids (including *Smilodon fatalis* and *Homotherium serum*) is that the dC of *Barbourofelis loveorum* and *B. morrisi* did not begin to erupt until after some of the permanent dentition had started to erupt (Smuts et al., 1978; Rawn-Schatzinger, 1983; Tejada-Flores and Shaw, 1984; Bryant, 1988, 1990). The pattern of delayed dC eruption is also apparent in *B. whitfordi* (Tseng et al., 2010). Unlike all of the felids for which there are available eruption sequences, initial C eruption in *B. morrisi* and *B. loveorum* occurred after all other teeth had started to erupt (Pocock, 1916; Smuts et al., 1978; Mazák, 1981; Currier, 1983; Rawn-Schatzinger, 1983; Tejada-Flores and Shaw, 1984; Jackson, 1988; Bryant, 1988, 1990). Bryant (1988) considers the delayed initial C eruption to be a consequence of the delayed dC eruption. *Barbourofelis loveorum* tooth eruption also differs from felid tooth eruption such that in *B. loveorum* c eruption was relatively delayed in its eruption sequence (Pocock, 1916; Smuts et al., 1978; Mazák, 1981; Rawn-Schatzinger, 1983; Tejada-Flores and Shaw, 1984; Bryant, 1988; Jackson, 1988). These delayed initial eruptions of the c and C in *B. loveorum* are striking when compared to the felid eruption sequences, including that of *S. fatalis* (Pocock, 1916; Smuts et al., 1978; Mazák, 1981; Rawn-Schatzinger, 1983; Tejada-Flores and Shaw, 1984; Bryant, 1988; Jackson, 1988). However, comparing the eruption sequences of *S. fatalis* and *B. loveorum* shows that these sabertooth species seem to share early permanent carnassial eruption relative to the eruption of other permanent teeth (i.e., the P^4 and M_1 begin eruption before the initial eruption of the P3, p3, p4, C, c, i3, and possibly additional permanent teeth) (see Table 9.1; Tejada-Flores and Shaw, 1984; Bryant, 1988).

Creodonts

Machaeroides and *Apataelurus*, two creodont genera within the tribe Machaeroidini, have sabertooth canines. The creodonts are an ancient group of carnivorous mammals, known from deposits dating from the Late Paleocene to the Miocene. The two sabertooth taxa occur from the Early to Middle Eocene (Gunnell, 1998). Family level assignment for *Machaeroides* and *Apataelurus* has been uncertain, with some researchers placing them in the oxyaenids

Table 9.1. Sabertooth carnivore tooth eruption sequences

Smilodon fatalis
dC^1 | I^1 P^4 | M^1 | I^2 | I^3 | C^1 P^3 |
   | I$_1$ M$_1$ | | I$_2$ | I$_3$ | C$_1$ P$_4$ | P$_3$

Smilodon populator
dC^1 | I^1 ↔ P^4 ↔ M^1 ↔ I^2 ↔ I^3 ↔ C^1 ↔ P^3
   | I$_1$ ↔ M$_1$ ↔ I$_2$ ↔ I$_3$ ↔ C$_1$ ↔ P$_4$ ↔ P$_3$

Homotherium serum
dI^1 dI^2 | dI^3 | dP^3 dP^4 | dC^1 | I^1 I^2 I^3 | C^1 ↔ P^3 ↔ P^4 ↔ M^1
dI$_1$ dI$_2$ | dI$_3$ | dP$_3$ dP$_4$ dC$_1$ | | I$_1$ I$_2$ I$_3$ | C$_1$ ↔ P$_3$ ↔ P$_4$ ↔ M$_1$

Machairodus
I$_1$ ↔ I$_2$ ↔ I$_3$ ↔ M$_1$ | C$_1$ ↔ P$_4$

Barbourofelis loveorum
P^4 | | dC^1 P^3 | | C^1
M$_1$ | I$_3$ P$_4$ | I$_2$?   P$_3$ | C$_1$ |

Barbourofelis morrisi
I^1 ↔ I^2 ↔ I^3 ↔ P^3 ↔ P^4 ↔ M^1 | dC^1 | C^1

Dinictis
I^1 | I^2 I^3 P^2 M^1 | P^4 | | P^3 | | C^1
I$_1$ | I$_2$ I$_3$ P$_2$ M$_1$ | | P$_4$ | P$_3$ M$_2$ | C$_1$ |

cf. Eusmilus bidentatus (FSP-ITD 342)
dI^1 ↔ dI^2 ↔ dI^3 ↔ dP^3 ↔ dP^4 | dC^1
dI$_1$ ↔ dI$_2$ ↔ dI$_3$ ↔ dP$_3$ ↔ dP$_4$ ↔ dC$_1$ |

Eusmilus cerebralis
dP^3 ↔ dP^4 | M^1 dC^1? |
dP$_3$ ↔ dP$_4$ | | M$_1$

Sequence of tooth eruption occurs from left to right.
| indicates separate stages of tooth eruption.
↔ indicates an unknown relationship for the timing of eruption between particular teeth.
? indicates tentative identification of a particular tooth.
Note: Data compiled from Schultz et al. (1970); Rawn-Schatzinger (1983); Tejada-Flores and Shaw (1984); Bryant (1988, 1990); Peigné and De Bonis (2003); Christiansen (2012), and personal observation.

(Gunnell, 1998; Antón 2013) and others placing them in the hyaenodontids (McKenna and Bell, 1997). Discussion of affinities to a particular higher taxon is beyond the scope of this study.

Unfortunately, at this writing, very little information is available regarding the tooth development of the sabertooth creodonts due to the limited number of both *Machaeroides* and *Apataelurus* specimens. However, it is clear that the sabertooth creodonts share a number of morphological characteristics with the sabertooth species of the Felidae, Barbourofelidae, and Nimravidae, including relatively well-developed carnassial teeth (Emerson and Radinsky, 1980; Polly, 1996). A notable distinction between the dentitions of these sabertooth taxa and sabertooth carnivorans is the particular teeth that function as the carnassials. For these creodonts, the M1 and the m2 serve as the carnassials, whereas the

P4 and the m1 are the carnassials in the carnivorans (Van Valen, 1969; Ewer, 1973; Rawn-Schatzinger, 1983; Tejada-Flores and Shaw, 1984; Bryant, 1988; Polly, 1996; Antón, 2013). The current lack of a representative ontogenetic sequence of specimens in either *Machaeroides* or *Apataelurus* makes the tooth eruption sequences unknown. In the event of auspicious discoveries within these sabertooth taxa, it would be valuable to examine the relative timing of *Machaeroides* and *Apataelurus* carnassial eruption within the context of the Creodonta as a whole. Because the sabertooth creodonts have different teeth that function as the carnassials (Van Valen, 1969; Ewer, 1973; Rawn-Schatzinger, 1983; Tejada-Flores and Shaw, 1984; Bryant, 1988; Polly, 1996; Antón, 2013), additional study in this area has the potential to enhance understanding of carnassial development and evolution in sabertooth carnivores overall and might provide insights pertinent to the relatively early permanent carnassial eruption of *Smilodon fatalis*.

Thylacosmilidae

The Thylacosmilidae is a family of sabertooth sparassodont metatherians. Although originally placed as a subfamily within the Borhyaenidae, the morphological characteristics are different enough for some researchers to raise the subfamily to family level, a designation utilized here (Marshall, 1976; Goin and Pascual, 1987). Three thylacosmilid genera—*Thylacosmilus*, *Anachlysictis*, and *Patagosmilus*—are known, occurring from the Miocene to the Pliocene of South America (Marshall, 1976; McKenna and Bell, 1997; Forasiepi and Carlini, 2010; Antón, 2013).

It appears that thylacosmilids have a pattern of tooth development that is considerably different from the tooth development of eutherian sabertooth carnivores. This occurs in ways that are consistent with the typical differences in tooth development patterns that exist between most metatherians and eutherians (Rawn-Schatzinger, 1983; Tejada-Flores and Shaw, 1984; Bryant, 1988; Luo et al., 2004; van Nievelt and Smith, 2005; Forasiepi and Sánchez-Villagra, 2014). Most metatherians lack functional replacement of the incisors, canines, and premolars except for the third premolars, whereas in many eutherians, deciduous incisors, canines, and premolars are replaced by permanent incisors, canines, and premolars (Luo et al., 2004; van Nievelt and Smith, 2005). The thylacosmilids *T. atrox* and *P. goini* appear to have exhibited a pattern of tooth development that deviates from those of most other metatherians in that these sabertooth species did not have third upper premolar replacement (Luo et al., 2004; van Nievelt and Smith, 2005; Forasiepi and Sánchez-Villagra, 2014). In addition, the tooth development of *T. atrox* deviates from that of many other metatherians and from other sabertooth carnivores because this species had hypselodont (i.e., ever-growing) upper canines (Riggs, 1934; Rawn-Schatzinger, 1983; Tejada-Flores and Shaw, 1984; Bryant, 1988; Forasiepi and Sánchez-Villagra, 2014).

Distinctive features of thylacosmilid tooth development notwithstanding, an intriguing parallel is apparent between some of the sabertooth carnivore species of the Felidae and the Thylacosmilidae. In a comparative analysis of molar wear, Forasiepi and Sánchez-Villagra (2014) conclude that *P. goini* had a faster upper canine growth rate than that of the fellow sparassodont *Borhyaena*. Similarly, data from analyses of stable oxygen isotope values in tooth enamel imply that C growth rate was faster in the dirk-toothed felids *Smilodon fatalis* (5.8 mm/month) and *S. gracilis* (6.3 mm/month) than in the scimitar-toothed felid, *Homotherium serum* (2.8 mm/month), or the conical-toothed felids *Panthera leo atrox* (2.6 mm/month) and modern lion, *P. leo* (2.9 mm/month) (Feranec, 2004, 2005, 2008). It appears that relatively accelerated upper canine growth rates might have occurred in some sabertooth carnivore taxa of the Felidae and the Thylacosmilidae.

Gorgonopsia

The Gorgonopsians were carnivorous, therapsid sabertooth taxa confined to the Late Permian, primarily in South Africa, but specimens have also been recovered from Russia. No specimens are known after the Permian–Triassic extinction (Kemp, 2012; Antón, 2013).

Although the sabertooth carnivores of the Gorgonopsia share a number of morphological features with mammalian sabertooth carnivores, tooth development, as might be expected, was markedly different in several ways (Kermack, 1956; Rawn-Schatzinger, 1983; Tejada-Flores and Shaw, 1984; Bryant, 1988; Forasiepi and Sánchez-Villagra, 2014). For instance, unlike *Smilodon fatalis*, which had some teeth that did not undergo replacement (i.e., the molars), every tooth appears to have been replaced in the gorgonopsians (Kermack, 1956; Tejada-Flores and Shaw, 1984). Moreover, the teeth of gorgonopsians were replaced a greater number of times than the teeth of mammals in general (Kermack, 1956; Rawn-Schatzinger, 1983; Tejada-Flores and Shaw, 1984; Bryant, 1988; Luo et al., 2004; van Nievelt and Smith, 2005; Forasiepi and Sánchez-Villagra, 2014). Furthermore, gorgonopsians, as well as some therocephalians (another therapsid taxon from the Permian and Triassic), had an upper canine replacement mechanism that was very different from the canine replacement of the felid *S. fatalis* (Kermack, 1956; Tejada-Flores and Shaw, 1984). Specifically, it appears that gorgonopsians had one functional upper canine pair at a time that was replaced via a functional tooth replacement process in which the functional upper canine alternated between two adjacent alveoli (i.e., one anterior alveolus and one posterior alveolus; Kermack, 1956). Kermack (1956) concludes that, at the end of this tooth replacement process, these sabertooth carnivores possessed a permanent set of functional upper canines within the anterior alveoli.

Interestingly, in a specimen of *Lycaenops ornatus*, it is apparent that the distal end of a partially formed, unerupted replacement canine is located dorsal to the functional canine and inside the alveolar border (Colbert, 1948). This is not inconsistent with the aforementioned pattern of upper canine replacement because the replacement canine described in Colbert (1948) is probably not the immediate functional replacement. Other gorgonopsian specimens exhibit a condition where a replacement upper canine is already present within the alveolus of the current, functional upper canine while having a further developed replacement upper canine within the adjacent upper canine alveolus (i.e., the immediate functional replacement; Kermack, 1956). Supporting this, a study by Kammerer (2014), analyzing the morphology of the *Eriphostoma microdon* holotype (AMNH FARB 5524) using CT scanning, shows that a functional upper canine replacement pattern is typical for gorgonopsians because this specimen has both a functional upper canine within the posterior upper canine alveolus and a replacement upper canine that is about to erupt from the anterior upper canine alveolus. In contrast, *S. fatalis* and several other mammalian sabertooth carnivores simply had a deciduous upper canine that was replaced by a permanent upper canine of the same tooth family (Rawn-Schatzinger, 1983; Tejada-Flores and Shaw, 1984; Bryant, 1988, 1990). Kermack (1956) states that gorgonopsian functional canine replacement was not an ancestral condition but a tooth replacement pattern that evolved because of competi-

tive advantages associated with having just one functional upper canine on either side of the skull. Despite the considerable overall differences between *S. fatalis* tooth development and gorgonopsian tooth development and the strikingly dissimilar tooth replacement mechanisms, particularly for the upper canine, both the gorgonopsians and *S. fatalis* had patterns of tooth development in which upper canine function was preserved throughout the canine replacement process (Kermack, 1956; Tejada-Flores and Shaw, 1984).

Discussion

Not surprisingly, sabertooth carnivores have similar morphological characteristics as well as certain common aspects of tooth development. Even in taxa as distantly related as *Smilodon fatalis* and the gorgonopsians, particular aspects of tooth development and functional morphology are shared (Kermack, 1956; Tejada-Flores and Shaw, 1984). It is apparent that despite having two very different upper canine replacement mechanisms, these taxa had tooth replacement processes that maintained functional upper canines throughout a portion of their ontogeny (Kermack, 1956; Tejada-Flores and Shaw, 1984). For *S. fatalis*, the dC remained in place for approximately 11 months of the C eruption process, which allowed for the C to erupt to about 65 to 80 mm (roughly 50–70% of its final crown height) by the time that the dC was shed (Tejada-Flores and Shaw, 1984; Wysocki et al., 2015). For gorgonopsians, a replacement functional upper canine developed in an adjacent alveolus and eventually became the functional canine when the prior functional canine was lost, a process that occurred multiple times (Kermack, 1956). This fact perhaps underscores the importance of functional upper canines for sabertooth carnivores and many hypercarnivores in general.

Early permanent carnassial eruption, which is apparent in both *Smilodon fatalis* and *Barbourofelis loveorum,* is another developmental characteristic that occurs in different sabertooth carnivore lineages (Tejada-Flores and Shaw, 1984; Bryant, 1988). Permanent carnassial eruption occurs earlier in the eruption sequence of *S. fatalis* than it does in the eruption sequence of any conical-toothed felid for which this information is available (Pocock, 1916; Smuts et al., 1978; Mazák, 1981; Tejada-Flores and Shaw, 1984; Jackson et al., 1988). In contrast to *S. fatalis*, Rawn-Schatzinger (1983) indicates that the permanent post-canines of *H. serum* did not appear until after the permanent incisors appeared. It seems that in *Machairodus* the i1, i2, i3, and m1 started erupting prior to initial eruption of the c and p4, but an analysis of additional specimens is necessary before any determination can be made regarding the precise timing of permanent carnassial eruption in *Machairodus*. The discovery of additional sabertooth carnivore tooth eruption sequences, particularly from the sabertooth creodont species in which different teeth function as the carnassials (Van Valen, 1967; Polly, 1996; Antón, 2013), has the potential to answer many questions regarding the carnassials of sabertooth carnivores and the relatively early permanent carnassial eruption that occurred in *S. fatalis* (Tejada-Flores and Shaw, 1984). Learning more about the timing of permanent carnassial eruption in other species of sabertooth carnivores will make it possible to discern whether early permanent carnassial eruption is primarily a product of developmental constraints or the forces of selection. Additionally, it may be possible to determine whether or not early permanent carnassial eruption was rare or common among sabertooth carnivores and if this pattern of tooth development was typical of just one particular type of sabertooth carnivore.

The relatively early permanent carnassial eruption of *Smilodon fatalis* and *Barbourofelis loveorum* might be related to common developmental constraints associated with the development of the extremely high-crowned upper canines of these two species compared to less high-crowned upper canines of other taxa (e.g., *Homotherium serum*) (Rawn-Schatzinger, 1983; Tejada-Flores and Shaw, 1984; Bryant, 1988). Alternatively, the early eruption of the permanent carnassials could have evolved as the result of competitive advantages associated with the larger shearing crests of the permanent carnassials as opposed to the relatively smaller shearing crests of the deciduous carnassials possessed earlier in life. Specifically, the larger shearing crests of the perma-

nent carnassials would allow individuals to process carcasses more efficiently, likely reducing the amount of time needed to remain at a carcass. In *S. fatalis*, early permanent carnassial eruption is not only evident in its tooth eruption sequence compared to the tooth eruption sequences of other felids, it is also apparent in the age when the permanent carnassials finished erupting (i.e., approximately 12 to 19 months old in *S. fatalis* versus 18 to 24 months old in *Panthera leo*) (Smuts et al., 1978; Wysocki et al., 2015). Tejada-Flores and Shaw (1984) state that the extinct American lion (*P. leo atrox*) tooth eruption sequence is identical to the modern lion (*P. leo*) tooth eruption sequence, and oxygen isotope analyses show that upper canine growth rates of *P. leo atrox* and *P. leo* are comparable (i.e., 2.6 mm/month in *P. leo atrox* and 2.9 mm/month in *P. leo*; Feranec, 2008). This implies that permanent carnassial eruption occurred at an earlier ontogenetic age in *S. fatalis* than it did in its potential competitor *P. leo atrox*. Antón (2013) suggests that the large carnassials of some sabertooth carnivore species, as well as the shearing potential of the post-canine teeth in *Thylacosmilus*, are indicative of ecosystems where these sabertooth carnivores had to feed very quickly because of substantial intra- or interspecific competition. Early permanent carnassial eruption might be fundamental for carnivore species in which juvenile survival is dependent on the capacity to process carcasses quickly. Whether or not this was the case for *S. fatalis*, it is evident that this species developed its permanent carnassials early in life (Tejada-Flores and Shaw, 1984; Wysocki et al., 2015) and it seems that young *S. fatalis* would have had enhanced carcass processing efficiency once the permanent carnassials finished replacing the deciduous carnassials.

Comparison of tooth development among various sabertooth taxa also indicates that some species evolved relatively rapid upper canine growth rates, notably *Smilodon* and the Thylacosmilidae (Feranec, 2008; Forasiepi and Sánchez-Villagra, 2014). However, it is apparent that rapid upper canine growth rates are not representative of all sabertooth carnivores as the permanent upper canine growth rate of the scimitar-toothed *Homotherium serum* is closer to the relatively slow permanent upper canine growth rates of conical-toothed felids (*Panthera leo atrox* and *P. leo*) than it is to the relatively fast permanent upper canine growth rates of the dirk-toothed felids (*S. fatalis* and *S. gracilis*; Feranec, 2004, 2005, 2008). Although they are both sabertooth species, *H. serum* (a scimitar-toothed felid) does not have extremely high-crowned permanent upper canines like *S. fatalis* (a dirk-toothed felid). Instead, *H. serum* has permanent upper canine crown heights that are similar to those of the conical-toothed felid *P. leo* (Rawn-Schatzinger and Collins, 1981; Rawn-Schatzinger, 1983; Tejada-Flores and Shaw, 1984; Rawn-Schatzinger, 1992). If an accelerated upper canine growth rate is not uncommon among sabertooth carnivores, it may be a developmental characteristic that is associated with extreme upper canine crown height. Theoretically, extremely high-crowned saberteeth that develop via faster growth rates may be more common than extremely high-crowned saberteeth that develop via slower growth rates because slower rates would necessitate a prolonged period of development that may not be adaptive when interspecific competition is great. Future research will further explore the relationship between relative upper canine crown height and upper canine growth rate in sabertooth carnivores, but it is apparent that *S. fatalis* was a species of sabertooth carnivore that had relatively rapid permanent upper canine growth and eruption rates (Feranec, 2004, 2008; Wysocki et al., 2015). These relatively fast rates of permanent upper canine growth and eruption in *S. fatalis* are notable because they made it possible for the much higher crowned permanent upper canine of *S. fatalis* to finish erupting at about 34 to 41 months old. This is an ontogenetic age that is not notably greater than the ontogenetic age at which the permanent upper canine of *P. leo* finishes erupting (24 to 36 months old) (Smuts et al., 1978; Tejada-Flores and Shaw, 1984; Haas et al., 2005; Feranec, 2008; Wysocki et al., 2015).

Smilodon fatalis shares many aspects of tooth development with other sabertooth species, but the differences in tooth developmental characteristics among sabertooth species are also ecologically and evolutionarily informative. One such developmental feature that occurred in sabertooth species of the Barbourofelidae and Nimravidae, but not in *S. fatalis*,

was delayed dC eruption (Tejada-Flores and Shaw, 1984; Bryant, 1988, 1990; Tseng et al., 2010; Peigné and de Bonis, 2003). Bryant (1988) suggests that *S. fatalis* did not have this delay because the dC crown height was not as extreme as the dC crown heights of the barbourofelid and nimravid species, and concludes that *S. fatalis* juveniles would have been able to start learning to hunt early in life. Considering that *Panthera leo* and *P. tigris* juveniles display hunting behaviors at relatively early ages and gradually develop their proficiencies as predators, it does seem plausible that *S. fatalis* juveniles could have started exhibiting hunting behaviors at an early ontogenetic age and progressively developed the ability to hunt independently (Schaller, 1967; Schaller 1972). Behavior is difficult to ascertain in ancient species, but it is apparent that *S. fatalis* juveniles younger than 4 to 7 months old were not frequently preserved at Rancho La Brea (Wysocki et al., 2015). This dearth of younger *S. fatalis* individuals at Rancho La Brea may be due to infrequent den site selection near Rancho La Brea, parental behavior that kept cubs from wandering away from den sites, and predator avoidance behavior that prevented cubs from leaving cover to directly feed on carcasses that were mired within the asphalt seeps (Tejada-Flores and Shaw, 1984; Shaw and Quinn, 2015; Wysocki et al., 2015). The recovery of individuals 4 to 7 months old and older at Rancho La Brea may indicate a behavioral shift in which *S. fatalis* juveniles increased direct utilization of food sources mired within the asphalt seeps (Tejada-Flores and Shaw, 1984; Shaw and Quinn, 2015; Wysocki et al., 2015). We suspect that these older individuals, which had more developed teeth (i.e., deciduous upper canine with an erupted length of 30 mm and all other deciduous teeth fully erupted), would have had greater carcass processing and defense capabilities than individuals of the younger age classes (Tejada-Flores and Shaw, 1984). Even though the relative dC crown height is not nearly as extreme as those of certain barbourofelid and nimravid species, from a functional standpoint it appears possible that the fully erupted dC of *S. fatalis* could still have been used in hunting considering that its crown height (60–65 mm) exceeds the crown height of the permanent upper canine of *P. leo* (43–52 mm) (Tejada-Flores and Shaw, 1984; Haas et al.,

Plate 1. "Rancho La Brea" (oil on canvas, Charles R. Knight, 1925)

Plate 2. *"Smilodon fatalis"* (gouache on board, John Dawson, 1977)

Plate 3. "A Time for Love" (gouache on board, Mark Hallett, 1987)

Plate 4. "La Brea Tar Pits" (gouache on board, John Sibbick, 1994)

Plate 5. "Group of *Smilodon* hunting a bison" (oil on canvas, Mauricio Anton, 1996)

Plate 6. "Standoff: Pleistocene Predators" (oil on canvas, William Stout, 2006)

Plate 7. *"Smilodon fatalis* in Snow" (photomanipulation and digital painting, Roman Yevseyev, 2012)

Plate 8. "The Sabertooth" (digital painting, Zubin Erik Dutta, 2016)

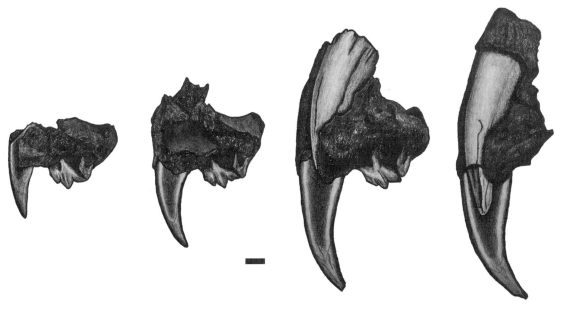

Figure 9.3. Ontogenetic sequence of *Smilodon fatalis* specimens illustrating eruption of the deciduous upper canine (dC) and initial eruption of the permanent upper canine (C). From left to right: LACMHC558 2001-579 lingual view; LACMHC7002 lingual view; UCMP152565 lingual view; UCMP152566 lingual view.

2005). Although it is not known at what ontogenetic age independent hunting occurred, it seems that the canine replacement pattern through which the function of the upper canine was preserved (Fig. 9.3) would have allowed *S. fatalis* juveniles to progressively become more proficient predators (Tejada-Flores and Shaw, 1984).

Conclusions

The sabertooth condition has evolved at least six times and includes some taxa that lived as long ago as the Late Permian, but there is no longer a living carnivore that has the sabertooth condition (Antón, 2013). Many questions remain about the ecology and evolution of sabertooth carnivores, including *Smilodon fatalis* despite its incredible fossil record (Stock, 1929; Akersten et al., 1983; Spencer et al., 2003). This investigation has explored patterns of tooth development in sabertooth carnivores to further understand the ecology and evolution of *S. fatalis*. Results suggest that there are some similar patterns of tooth development among sabertooth carnivores, as well as several notable differences. Early permanent carnassial eruption is apparent in both *S. fatalis* and *Barbourofelis loveorum*, which is

intriguing because each of these species is among the more derived sabertooth taxa of its respective lineage (Tejada-Flores and Shaw, 1984; Bryant, 1988; Slater and Van Valkenburgh, 2008; Tseng et al., 2010; Meachen-Samuels, 2012). Another similarity in sabertooth carnivore tooth development is that some taxa of the Felidae (i.e., *S. fatalis* and *S. gracilis*) and Thylacosmilidae (i.e., *Patagosmilus goini*) had relatively rapid upper canine growth rates, but fast rates of upper canine growth are not necessarily characteristic of all sabertooth carnivores (e.g., the scimitar-toothed *Homotherium serum*, which had a relatively slow upper canine growth rate) (Feranec, 2004, 2005, 2008; Forasiepi and Sánchez-Villagra, 2014). Two markedly dissimilar upper canine replacement mechanisms are apparent in the distantly related *S. fatalis* and gorgonopsians, yet their tooth replacement processes are similar in that both maintained functional upper canines through a portion of ontogeny (Kermack, 1956; Tejada-Flores and Shaw, 1984). An additional common pattern in sabertooth carnivore tooth development is that several species of the Nimravidae and the Barbourofelidae had delayed eruption of the deciduous upper canine, whereas the felid *S. fatalis* did not (Tejada-Flores and

Shaw, 1984; Bryant, 1988, 1990; Peigné and De Bonis, 2003; Tseng et al., 2010). As Bryant (1988) indicates, this difference in the timing of deciduous upper canine eruption is likely because the deciduous upper canine has a more extreme crown height in each of these nimravid and barbourofelid species than it does in *S. fatalis*. This dissimilarity in deciduous upper canine eruption appears to be responsible for another difference in tooth development: the delayed permanent upper canine eruption that is evident in species of the Nimravidae and the Barbourofelidae, but not in *S. fatalis* (Tejada-Flores and Shaw, 1984; Bryant, 1988, 1990). This absence of a delay to initial permanent canine eruption, in combination with the relatively fast rates of permanent upper canine growth and eruption, results in the permanent dentition of *S. fatalis* completing eruption at an age not much older than the age at which the permanent dentition is fully erupted in the conical-toothed felid *P. leo* (Smuts et al., 1978; Tejada-Flores and Shaw, 1984; Feranec, 2004, 2008; Wysocki et al., 2015). Overall, it appears that even though the specialized sabertooth dentition of *S. fatalis* includes extremely high-crowned permanent upper canines, several characteristics of its tooth development made it possible for this sabertooth carnivore to develop its formidable permanent dentition without necessarily incurring the competitive disadvantages of requiring an especially prolonged period of development.

ACKNOWLEDGEMENTS

The authors wish to thank H. G. McDonald, C. A. Shaw, and L. Werdelin for providing suggestions that improved this manuscript. The authors also thank the American Museum of Natural History, University of California Museum of Paleontology, and George C. Page Museum for access to specimens. Some logistical support for this study was provided by the New York State Museum.

REFERENCES

Akersten, W. A. 1985. Canine function in *Smilodon* (Mammalia, Felidae, Machairodontinae). Natural History Museum of Los Angeles County, Contributions in Science 356:1-22.

Akersten, W. A., C. A. Shaw, and G. T. Jefferson. 1983. Rancho La Brea: status and future. Paleobiology 9:211-217.

Antón, M. 2013. Sabertooth. Indiana University Press, Bloomington, Indiana, 243 pp.

Bryant, H. N. 1988. Delayed eruption of the deciduous upper canine in the saber-toothed carnivore *Barbourofelis lovei* (Carnivora, Nimravidae). Journal of Vertebrate Paleontology 8:295-306.

Bryant, H. N. 1990. Implications of the dental eruption sequence in *Barbourofelis* (Carnivora, Nimravidae) for the function of upper canines and the duration of parental care in sabretoothed carnivores. Journal of Zoology 222:585-590.

Christiansen, P. 2012. The making of a monster: postnatal ontogenetic changes in craniomandibular shape in the great sabercat *Smilodon*. PLoS One 7:e29699.

Colbert, E. H. 1948. The mammal-like reptile *Lycaenops*. Bulletin of the American Museum of Natural History 89:353-404.

Cope, E. D. 1880. On the extinct cats of America. The American Naturalist 14:832-858.

Crowe, D. M. 1975. Aspects of ageing, growth, and reproduction of bobcats from Wyoming. Journal of Mammalogy 56:177-198.

Currier, M. J. P. 1983. *Felis concolor*. Mammalian Species 200:1-7.

Emerson, S. B., and L. Radinsky. 1980. Functional analysis of sabertooth cranial morphology. Paleobiology 6:295-312.

Ewer, R. F. 1973. The Carnivores. Weidenfeld and Nicolson, London, 494 pp.

Feranec, R. S. 2004. Isotopic evidence of saber-tooth development, growth rate, and diet from the adult canine of *Smilodon fatalis* from Rancho La Brea. Palaeogeography, Palaeoclimatology, Palaeoecology 206:303-310.

Feranec, R. S. 2005. Growth rate and duration of growth in the adult canine of *Smilodon gracilis*, and inferences on diet through stable isotope analysis. Bulletin of the Florida Museum of Natural History 45:369-377.

Feranec, R. S. 2008. Growth differences in the saber-tooth of three felid species. Palaios 23:566-569.

Forasiepi, A. M., and A. A. Carlini. 2010. A new thylacosmilid (Mammalia, Metatheria, Sparassodonta) from the Miocene of Patagonia, Argentina. Zootaxa 2552:55-68.

Forasiepi, A. M., and M. R. Sánchez-Villagra. 2014. Heterochrony, dental ontogenetic diversity, and the circumvention of constraints in marsupial mammals and extinct relatives. Paleobiology 40:222-237.

Goin, F. J., and R. Pascual. 1987. News on the biology and taxonomy of the marsupials Thylacosmilidae (Late Tertiary of Argentina). Anales del Academia Nacional de Ciencas Exactas, Físicas y Naturales, Buenos Aires 39:219-246.

Gonyea, W. J. 1976. Behavioral implications of saber-toothed felid morphology. Paleobiology 2:332-342.

Gunnell, G. F. 1998. Creodonta; pp. 91-109 in C. M. Janis, K. M. Scott, and L. L. Jacobs (eds.), Evolution of Tertiary Mammals of North America. Volume 1, Terrestrial Carnivores, Ungulates, and Ungulatelike Mammals. Cambridge University Press, Cambridge, UK.

Haas, S. K., V. Hayssen, and P. R. Krausman. 2005. *Panthera leo*. Mammalian Species 762:1-11.

Jackson, D. L., E. A. Gluesing, and H. A. Jacobson. 1988. Dental eruption in bobcats. Journal of Wildlife Management 52:515-517.

Kammerer, C. F. 2014. A redescription of *Eriphostoma microdon* Broom, 1911 (Therapsida, Gorgonopsia) from the *Tapinocephalus* assemblage zone of South Africa and a review of Middle Permian Gorgonopsians; pp. 171-184 in C. F. Kammerer, K. D. Angielczyk, and J. Fröbisch (eds.), Early Evolutionary History of the Synapsida. Springer, New York.

Kemp, T. S. 2012. The origin and radiation of therapsids; pp. 3-28 in A. Chinsamy-Turan (ed.), Forerunners of Mammals. Indiana University Press, Bloomington, Indiana.

Kermack, K. A. 1956. Tooth replacement in mammal-like reptiles of the suborders Gorgonopsia and Therocephalia. Philosophical Transactions of the Royal Society of London, Biological Sciences 240:95-133.

Luo, Z.-X., Z. Kielan-Jaworowska, and R. L. Cifelli. 2004. Evolution of dental replacement in mammals. Bulletin of Carnegie Museum of Natural History 36:159-175.

Marshall, L. G. 1976. Evolution of Thylacosmilidae, extinct saber-tooth marsupials of South America. PaleoBios 23:1-30.

Martin, L. D. 1980. Functional morphology and the evolution of cats. Transactions of the Nebraska Academy of Sciences 8:141-154.

Martin, L. D. 1998. Nimravidae; pp. 228-235 in C. M. Janis, K. M. Scott, and L. L. Jacobs (eds.), Evolution of Tertiary Mammals of North America. Volume 1: Terrestrial carnivores, ungulates, and ungulatelike mammals. Cambridge University Press, Cambridge.

Mazák, V. 1981. *Panthera tigris*. Mammalian Species 152:1-8.

McHenry, C. R., S. Wroe, P. D. Clausen, K. Moreno, and E. Cunningham. 2007. Supermodeled sabercat, predatory behavior in *Smilodon fatalis* revealed by high-resolution 3D computer simulation. Proceedings of the National Academy of Sciences of the United States of America 104:16010-16015.

McKenna, M. C., and S. Bell. 1997. Classification of Mammals above the Species Level. Columbia University Press, New York, 631 pp.

Meachen-Samuels, J. 2012. Morphological convergence of the prey-killing arsenal of sabertooth predators. Paleobiology 38:715-728.

Morlo, M., S. Peigné, and D. Nagel. 2004. A new species of *Prosansanosmilus*: implications for the systematic relationships of the family Barbourofelidae new rank (Carnivora, Mammalia). Zoological Journal of the Linnean Society 140:43-61.

Peigné, S., and L. de Bonis. 2003. Juvenile cranial anatomy of Nimravidae (Mammalia, Carnivora): biological and phylogenetic implications. Zoological Journal of the Linnean Society 138:477-493.

Pocock, R. I. 1916. On the tooth-change, cranial characters, and classification of the snow-leopard or ounce (*Felis uncia*). Journal of Natural History Series 8 18:306-316.

Polly, P. D. 1996. The skeleton of *Gazinocyon vulpeculus* gen. et comb. nov. and the cladistic relationships of Hyaenodontidae (Eutheria, Mammalia). Journal of Vertebrate Paleontology 16:303-319.

Rawn-Schatzinger, V. M. 1983. Development and eruption sequence of deciduous and permanent teeth in the saber-tooth cat *Homotherium serum* Cope. Journal of Vertebrate Paleontology 3:49-57.

Rawn-Schatzinger, V. M. 1992. The scimitar cat *Homotherium serum* Cope: osteology, functional morphology, and predatory behavior. Illinois State Museum Reports of Investigations 47:1-80.

Rawn-Schatzinger, V. M., and R. L. Collins. 1981. Scimitar cats, *Homotherium serum* Cope from Gassaway Fissure, Cannon County, Tennessee and the North American distribution of *Homotherium*. Journal of the Tennessee Academy of Science 56:15-19.

Riggs, E. S. 1934. A new marsupial saber-tooth from the Pliocene of Argentina and its relationships to other South American predacious marsupials. Transactions of the American Philosophical Society, new series 24:1-32.

Schaller, G. B. 1967. The Deer and the Tiger: A Study of Wildlife in India. University of Chicago Press, Chicago, Illinois, 384 pp.

Schaller, G. B. 1972. The Serengeti Lion: A Study of Predator-Prey Relations. University of Chicago Press, Chicago, Illinois, 504 pp.

Schneider, K. M. 1959. Zum zahndurchbruch des löwen (*Panthera leo*) nebst bemerkungen über das zahnen einiger anderer grosskatzen und der hauskatze (*Felis catus*). Zoologische Garten 22:240-361.

Schultz, C. B., M. R. Schultz, and L. D. Martin. 1970. A new tribe of saber-toothed cats (Barbourofelini) from the Pliocene of North America. Bulletin of the University of Nebraska State Museum 9:1-31.

Shaw, C. A., and J. P. Quinn. 2015. The addition of *Smilodon fatalis* (Mammalia, Carnivora, Felidae) to the biota of the Late Pleistocene Carpinteria asphalt deposits in California, with ontogenetic and ecologic implications for the species; pp. 91-95, in J. M. Harris (ed.), La Brea and Beyond: The Paleontology of Asphalt-Preserved Biotas. Natural History Museum of Los Angeles County, Science Series 42.

Slater, G. J., and B. Van Valkenburgh. 2008. Long in the tooth: evolution of sabertooth cat cranial shape. Paleobiology 34:403-419.

Smith, J. L. D. 1993. The role of dispersal in structuring the Chitwan tiger population. Behaviour 124:165-195.

Smuts, G. L., J. L. Anderson, and J. C. Austin. 1978. Age determination of the African lion (*Panthera leo*). Journal of Zoology 185:115-146.

Spencer, L. M., B. Van Valkenburgh, and J. M. Harris. 2003. Taphonomic analysis of large mammals recovered from the Pleistocene Rancho La Brea tar seeps. Paleobiology 29:561-575.

Stander, P. E. 1997. Field age determination of leopards by tooth wear. African Journal of Ecology 35:156-161.

Stock, C. 1929. A census of the Pleistocene mammals of Rancho La Brea, based on the collections of the Los Angeles Museum. Journal of Mammalogy 10:281-289.

Sunquist, M., and F. Sunquist. 2002. Wild Cats of the World. University of Chicago Press, Chicago, Illinois, 452 pp.

Tejada-Flores, A. E., and C. A. Shaw. 1984. Tooth replacement and skull growth in *Smilodon* from Rancho La Brea. Journal of Vertebrate Paleontology 4:114-121.

Tseng, Z. J., G. T. Takeuchi, and X. Wang. 2010. Discovery of the upper dentition of *Barbourofelis whitfordi* (Nimravidae, Carnivora) and an evaluation of the genus in California. Journal of Vertebrate Paleontology 30:244-254.

Tummers, M., and I. Thesleff. 2003. Root or crown: a developmental choice orchestrated by the differential regulation of the epithelial stem cell niche in the tooth of two rodent species. Development 130:1049-1057.

van Nievelt, A. F. H., and K. K. Smith. 2005. To replace or not to replace: the significance of reduced functional tooth replacement in marsupial and placental mammals. Paleobiology 31:324-346.

Van Valen, L. 1969. Evolution of dental growth and adaptation in mammalian carnivores. Evolution 23:96-117.

Wang, X. P., M. Suomalainen, S. Felszeghy, L. C. Zelarayan, M. T. Alonso, M. V. Plikus, R. L. Maas, C. M. Chuong, T. Schimmang, and I. Thesleff. 2007. An integrated gene regulatory network controls stem cell proliferation in teeth. PLoS Biology 5:e159.

Wroe, S., U. Chamoli, W. C. Parr, P. Clausen, R. Ridgely, and L. Witmer. 2013. Comparative biomechanical modeling of metatherian and placental saber-tooths: a different kind of bite for an extreme pouched predator. PLoS One 8:e66888.

Wysocki, M. A., R. S. Feranec, Z. J. Tseng, and C. S. Bjornsson. 2015. Using a novel absolute ontogenetic age determination technique to calculate the timing of tooth eruption in the saber-toothed cat, *Smilodon fatalis*. PLoS One 10:e0129847.

Yokohama-Tamaki, T., H. Ohshima, N. Fujiwara, Y. Takada, Y. Ichimori, S. Wakisaki, H. Ohuchi, and H. Harada. 2006. Cessation of Fgf10 signaling, resulting in a defective dental epithelial stem cell compartment, leads to the transition from crown to root formation. Development 133:1359-1366.

10 Dietary Ecology of *Smilodon*

LARISA R. G. DESANTIS

Introduction

Clarifying the dietary ecology of sabertooth cats of the genus *Smilodon* is necessary to understanding their paleobiology and may reveal potential reasons for their extinction. While their pronounced saber-like canines and the large shearing facets on their carnassial teeth undeniably attest to their placement as meat-eating carnivores, significant research has focused on better understanding potential prey resources, hunting modes (including hunting socially), and the degree to which they scavenged for food. Dietary behavior of *Smilodon* is possible to investigate using a diversity of proxy methods, including studies of morphology in other sabertooth cat taxa, tooth breakage, dental microwear (including microwear texture analysis), and stable isotope analyses of bone and tooth enamel. This chapter reviews dietary studies of *Smilodon*, including the best-known Pleistocene taxon, *Smilodon fatalis*; the smallest of the three species (occurring primarily in eastern North America), *Smilodon gracilis*; and the largest of the three species (occurring in much of South America), *Smilodon populator*. Further, it presents and discusses novel stable isotope data for *Smilodon fatalis* and novel dental microwear data for *S. fatalis*, *S. gracilis*, and *S. populator*.

Diet and Hunting Inferred from Morphology

Modern felids (i.e., cats and their relatives) hunt prey using a diversity of strategies, ranging from ambush predation (e.g., cougars and jaguars) to pursuit predation (e.g., cheetahs). Pursuit predation may be fairly novel, evolving sometime during the past few million years (Janis and Wilhelm, 1993); the majority of felids stay concealed when hunting and approach their prey from dense cover (Elliot et al., 1977). Whether felids stalk and then chase prey or ambush prey, either mode benefits from some form of concealment. Further, cats can vary their hunting strategies depending on whether they hunt individually or in pairs or groups. The majority of large cats are solitary hunters (e.g., the tiger, *Panthera tigris,* and the jaguar, *Panthera onca*); African lions (*Panthera leo*) hunt cooperatively, with females often working in pairs to take down prey (Nowak, 2005). As one would expect, social felids occur in higher densities compared to solitary species, and multiple individuals will congregate around fresh kills. Thus, the relatively high abundance of *Smilodon fatalis* and *Canis dirus* (dire wolf)—the two most abundant carnivorans at Rancho La Brea—suggests that these carnivorans were social (Carbone et al., 2009). Additionally, there is evidence that *S. fatalis* exhibited some sexual dimorphism, although much less so than African lions (Van Valkenburgh and Sacco,

2002; Christiansen and Harris, 2012). Thus, sexual dimorphism patterns match other, less social felids and suggest that *S. fatalis* could have been social, solitary, and polygamous (similar to most living felids), have cohabited in groups composed of monogamous pairs, or have lived in unisexual groups similar to those of the Asiatic lion, *Panthera leo persica* (Van Valkenburgh and Sacco, 2002; Christiansen and Harris, 2012).

Early work on *Smilodon fatalis* suggested that it consumed "large, slow-footed animals" (Kurtén and Anderson, 1980). However, an analysis of felid body proportions demonstrates that *Smilodon* was most similar to modern forest felids—"dwellers of high structured dense forest" (Gonyea, 1976:332). The similarity of *Smilodon* limb proportions to other forest felids that usually ambush or stalk exclusively suggests that *Smilodon* likely hunted in a similar way, while the size of the limbs suggests that they preyed on large animals (Gonyea, 1976).

More recently, a study of the postcranial morphology—specifically cortical thickening on the humerus (which is greater than in any extant felid)—suggests that *Smilodon fatalis* had a greater ability than modern large felids to subdue prey using its forelimbs (Meachen-Samuels and Van Valkenburgh, 2010). This was likely a necessary adaption to reduce forces exerted by struggling prey and to protect the elongated canines, which otherwise would have been more vulnerable to breakage (Meachen-Samuels and Van Valkenburgh, 2010). The robust forelimbs in *S. fatalis* may have resulted in a tradeoff, making this sabertooth cat better suited for ambush hunting as opposed to any pursuit hunting. Further, this adaptation may have contributed to extinction if the pronounced forelimbs resulted in a diminished ability to hunt smaller and more agile prey—the size class of prey that survived the Late Pleistocene megafaunal extinction (e.g., Van Valkenburgh and Hertel, 1998; Meachen-Samuels and Van Valkenburgh, 2010).

Cranial morphology documents that both the elongated canines and a battery of other features made *Smilodon* a formidable predator (e.g., Merriam and Stock, 1932; Meachen-Samuels, 2012; chapter 8, this volume). More recent cranial studies using morphological modeling demonstrate that *Smilodon fatalis* had a relatively weak bite, approximately one-third that of a comparably sized African lion (Wroe et al., 2005; McHenry et al., 2007; chapter 6, this volume). Finite element analysis also demonstrates that sabertooth cat skulls were not well suited to resist high, multidimensional loads caused by unrestrained prey (McHenry et al., 2007), consistent with the morphology of the mandible (Therrien, 2005). Collectively, these data are in agreement with a canine-shear bite model (Akersten, 1985), where the elongated canines penetrate prey with significant input from jaw adductors in conjunction with cervical musculature (Wroe et al., 2005; chapter 6, this volume) and forelimb morphology suggests that *S. fatalis* subdued its prey with its forelimbs and not its canines (Meachen-Samuels and Van Valkenburgh, 2010) (see Fig. 10.1).

Dental Macrowear and Microwear

Hunting and Defensive Injuries Inferred from Teeth

Dental macrowear and microwear have the potential to clarify the degree of bone processing and/or hunting behavior exhibited by *Smilodon*. Early work by Van Valkenburgh and Hertel (1993) demonstrates a higher incidence of broken teeth in Rancho La Brea carnivorans, including both felids and canids, in contrast to modern assemblages of carnivorans. Specifically, the American lion, *Panthera atrox*, the largest Rancho La Brea felid, had the highest incidence of broken canines (36%) while *S. fatalis* had the lowest incidence among extinct Pleistocene taxa (11.2%). However, these breakage indices are greater than the average canine breakage rate of ~7% found in extant felids, which range between 3.2% and 9.8% in the jaguar (*Panthera onca*) and leopard (*Panthera pardus*), respectively. These results, in addition to follow-up work that expanded the extant carnivoran baseline (Van Valkenburgh, 2009), led to the suggestion that 'tough times' prior to the extinction of the Late Pleistocene megafauna and sabertooth cats necessitated increased carcass consumption. This was also suggested to occur in other carnivorans at Rancho La Brea, including the American lion (*Panthera atrox*) and the dire wolf (*Canis dirus*). Tooth breakage data further suggest that large carnivorans at Rancho La Brea—including *S. fatalis*—were

Figure 10.1. Images of juvenile and adult *Smilodon* sp. demonstrating the presence of elongated canines, even as juveniles, and illustrating skull proportions of *Smilodon* species. A = juvenile *S. populator* (Swedish Museum of Natural History, Stockholm NRM-PZ M); B = adult *S. populator* (Natural History Museum, London, cast); C = juvenile *S. fatalis* (from Dinocasts); D = adult *S. fatalis* (Natural History Museum of Los Angeles County, Hancock Collection, LACMHC 2001-173). Scale bars equal 5 cm. Image from Christiansen (2012).

potentially more vulnerable to extinction due to increased competition for prey (Van Valkenburgh and Hertel, 1993).

The idea of tough times (i.e., increased carcass consumption and/or declining health of carnivorans prior to their extinction) has been challenged by dental microwear texture studies that question significant carcass consumption by Rancho La Brea carnivorans (e.g., DeSantis et al., 2012; Donohue et al., 2013; DeSantis et al., 2015; DeSantis and Haupt, 2014). Additionally, studies of Harris line frequency (an indicator of stress in carnivorans as well as in herbivores and humans) challenge the idea of

stressful conditions at Rancho La Brea (Duckler and Van Valkenburgh, 1998). Specifically, *Smilodon fatalis* had a lower frequency of Harris lines (6.2%) than both healthy cougars (*Puma concolor*, 12.3%) and 'stressed' Florida panthers (*Puma concolor coryi*; 28.7%). In fact, data from all Rancho La Brea taxa studied, including the most abundant herbivores and carnivores, indicate that the "late Pleistocene species were not suffering from unusual levels of poor health, even as they approached the extinction horizon" (Duckler and Van Valkenburgh, 1998:180).

Instead, the indisputable high incidence of broken teeth during the Pleistocene may have resulted from the greater forces exerted when taking down larger prey. Larger teeth (including canines) are more vulnerable to fracture than smaller teeth and support a decreasing proportion of a predator's weight with increased tooth size (Freeman and Lemen, 2007). For example, a fox-sized predator can support prey ~7.3 times its body weight, in contrast to ~4.4 times and ~2.2 times for a lion-sized predator or *Smilodon* with elongated canines, respectively (Freeman and Lemen, 2007). The idea that teeth are most likely to break during prey acquisition is also supported by tooth breakage data on modern lions, where females have significantly greater tooth breakage than males (Van Valkenburgh, 1988). Furthermore, the highest incidence of broken canines occurred in the largest felid (the American lion, *Panthera atrox*, ~36%; Van Valkenburgh, 2009), which likely took down the largest prey by itself; the theory that it used a solitary hunting strategy is based on the rarity of specimens at Rancho La Brea (Carbone et al., 2009; Van Valkenburgh, 2009). In contrast, the lowest incidence of canine breakage occurred in *S. fatalis* (~10-11.2%), which is thought to be social and therefore potentially to incur less damage during prey-acquisition (Carbone et al., 2009; Van Valkenburgh, 2009). Tooth breakage data are likely indicative of hunting behavior and do suggest that large carnivorans may have been more vulnerable to injury from prey acquisition during the Pleistocene.

A more detailed temporal analysis of tooth breakage and tooth macrowear (i.e., visible wear occurring on the occlusal surface of teeth, affected by diet and age) was performed on *Smilodon fatalis* from depositional units representing three time periods at Rancho La Brea (Binder and Van Valkenburgh, 2010). Unlike dire wolves, which exhibit increased tooth breakage during Pit 13 (~16,000 Ka; O'Keefe et al., 2009), *S. fatalis* has a greater incidence of broken teeth during Pit 3 (~18,500; O'Keefe et al., 2009; in contrast to Pit 61/67 ~11,500 Ka; Binder and Van Valkenburgh, 2010). Further, *S. fatalis* does exhibit greater wear on their teeth than co-occurring dire wolves during the most recent time period studied (Pit 61/67, ~11,500 Ka; O'Keefe et al., 2009). Collectively, tooth breakage and macrowear data suggest that the acquisition of prey and hunting efficiencies of *S. fatalis* may have varied over time.

Times were in fact 'tough' in terms of damage incurred during prey capture, although less so in *S. fatalis* than in most other co-occurring carnivorans (Van Valkenburgh and Hertel, 1993; Van Valkenburgh, 2009).

Dental Microwear Suggests Flesh and Bone Consumption

Dental microwear, the microscopic wear features found on the surface of teeth that are due to the processing of food, has long been used to infer diets in extinct species (e.g., Walker et al., 1978; see review by DeSantis, 2016). As opposed to cranial dental morphology and dental macrowear, which reveal potential and average diets, respectively, dental microwear helps identify the diet of individual organisms shortly before their death (Grine, 1986). Early dental microwear analyses using scanning electron microscopy revealed distinct dental microwear features in *Smilodon fatalis* as compared to all extant carnivorans examined (Van Valkenburgh et al., 1990). Specifically, *S. fatalis* has relatively narrow, long features and a low incidence of pits. This indicates that *S. fatalis* likely consumed even less bone than the cheetah (*Acinonyx jubatus*), a felid observed to consume largely flesh and a low incidence of bone (Van Valkenburgh, 1996). *Smilodon* may have left behind significant carrion for scavengers (Van Valkenburgh et al., 1990).

Dental microwear studies of feliforms have typically focused on analyzing the carnassial shearing facet that is known to interact with bone in extant taxa (e.g., Van Valkenburgh et al., 1990; Schubert et al., 2010; DeSantis et al., 2012). However, early work by Anyonge (1996) examined dental microwear features on the canines of *Smilodon fatalis* compared to six extant carnivorans (including leopard, *Panthera pardus*, cheetah, *Acinonyx jubatus*, African lion, *Panthera leo*, spotted hyena, *Crocuta crocuta*, African wild dog, *Lycaon pictus*, and gray wolf, *Canis lupus*). His dental microwear data demonstrate that *S. fatalis* specimens had fewer features on their canines as compared to all other carnivorans examined. Further, *S. fatalis* showed no consistent similarity to any of the extant carnivorans studied (Anyonge, 1996). These data suggest that *S. fatalis* exhibited different killing behaviors than the other carnivorans studied and/or avoided canine contact with bone (Anyonge, 1996), all consistent with the 'canine-shear bite' model (Akersten, 1985; Wroe et al., 2005).

Dental microwear texture analysis (DMTA), a dental microwear method that analyzes wear surfaces in three dimensions using scale-sensitive fractal analysis (e.g., Ungar et al., 2003; Scott et al., 2005, 2006), is useful for inferring the likelihood of durophagy (i.e., bone processing; Schubert et al., 2010; DeSantis et al., 2012, 2013, 2015; DeSantis, 2016; DeSantis and Patterson, 2017). Specifically, DMTA can help distinguish between hard-, soft-, or tough-object consumption (DeSantis, 2016). In carnivorans (and more specifically feliforms), taxa most often observed to avoid complete carcass utilization (including shunning bone, like cheetahs) are distinct from those that are more generalized (e.g., African lions) and those that more fully utilize carcasses, as hyenas will (Schubert et al., 2010; DeSantis et al., 2012, 2013, 2015, 2017). Dental microwear textures of *Smilodon* are more generalized, much like extant African lions, which suggests that these carnivores consumed both flesh and bone (DeSantis et al., 2012, 2015) (Fig. 10.2). Further, dental microwear texture attributes of *S. fatalis* were distinct from those of the co-occurring American lion, *Panthera atrox*, which largely avoided bone and has dental microwear textures indistinguishable from extant cheetahs (DeSantis et al., 2012). Collectively, data from DMTA and dental macrowear suggest that *S. fatalis* had a fairly generalized diet of flesh and bone and incurred fewer injuries than co-occurring *P. atrox*. The higher incidence of broken canines and less generalized microwear suggests that *P. atrox* consumed a greater proportion of flesh from solitary kills, but likely paid a price for using this hunting technique. *Smilodon* at Rancho La Brea was probably very similar to African lions in modes of hunting, including having access to carcasses even when not directly killing the prey (see summary by Yamaguchi et al., 2004) and engaging in moderate levels of muscle and bone consumption (Van Valkenburgh, 1996).

Unlike the 'tough times' scenario in which scavenging of carcasses may have resulted from desperate times and increased prey competition with humans and/or other apex predators, more generalized

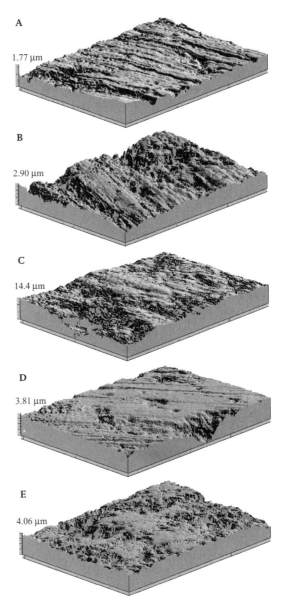

Figure 10.2. Three-dimensional photo simulations of microwear surfaces of the extant cheetah, A = *Acinonyx jubatus* (American Museum of Natural History, AMNH 161139); the extant lion, B = *Panthera leo* (US National Museum, USNM 236919); the extant spotted hyena, C = *Crocuta crocuta* (AMNH 83592); the extinct American lion from Rancho La Brea, D = *Panthera atrox* (LACMHC 6996); and the sabertooth cat from Rancho La Brea, E = *Smilodon fatalis* (LACMHC 2002-298). Image from DeSantis et al. (2012).

dietary behavior including scavenging and/or more fully consuming smaller carcasses may have been a useful strategy and key to the survival of the cougar, *Puma concolor* (DeSantis and Haupt, 2014). While the DMTA attribute indicative of hard-object feeding (i.e., complexity, *Asfc*) is indistinguishable between cougars and *Smilodon fatalis*, the former have more variable complexity values than *S. fatalis* and may have demonstrated more variable dietary behavior than the now-extinct sabertooth cats.

Beyond La Brea: Similarity in Dental Microwear Textures across Species

In addition to the examination of *Smilodon fatalis* at Rancho La Brea, this chapter presents data on the dental microwear textures of *S. fatalis* from the Late Pleistocene Friesenhahn Cave (n = 4) in Texas (Evans, 1961), *Smilodon gracilis* from fossil assemblages deposited during a glacial (Inglis 1A, n = 4) and an interglacial period (Leisey Shell Pit 1A, n = 7) (DeSantis et al., 2009), and a number of isolated specimens of *S. populator* from South America (n = 3). Dental microwear analyses follow the same methods as those presented in DeSantis et al. (2012, 2015), including the analysis of each wear surface in three dimensions and in four areas (for a total sample area of 204×276 µm²). All scans were analyzed using scale-sensitive fractal analysis (SSFA) software (Toothfrax and Sfrax, Surfract Corp., www.surfrait .com) to characterize tooth surfaces according to the following variables: (1) complexity (*Asfc*), a scale-sensitive measure of roughness used to distinguish taxa that consume hard, brittle foods (i.e., high *Asfc*) from those that eat softer food items, and (2) anisotropy (*epLsar*), the degree to which surface textures show a preferred orientation, such as the dominance of parallel striations that might be formed by carnassial action in meat slicing given constraints on tooth-on-tooth movement during occlusion (Ungar et al., 2003; Scott et al., 2005, 2006; DeSantis, 2016). Because the majority of DMTA variables are not normally distributed, non-parametric statistical tests (Mann-Whitney U tests) were used to compare *S. gracilis* specimens between glacial and interglacial sites in Florida, and to *S. fatalis* from Rancho La Brea and Friesenhahn Cave (collectively). *Smilodon populator* specimens from

South America were independently compared to different species. Results should be viewed cautiously, however, due to the small sample size of *S. populator*. Further comparisons of *S. gracilis* specimens between glacial and interglacial sites in Florida are also hampered by low sample sizes and may be considered suggestive or preliminary.

Dental microwear texture attributes of all *Smilodon* taxa are indistinguishable from one another in both complexity and anisotropy (p > 0.05) (see Tables 10.1 and 10.2). All *Smilodon* specimens suggest some mix of flesh and bone consumption.

'Hard Times' during Glacial Periods in Florida

Dental microwear texture analysis of *Smilodon gracilis*, while indistinguishable from other *Smilodon* species, does demonstrate a significant decline in hard-object feeding (and potentially scavenging) during the interglacial period in Florida (Fig. 10.3). *Smilodon gracilis* complexity values suggest that increased durophagy occurred during colder times. Dietary shifts coincident with changing climates, floral habitats, and changing diets in prey species are also consistent with and inferred from changing stable isotope values in *S. gracilis* between the Florida sites (see below). Similar work aimed at assessing temporal variability in DMTA attributes in *S. fatalis* coincident with changing climates at Rancho La Brea is ongoing. However, preliminary data suggest that sabertooth cats may have altered their dietary behavior coincident with changing climates.

Potential Prey Resources

Stable carbon and nitrogen isotopes are useful for resolving potential prey resources of predators, including assessing the degree to which predators consumed primary consumers. Stable carbon isotopes (δ^{13}C values) in tooth enamel reflect whole diet (protein and carbohydrates) while bone collagen and tooth dentine collagen reflect the protein component of an individual's diet (Ambrose and Norr, 1993). Carbon isotopes in herbivores record changes in vegetation consumed via the relative abundance of C_3 versus C_4 plants, while carnivores track changes in prey preferences (e.g., DeNiro and Epstein, 1978; Cerling et al., 1997; Badgley et al., 2008; DeSantis et al., 2009; Feranec and DeSantis, 2014).

Table 10.1. All new *Smilodon* specimens examined and dental microwear characters

Taxon	Site	Collection	ID	Tooth	*Asfc*	*epLsar*
Smilodon fatalis[†]	Friesenhahn Cave, Texas	TMM	933-897	rml	1.856	0.0046
		TMM	933-1314	lml	2.172	0.0031
		TMM	933-2690	LP4	2.776	0.0019
		TMM	933-2691	RP4	3.747	0.0018
Smilodon gracilis[†]	Inglis 1A, Florida	UF/TRO	1665	lml	3.809	0.0062
		UF	18102	rml	8.416	0.0026
		UF	20065	rml	3.201	0.0044
		UF	uncatalogued	lml	15.578	0.0017
	Leisey Shell Pit 1A, Florida	UF	63656	lml	2.350	0.0027
		UF	63657	rml	1.382	0.0031
		UF	81723	rml	2.786	0.0033
		UF	81724	lml	4.420	0.0025
		UF	82529	rml	1.938	0.0025
		UF	87241	lml	3.606	0.0021
		UF	217423	lml	3.196	0.0030
Smilodon populator[†]	Quequén River, Buenos Aires, Argentina	FMNH	14271	lml	5.141	0.0035
		FMNH	14279	lml	2.096	0.0015
	Arroyo Quequén Salado, Buenos Aires, Argentina	FMNH	14294	lml	2.600	0.0048

Abbreviations: Collections, TMM, Texas Memorial Museum; UF/TRO, University of Florida/TRO; UF, University of Florida/ Florida Museum of Natural History; FMNH, Field Museum of Natural History; *Asfc*, area-scale fractal complexity; *epLsar*, anisotropy

Note: Tooth positions noted with lower case noting mandibular teeth and upper case noting maxillary teeth, and tooth positions noted; [†] denotes extinct taxon.

Signatures of $\delta^{13}C$ from C_3 versus C_4 plants are incorporated into herbivore bone collagen and enamel apatite of mammals with an enrichment factor of ~5‰ and ~14‰, respectively (Ambrose and Norr, 1993; Cerling and Harris, 1999). Further, to convert $\delta^{13}C$ values to local plant values the above-mentioned values must also be reduced by 1.5‰ due to the Suess effect (increased CO_2 emissions over the past few centuries have reduced atmospheric and subsequent floral and faunal $\delta^{13}C$ values; e.g., Friedli et al., 1986). Subsequently, carnivore carbon isotope values reflect prey consumed, with an additional depletion of 1.3‰ between predator and prey $\delta^{13}C$ bioapatite values (Clementz et al., 2009), and enrichment of approximately 1.1-1.3‰ in collagen of predators relative to their prey (Fox-Dobbs et al., 2007; Bocherens et al., 2015).

Isotopic offsets between predator and prey are the source of much debate and vary from 0 to 1.3‰ in bioapatite (Lee-Thorp et al., 1989; Clementz et al., 2009) and ~1-2.5‰ in collagen (Lee-Thorp et al., 1989;

Fox-Dobbs et al., 2007; Bocherens et al., 2015). For example, $\delta^{13}C$ enamel values in medium- to large-bodied herbivores of less than −9‰ reflect a predominantly C_3 diet, whereas values greater than −2‰ indicate a primarily C_4 diet (Cerling et al., 1997). For bone collagen, $\delta^{13}C$ values in fossil carnivores of less than −16.8‰ reflect a predominantly C_3 diet, whereas values greater than −9.8‰ indicate a primarily C_4 diet. When converting $\delta^{13}C$ from collagen to modern plant values, an offset of −7.7‰ is applied to published values. This includes an offset of −5.0‰ between herbivores and plants (Lee-Thorp et al., 1989), an offset of −1.2‰ between predators and prey (Bocherens et al., 2015), and −1.5‰ for the Suess effect (Friedli et al., 1986). When converting $\delta^{13}C$ from enamel to modern plant values, an offset of −14.3‰ is applied to published values; this includes an offset of −14.1‰ between herbivores and plants (Cerling and Harris, 1999), an offset of +1.3‰ between predators and prey (Clementz et al., 2009), and −1.5‰ for the Suess effect (Friedli et al., 1986).

Table 10.2. Descriptive statistics for each DMTA variable by feliform species, summarized from published data or newly noted here

	Taxon Statistic	*n*	*Asfc*	*epLsar*
Acinonyx jubatus[a]	Mean	9	1.590	0.0049
(modern)	Median		1.767	0.0047
	Standard deviation		0.737	0.0011
	Minimum		0.759	0.0030
	Maximum		2.674	0.0064
	Total range		1.915	0.0034
Crocuta crocuta[a]	Mean	12	9.315	0.0031
(modern)	Median		7.070	0.0034
	Standard deviation		6.708	0.0011
	Minimum		2.280	0.0012
	Maximum		23.864	0.0052
	Total range		21.584	0.0040
Panthera leo[a]	Mean	15	4.616	0.0031
(modern)	Median		4.690	0.0033
	Standard deviation		1.729	0.0017
	Minimum		1.807	0.0009
	Maximum		7.354	0.0075
	Total range		5.547	0.0066
Puma concolor[b]	Mean	38	4.086	0.0035
(modern)	Median		4.039	0.0034
	Standard deviation		1.929	0.0017
	Minimum		1.075	0.0004
	Maximum		9.505	0.0071
	Total range		8.430	0.0067
Smilodon fatalis[†]	Mean	19	2.845	0.0026
(La Brea[a] and Friesenhahn Cave)	Median		2.808	0.0023
	Standard deviation		0.826	0.0013
	Minimum		1.173	0.0009
	Maximum		4.590	0.0054
	Total range		3.417	0.0045
Smilodon gracilis[†]	Mean	11	4.607	0.0031
(Inglis 1A and Leisey Shell Pit 1A, FL)	Median		3.201	0.0027
	Standard deviation		4.080	0.0012
	Minimum		1.382	0.0017
	Maximum		15.578	0.0062
	Total range		14.195	0.0045
Smilodon populator[†]	Mean	3	3.279	0.0033
(Buenos Aires, Argentina)	Median		2.600	0.0035
	Standard deviation		1.632	0.0017
	Minimum		2.096	0.0015
	Maximum		5.141	0.0048
	Total range		3.045	0.0033

Abbreviations: *n* = number of individuals sampled or summarized; *Asfc* = area-scale fractal complexity; *epLsar* = anisotropy; † = denotes extinct taxon

[a] = some or all taken from DeSantis et al. (2012); [b] = all data taken from DeSantis and Haupt (2014).

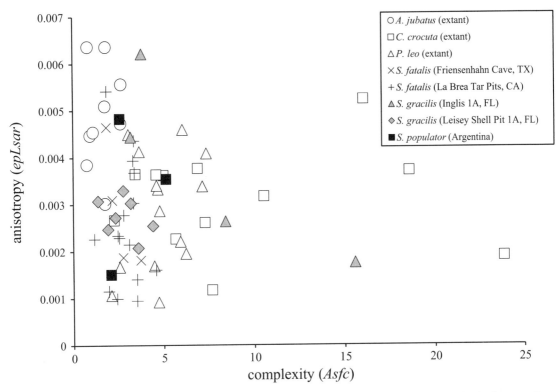

Figure 10.3. Bivariate plot of anisotropy and complexity of extant carnivorans and sabertooth cats (*Smilodon fatalis, S. gracilis,* and *S. populator*) throughout the Americas. Data on extant carnivorans and *S. fatalis* from Rancho La Brea are from DeSantis et al. (2012). All other data are newly presented here (see Tables 10.1 and 10.2).

Stable carbon isotope data of extinct and extant taxa can also be used to identify whether mammalian taxa consumed resources in denser canopy forests, due to greater ^{13}C discrimination in dense closed canopies as compared to more open C_3 environments (van der Merwe and Medina, 1989, 1991; Cerling et al., 2004; DeSantis and Wallace, 2008; DeSantis, 2011). Because $\delta^{13}C$ values increase with decreasing canopy density and increasing distance from dense forests, lower $\delta^{13}C$ herbivore enamel values reflect the consumption of browse in forests with denser canopies. Thus, stable carbon isotopes of carnivores can differentiate whether *Smilodon* species were selectively consuming prey in forests or in more open grasslands.

Nitrogen isotopes are useful for paleoecological reconstructions because their fractionation provides a proxy for trophic level in mammals (Schoeninger et al., 1983; Bocherens et al., 1994; Fox-Dobbs et al., 2007). Furthermore, because marine food chains

typically contain more tiers than terrestrial ones, $\delta^{15}N$ can be a sensitive proxy for identifying marine dietary components in terrestrial carnivores (e.g., Reid and Koch, 2013). Higher $\delta^{15}N$ values are found in carnivorous animals, including those consuming a greater proportion of marine resources, as compared to more herbivorous prey with lower $\delta^{15}N$ values (Schoeninger et al., 1983; Kelly, 2000). Thus, nitrogen isotopes can help reveal dietary variability and have the potential to identify more omnivorous taxa and/ or those utilizing marine resources. Trophic-level enrichment in $\delta^{15}N$ values is approximately 4.6‰ in carnivorans from Isle Royal in Lake Superior, Michigan—a simple island system where moose are the dominant prey of wolves (Fox-Dobbs et al., 2007).

Although both carbon and nitrogen from bone or dentine collagen are useful for resolving diet in extinct organisms, the ability to use nitrogen and carbon from collagen is limited by preservation of organic material. Fortunately, the latest Pleistocene

Smilodon species (*S. fatalis* and *S. populator*) have remains with organic molecules present, which allows stable isotope analyses of collagen (e.g., Coltrain et al., 2004; Barnett et al., 2005; Hubbe et al., 2009; Bocherens et al., 2016). However, studies of *S. gracilis* require the analysis of tooth enamel because collagen is not preserved and bone apatite is prone to postmortem alteration (Wang and Cerling, 1994).

Smilodon fatalis: Generalist or Forest Specialist?

Early stable isotope studies of *Smilodon fatalis* demonstrated that the most abundant carnivorans, *S. fatalis* and *Canis dirus*, and the largest felid, *Panthera atrox*, all competed for similar prey and consumed both ruminant and nonruminant herbivores. Nitrogen isotopes, not surprisingly, support the consumption of primary consumers (i.e., *S. fatalis* was carnivorous) and that these carnivorans did not consume a significant proportion of marine resources (Coltrain et al., 2004). Specifically, $\delta^{13}C$ values averaged −18.7‰ (ranging from −19.7‰ to −16.5‰), which corresponds to modern plant values of ∼−26.4‰ and is suggestive of the consumption of prey in primarily C_3 environments, potentially shrubland C_3 environments. *Smilodon* carbon isotope values are not significantly different from those of any of the above-mentioned carnivorans and all carnivoran $\delta^{13}C$ values of the above-mentioned taxa range from ∼−20.0‰ to −16.4‰ in *C. dirus* (the total range of all carnivoran $\delta^{13}C$ values) and from ∼−20.1‰ to −17.9‰ in *P. atrox*. A handful of additional collagen values from *Smilodon* also fall within this range (Fuller et al., 2014) (see Table 10.3).

Stable carbon isotope values have also been analyzed on the canines of *Smilodon fatalis* (and also *S. gracilis*) by Feranec (2004, 2005). In summary, these carbon values average −10.7‰ (average of two canines, each sampled multiple times perpendicular to the tooth's growth axis; Feranec, 2005) in *S. fatalis* and an average of −8.8‰ (average from one canine) in *S. gracilis* from Leisey Shell Pit 1A in Florida, which translates to prey-consuming plants with $\delta^{13}C$ values of −26.3‰ in California and −24.4‰ in Florida. This suggests they primarily hunted in C_3 forest and/or shrubland environments. Further, serial samples of *Smilodon* taxa demonstrate that the

diets of both *S. gracilis* and *S. fatalis* were fairly consistent seasonally, although much less variable in *S. gracilis* than in *S. fatalis* (Feranec, 2004, 2005).

In addition to prior studies, included here are additional stable isotope data from tooth enamel of *Smilodon fatalis* and *Canis dirus* from Pits 61/67 (dated to ∼11,500 Ka; O'Keefe et al., 2009), provided in Table 10.3 along with relevant herbivore data from Jones and DeSantis (2017) (see Fig. 10.4). In contrast to Coltrain et al. (2004), $\delta^{13}C$ values of *S. fatalis* and *C. dirus* from Pits 61/67 are significantly different from one another (p < 0.0001) with $\delta^{13}C$ values averaging −10.9‰ (ranging from −12.6‰ to −9.9‰) in *S. fatalis* (note, these values and all reported enamel values hereafter are adjusted by +1.3‰ as per Clementz et al., 2009, i.e., $\delta^{13}C_{prey}$ values noted in Table 10.3) and $\delta^{13}C$ values averaging −7.1‰ in *C. dirus* (ranging from −8.3‰ to −4.9‰) (Table 10.3 and Fig 10.4). The discrepancy in $\delta^{13}C$ enamel and collagen values is not apparent when one looks at only *S. fatalis* from Rancho La Brea (Fig. 10.5). Instead, when bone collagen $\delta^{13}C$ values are converted to modern plant values (+7.7‰, including +5.0‰ herbivore-vegetation trophic fractionation, +1.2‰ predator-prey trophic fractionation, and +1.5 Suess effect) (see Friedli et al., 1986; Lee-Thorp et al., 1989; Bocherens et al., 2015), these data suggest that *S. fatalis* was reliant on more wooded environments than *C. dirus*.

While analyses are underway to expand on the potential carnivoran prey base, including the incorporation of closed-canopy dwelling tapirs and deer in Bayesian mixing models (e.g., Stable Isotope Analysis in R-package, SIAR, similar to Feranec and DeSantis, 2014), these data suggest that *Smilodon fatalis* and *Canis dirus* consumed different prey and contradicts the suggestion that the most abundant carnivorans at Rancho La Brea were competing for similar prey (Fig. 10.4). Further, these data in combination with postcranial morphology suggest that *S. fatalis* specialized on taking down prey via ambush hunting using forest or shrub cover but were generalized in the degree of flesh and bone incorporated in their diet.

It is surprising that enamel $\delta^{13}C$ values of extinct *Canis dirus* yield different dietary interpretations than those from bone collagen. These differences are unlikely to stem from differences in the incorporation of protein (from collagen) and protein and carbohydrates (from enamel carbonate) because hypercarnivorous mammals, unlike more omnivorous taxa, are unlikely to consume carbohydrates from nonmeat sources. Second, differences in isotopic values could result from a change in diet over time, as bone collagen reflects a more recent average than tooth enamel. Third, one must evaluate whether sample preparation and treatment (e.g., the boiling of bones and teeth in kerosene to clean bones recovered during early excavations at Rancho La Brea) has affected bone collagen differently from enamel. Finally, as herbivorous *Bison antiquus* and *Equus occidentalis* at Rancho La Brea also seem to demonstrate this discrepancy, the increased incorporation of C_4 plants (and/or predators consuming prey that consumed C_4 plants) may affect isotopic fractionation, as suggested by Lee-Thorp et al. (1989).

The idea that *Smilodon fatalis* preferred prey living in a mixed grassland and/or forest is not new. Kohn et al. (2005) were the first to demonstrate low $\delta^{13}C$ values for *S. fatalis* (and another canid, *Canis armbrusteri*) even when occurring in regions with C_4 grasslands. Although Kohn et al. (2005) applied a smaller correction to the carnivoran data (+0.5 rather than +1.3‰, per Clementz et al. 2009), this minor difference would not change the overall interpretation that *S. fatalis* had a preference for forest and/or woodland herbivores in South Carolina during the Late Irvingtonian (~400 Ka), much like at Rancho La Brea in southern California during the latest Pleistocene (~11,500 Ka).

Smilodon gracilis: Changing Diets with Changing Climates

Stable isotopes from bone collagen could not be obtained for *Smilodon gracilis* due to its older geologic age, occurring during the Early to Middle Pleistocene, ~2.5 Ma to 500,000 Ka. Instead, stable carbon isotopes from tooth enamel were analyzed in both *S. gracilis* and *Canis edwardii* (an Irvingtonian wolf) from an Early Pleistocene glacial site (Inglis 1A) and a Middle Pleistocene interglacial site (Leisey Shell Pit 1A) in Florida (Feranec and DeSantis, 2014). When compared to $\delta^{13}C$ values in prey that largely switch from eating mainly forested C_3 resources during the glacial period to mixed feeding and/or grazing under

interglacial conditions (DeSantis et al., 2009), *S. gracilis* shows the opposite pattern, with lower $\delta^{13}C$ values occurring during the interglacial period. *Smilodon fatalis* was more generalized when environmental conditions were largely forested but specialized a bit more on forest-dwelling taxa when floral conditions were more open and contained forests and grasslands, based on mammalian herbivore carbon isotope values (DeSantis et al., 2009; Feranec and DeSantis, 2014).

In contrast to *Smilodon fatalis* $\delta^{13}C$ enamel values from Rancho La Brea, *S. gracilis* mean $\delta^{13}C$ values were more enriched in ^{13}C relative to ^{12}C, with $\delta^{13}C$ enamel values of −8.0‰ (ranging from −11.7‰ to −8.4‰; n = 13) at Leisey Shell Pit 1A and −7.8‰ (ranging from −3.1‰ to −12‰; n = 7) at Inglis 1A (Table 10.3, Fig 10.5). Further, *S. gracilis* is indistinguishable in mean values and variance from *Canis edwardii* at Inglis 1A, but *S. gracilis* has significantly lower $\delta^{13}C$ values and lower variance than *C. edwardii* at Leisey Shell Pit 1A (Feranec and DeSantis, 2014). Collectively, when considering both stable isotope and DMTA data, *S. gracilis* fluctuated in its diet and degree of bone processing, likely consuming prey in more open regions and engaging in more scavenging during glacial periods in Florida.

Smilodon populator: A Specialist of C_3 Consumers

Stable isotope analyses of *Smilodon populator* from southern Patagonia, Chile, were conducted by Barnett et al. (2005) and discussed by Prevosti and Martin (2013). One value for *S. populator* in Brazil was published by Hubbe et al. (2009), and *S. populator* from Argentina was recently studied by Bocherens et al. (2016). Because *S. populator* was present during the latest Pleistocene in South America, bone collagen carbon and nitrogen isotopes can be assessed, although only a third of bones sampled yielded collagen (Bocherens et al., 2016).

In southern Chile (Patagonia), carbon isotope values from bone collagen averaged −19.1‰ (n = 2) (see Table 10.3) (Barnett et al., 2005; Prevosti and Martin, 2013). When converted to modern plant values (Fig. 10.5), these values suggest that *Smilodon populator* hunted in a primarily C_3 environment, but potentially more open. These data are not surprising

Table 10.3. Summary of all stable carbon isotope data of *Smilodon* species known to date

Taxon	Site	Specimen ID	Material	δ¹³C (prey)	δ¹³C (modern plant)	Ref.
S. gracilis	Inglis 1A, FL	UF1664	enamel	−10.8	−26.4	a
		UF1707	enamel	−10.4	−26.0	a
		UF12929	enamel	−10.3	−25.9	a
		UF12933	enamel	−7.9	−23.5	a
		UF18103	enamel	−12.0	−27.6	a
		UF18104	enamel	−3.1	−18.7	a
		UF18105	enamel	−9.5	−25.1	a
	Leisey Shell Pit 1A, FL	UF63656	enamel	−11.7	−27.3	a
		UF63657	enamel	−9.9	−25.5	a
		UF80221	enamel	−9.1	−24.7	a
		UF81724	enamel	−9.3	−24.9	a
		UF83087	enamel	−9.1	−24.7	a
		UF83088	enamel	−8.5	−24.1	a
		UF83089	enamel	−8.4	−24.0	a
		UF86692	enamel	−9.7	−25.3	a
		UF87236	enamel	−9.5	−25.1	a
		UF87241	enamel	−8.8	−24.4	a
		UF87242	enamel	−8.9	−24.5	a
		UF87246	enamel	−9.3	−24.9	a
		UF87259	enamel	−8.8	−24.4	a, b
S. fatalis	Camelot Local Fauna, SC	SC2003.75.12	enamel	−8.7	−24.3	c
		SC2003.75.13	enamel	−8.4	−24.0	c
	Rancho La Brea, CA (Pits 61/67)	LACMHC141861	enamel	−9.9	−25.5	d
		LACMHC141862	enamel	−11.4	−27.0	d
		LACMHC141863	enamel	−10.0	−25.6	d
		LACMHC141864	enamel	−10.9	−26.5	d
		LACMHC141865	enamel	−12.6	−28.2	d
		LACMHC141866	enamel	−10.3	−25.9	d
		LACMHC141867	enamel	−10.9	−26.5	d
		LACMHC141868	enamel	−10.7	−26.3	d
		LACMHC141869	enamel	−12.0	−27.6	d
		LACMHC141870	enamel	−10.6	−26.2	d
		LACMHC141875	enamel	−10.9	−26.5	d
	Rancho La Brea, CA (UCMP Collection)	UCMP158250	enamel	−10.0	−25.6	e
		UCMP173179	enamel	−8.8	−24.4	e
	Fairmead, CA	MCPCA2210	enamel	−12.1	−27.7	f
		MCPCA2201	enamel	−10.5	−26.1	f
		MCPCA1757	enamel	−9.9	−25.5	f
	Rancho La Brea, CA (Pit 3)	LACMHC35349	collagen	−20.3	−26.8	g
		LACMHC35350	collagen	−20.1	−26.6	g
		LACMHC35352	collagen	−20.0	−26.5	g
		LACMHC35353	collagen	−20.2	−26.7	g
		LACMHC35354	collagen	−20.0	−26.5	g
	Rancho La Brea, CA (Pit 60)	LACMHC35770	collagen	−19.7	−26.2	g
		LACMHC35771	collagen	−19.9	−26.4	g
		LACMK2250	collagen	−19.6	−26.1	g
		LACMK2254	collagen	−19.5	−26.0	g
		LACMK5525	collagen	−17.7	−24.2	g

Table 10.3. (continued)

Taxon	Site	Specimen ID	Material	δ¹³C (prey)	δ¹³C (modern plant)	Ref.
	Rancho La Brea, CA (Pit 91)	LACMR12485	collagen	−19.7	−26.2	g
		LACMR16695	collagen	−19.2	−25.7	h
		LACMR17668	collagen	−19.4	−25.9	g
		LACMR18585	collagen	−18.4	−24.9	g
		LACMR34802	collagen	−19.4	−25.9	h
		LACMR42645	collagen	−20	−26.5	h
	Rancho La Brea, CA (Pits 61/67)	LACMHC35846	collagen	−20.0	−26.5	g
		LACMHC35847	collagen	−19.9	−26.4	g
		LACMHC35849	collagen	−20.9	−27.4	g
		LACMHC35850	collagen	−19.7	−26.2	g
		LACMHC35905	collagen	−20.3	−26.8	g
		LACMHC35907	collagen	−20.9	−27.4	g
		LACMHC35912	collagen	−20.6	−27.1	g
		LACMHC35914	collagen	−19.5	−26.0	g
		LACMHC35915	collagen	−20.3	−26.8	g
		LACMK1807	collagen	−20.7	−27.2	g
		LACMK3569	collagen	−20.6	−27.1	h
		LACMK3593	collagen	−19.9	−26.4	h
	Rancho La Brea, CA (Project 23)	LACMP23-967	collagen	−19.2	−25.7	h
		LACMP23-6185	collagen	−19.1	−25.6	h
		LACMP23-12102	collagen	−17.8	−24.3	h
		LACMP23-15065	collagen	−20.1	−26.6	h
S. populator	General Belgrano, Argentina	MHMP31	collagen	−18.9	−25.4	i
		MHMP53	collagen	−18.9	−25.4	i
		MHMP71	collagen	−18.9	−25.4	i
		MHMP85	collagen	−18.9	−25.4	i
	Arroyo Tapalque, Argentina	MLP98-II-20-1	collagen	−18.4	−24.9	i
	Camet Norte North, level B, Argentina	Santa Clara Lab#ARGC-44	collagen	−18.6	−25.1	i
		Santa Clara Lab#ARGC-45	collagen	−18.8	−25.3	i
	Mylodon Cave, Chile	OXA-13717	collagen	−20.4	−26.9	j
		OXA-14457	collagen	−20.1	−26.6	j
	Cuvieri Cave, Minas Gerais, Brazil	CV-L2-13122	collagen	−16.2	−22.7	k

Note: Data were gathered from the following references: new data from *S. fatalis* from Pits 61/67 at Rancho La Brea are also noted. δ¹³C (prey), all δ¹³C values were converted to equivalent to δ¹³C (prey) values by adding 1.3‰ for enamel (per Clementz et al., 2009) or subtracting 1.2‰ for collagen (per Bocherens et al., 2015), if not done so as per the relevant reference; δ¹³C (modern plant), all δ¹³C values were converted to equivalent δ¹³C (modern plant) values by further reducing all δ¹³C (prey) enamel values by −15.6‰ (−14.1‰ for enamel-plant fractionation; Cerling and Harris, 1999; and −1.5‰ due to the Suess effect; Friedli et al., 1986) and collagen values by −6.5‰ (−5.0‰ for herbivore collagen-plant fractionation; Lee-Thorp et al., 1989; and −1.5‰ due to the Suess effect; Friedli et al., 1986). References: a = Feranec and DeSantis, 2014; b = Feranec, 2005; c = Kohn et al., 2005; d = new data here reported; e = Feranec, 2004; f = Trayler et al., 2015; g = Coltrain et al., 2004; h = Fuller et al., 2014; i = Bocherens et al., 2016; j = Barnett et al., 2005; k = Hubbe et al., 2009.

when considering the location of the fossil specimens, occurring near the southernmost point of South America (dominated by C₃ vegetation) and in a region that is known today to be dominated by open-landscape vegetation. Nitrogen isotope values were similar to other carnivorans (mean δ¹⁵N = 10.0‰, values of 10.4‰ and 9.6‰), although slightly more elevated than other taxa, including *Panthera onca* (mean δ¹⁵N = 7.5‰) (Barnett et al., 2005; Prevosti and Martin, 2013). A more detailed dietary analysis based

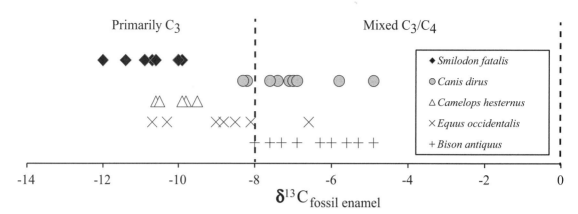

Figure 10.4. Stable carbon isotope data from tooth enamel of the most abundant carnivorous (*Smilodon fatalis* and *Canis dirus*) and herbivorous (*Bison antiquus*, *Camelops hesternus*, and *Equus occidentalis*) species from Pit 61/67 at Rancho La Brea (1.3‰ added to original carnivore δ¹³C values to account for trophic enrichment; Clementz et al., 2009; all data reported in Table 3). Unlike data from Coltrain et al. 2004, these data show separation in diet between *Bison* and other herbivore taxa, and between dire wolves and *S. fatalis*.

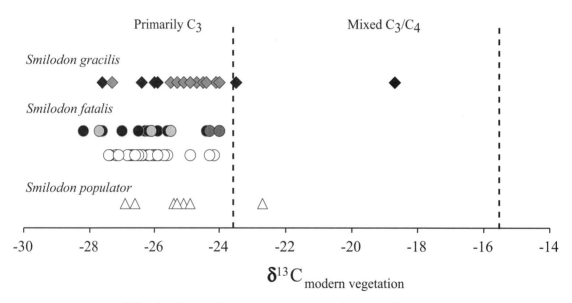

Figure 10.5. A summary of δ¹³C values from *Smilodon* species as noted in the literature and reported here (also see Table 10.4). White values denote collagen, and black and gray values (representing different fossil sites, see Table 10.3) denote enamel values. *S. fatalis* values are slightly offset on the y-axis so that the majority of isotopic values can be seen. Note: All enamel and collagen values were converted to modern plant δ¹³C values per the methods described in the text.

on Bayesian mixing models and body size analysis conducted by Prevosti and Martin (2013) suggests that *S. populator* likely focused on the larger prey such as the sloth, *Mylodon,* and the extinct ungulate, *Macrauchenia.*

An isolated carbon isotope value for bone collagen from a specimen in Brazil was elevated compared to

other *Smilodon populator* specimens (δ¹³C= −15.0‰) (see Table 10.3) (Hubbe et al., 2009). While this is the highest δ¹³C noted for *S. populator*, it is from a region that today contains open cerrado environments (a tropical savanna ecoregion of Brazil, similar to North American grassland/open woodland environments).

In Argentina, *Smilodon populator* had nitrogen and carbon isotope values that were slightly elevated as compared to individuals from southern Patagonia (p=0.01 and p < 0.0001 for nitrogen and carbon, respectively) although statistical analyses should be considered highly speculative considering the small sample sizes (n = 7 in Argentina and n = 2 in southern Patagonia). Further, *S. populator* from Buenos Aires Province of Argentina has elevated $\delta^{15}N$ values that range from 11.2‰ to 15.4‰, as compared to herbivores, as well as values for the jaguar *Panthera onca* and canid *Protocyon* (Bocherens et al., 2016).

Stable carbon isotopes from bone collagen of *Smilodon populator* from Argentina yielded an average $\delta^{13}C$ value of −17.6‰, with a narrow range of only 0.5‰ (from −17.2 to −17.7‰, n = 7) (Table 10.3). As mentioned by Bocherens et al. (2016), the narrow range of $\delta^{13}C$ values could result from specialization on a narrow range of prey or could result from the averaging of different types of prey. When predator $\delta^{13}C$ values are converted to prey $\delta^{13}C$ values (by subtracting 1.2‰) (Bocherens et al., 2015), potential prey $\delta^{13}C$ values are −18.8‰, which translates to ∼ −25.3‰ in modern plant values. Values in this range suggest that *S. populator* largely consumed prey in C_3 environments, similar to other sabertooth cats. The consumption of prey in more open environments by *S. populator* is particularly interesting because prey occupying closed forests were common in the regions studied. Although speculative, the "unusual pectoral girdle morphology" may contribute to the ability of *S. populator* to stalk prey even in more open landscapes (Bocherens et al., 2016:472).

Summary and Future Directions

Morphological data (including craniodental and postcranial) suggest that *Smilodon* was an ambush hunter and made use of its robust forelimbs to subdue its prey and its elongated canines to quickly kill it. Dental microwear analysis of all *Smilodon* species suggests that these sabertooth cats were fairly generalized in consuming both bone and flesh but were not hyena-like and fully consuming carcasses during the Late Pleistocene. However, dental microwear analysis of *S. gracilis* demonstrates a fluctuating diet consisting of harder objects during glacial times. Stable isotopes of *Smilodon* suggest that

it largely preferred prey residing in C_3 environments (even when C_4 environments such as grasslands were present). Future work is needed to assess whether and how *Smilodon* varied its diet over time in response to fluctuating climates. Additionally, with technological advances assisting our ability to study fossil carnivores using a diversity of proxy methods (including stable isotopes, dental microwear texture analysis, morphometrics, and finite element analysis), the integration of these analytical methods can here and in the future help clarify the dietary behavior of sabertooth cats.

ACKNOWLEDGMENTS

The author would like to thank the following institutions for allowing access to and use of their collections: Natural History Museum of Los Angeles County; George C. Page Museum of La Brea Discoveries (recently renamed the La Brea Tar Pits and Museum), Los Angeles; Field Museum of Natural History, Chicago; Florida Museum of Natural History, Gainesville; Santa Barbara Museum of Natural History; Iziko South African Museum, Cape Town; Texas Memorial Museum, Austin; American Museum of Natural History, New York City; the National Museum of Natural History, Washington, D.C.; and Swedish Museum of Natural History, Stockholm.

The author also thanks the National Science Foundation and Vanderbilt University for funding this research.

REFERENCES

Akersten, W. A. 1985. Canine function in *Smilodon* (Mammalia, Felidae, Machairodontinae). Natural History Museum of Los Angeles County, Contributions in Science 356:1-22.

Ambrose, S. H., and L. Norr. 1993. Experimental evidence for the relationship of the carbon isotope ratios of whole diet and dietary protein to those of bone collagen and carbonate; pp. 1-37 in J. B. Lambert and G. Grupe (eds.), Prehistoric Human Bone: Archaeology at the Molecular Level. Springer, Berlin.

Anyonge, W. 1996. Microwear on canines and killing behavior in large carnivores: saber function in *Smilodon fatalis*. Journal of Mammalogy 77:1059-1067.

Badgley, C., J. C. Barry, M. E. Morgan, S. V. Nelson, A. K. Behrensmeyer, T. E. Cerling, and D. Pilbeam. 2008.

Ecological changes in Miocene mammalian record show impact of prolonged climatic forcing. Proceedings of the National Academy of Sciences of the United States of America 105:12145-12149.

Barnett, R., I. Barnes, M. J. Phillips, L. D. Martin, C. R. Harington, J. A. Leonard, and A. Cooper. 2005. Evolution of the extinct sabretooths and the American cheetah-like cat. Current Biology 15:R589-R590.

Binder, W. J., and B. Van Valkenburgh. 2010. A comparison of tooth wear and breakage in Rancho La Brea saber-tooth cats and dire wolves across time. Journal of Vertebrate Paleontology 30:255-261.

Bocherens, H., M. Fizet, and A. Mariotti. 1994. Diet, physiology and ecology of fossil mammals as inferred from stable carbon and nitrogen isotope biogeochemistry: implications for Pleistocene bears. Palaeogeography, Palaeoclimatology, Palaeoecology 107:213-225.

Bocherens, H., M. Cotte, R. Bonini, D. Scian, P. Straccia, L. Soibelzon, and F. J. Prevosti. 2016. Paleobiology of sabretooth cat *Smilodon populator* in the Pampean region (Buenos Aires Province, Argentina) around the Last Glacial Maximum: insights from carbon and nitrogen stable isotopes in bone collagen. Palaeogeography, Palaeoclimatology, Palaeoecology 449:463-474.

Bocherens, H., D. G. Drucker, M. Germonpré, M. Lázničková-Galetová, Y. Naito, C. Wissing, J. Brůžek, and M. Oliva. 2015. Reconstruction of the Gravettian food-web at Předmostí I using multi-isotopic tracking ^{13}C, ^{15}N, ^{34}S) of bone collagen. Quaternary International 359-360:211-228.

Carbone, C., T. Maddox, P. J. Funston, M. G. L. Mills, G. F. Grether, and B. Van Valkenburgh. 2009. Parallels between playbacks and Pleistocene tar seeps suggest sociality in an extinct sabretooth cat, *Smilodon*. Biology Letters 5:81-85.

Cerling, T. E., and J. M. Harris. 1999. Carbon isotope fractionation between diet and bioapatite in ungulate mammals and implications for ecological and paleoecological studies. Oecologia 120:347-363.

Cerling, T. E., J. A. Hart, and T. B. Hart. 2004. Stable isotope ecology in the Ituri Forest. Oecologia 138:5-12.

Cerling, T. E., J. M. Harris, B. J. MacFadden, M. G. Leakey, J. Quade, V. Eisenmann, and J. R. Ehleringer. 1997. Global vegetation change through the Miocene/Pliocene boundary. Nature 389:153-158.

Christiansen, P. 2012. The making of a monster: postnatal ontogenetic changes in craniomandibular shape in the great sabercat *Smilodon*. PloS One 7:e29699.

Christiansen, P., and J. M. Harris. 2012. Variation in craniomandibular morphology and sexual dimorphism

in pantherines and the sabercat *Smilodon fatalis*. PLoS One 7(10):e48352.

Clementz, M. T., K. Fox-Dobbs, P. V. Wheatley, P. L. Koch, D. F. Doak. 2009. Revisiting old bones: coupled carbon isotope analysis of bioapatite and collagen as an ecological and palaeoecological tool. Geological Journal 44:605-620.

Coltrain, J. B., J. M. Harris, T. E. Cerling, J. R. Ehleringer, M.-D. Dearing, J. Ward, and J. Allen. 2004. Rancho La Brea stable isotope biogeochemistry and its implications for the paleoecology of the Late Pleistocene, coastal southern California. Palaeogeography, Palaeoclimatology, Palaeoecology 205:199-219.

DeNiro, M. J., and S. Epstein. 1978. Influence of diet on distribution of carbon isotopes in animals. Geochimica et Cosmochimica Acta 42:495-506.

DeSantis, L. R. G. 2011. Stable isotope ecology of extant tapirs from the Americas. Biotropica 43:746-754.

DeSantis, L. R. G. 2016. Dental microwear textures: reconstructing diets of fossil mammals. Surface Topography: Metrology and Properties 4:023002.

DeSantis, L. R. G., and R. J. Haupt. 2014. Cougars' key to survival through the Late Pleistocene extinction: insights from dental microwear texture analysis. Biology Letters 10:20140203.

DeSantis, L. R. G., and B. D. Patterson. 2017. Dietary behaviour of man-eating lions as revealed by dental microwear textures. Scientific Reports 7:904.

DeSantis, L. R. G., and S. C. Wallace. 2008. Neogene forest from the Appalachians of Tennessee, USA: geochemical evidence from fossil mammal teeth. Palaeogeography, Palaeoclimatology, Palaeoecology 266:59-68.

DeSantis, L. R. G., R. S. Feranec, and B. J. MacFadden. 2009. Effects of global warming on ancient mammalian communities and their environments. PLoS One 4:e5750.

DeSantis, L. R. G., B. W. Schubert, J. R. Scott, and P. S. Ungar. 2012. Implications of diet for the extinction of saber-toothed cats and American lions. PLoS One 7:e52453.

DeSantis, L. R. G., B. W. Schubert, E. Schmitt-Linville, P. Ungar, S. Donohue, and R. J. Haupt. 2015. Dental microwear textures of carnivorans from the La Brea Tar Pits, California and potential extinction implications; pp. 37-52 in J. M. Harris (ed.), La Brea and Beyond: The Paleontology of Asphalt Preserved Biotas. Natural History Museum of Los Angeles County, Science Series 42.

DeSantis, L. R. G., Z. J. Tseng, J. Liu, A. Hurst, B. W. Schubert, and Q. Jiangzuo. 2017. Assessing niche conservatism using a multiproxy approach: dietary ecology of extinct and extant spotted hyenas. Paleobiology 43:286-303.

DeSantis, L. R. G., J. R. Scott, B. W. Schubert, S. L. Donohue, B. M. McCray, C. A. Van Stolk, A. A. Winburn, M. A. Greshko, and M. C. O'Hara. 2013. Direct comparisons of 2D and 3D dental microwear proxies in extant herbivorous and carnivorous mammals. PLoS One 8:e71428.

Donohue, S. L., L. R. G. DeSantis, B. W. Schubert, and P. S. Ungar. 2013. Was the giant short-faced bear a hyper-scavenger? a new approach to the dietary study of ursids using dental microwear textures. PLoS One 8:e77531.

Duckler, G., and B. Van Valkenburgh. 1998. Exploring the health of Late Pleistocene mammals: the use of Harris lines. Journal of Vertebrate Paleontology 18:180-188.

Elliot, J. P., I. M. Cowan, and C. S. Holling. 1977. Prey capture in the African lion. Canadian Journal of Zoology 55:1811-1828.

Evans, G. L. 1961. The Friesenhahn Cave. Bulletin of the Texas Memorial Museum 2:3-22.

Feranec, R. S. 2004. Isotopic evidence of saber-tooth development, growth rate, and diet from the adult canine of *Smilodon fatalis* from Rancho La Brea. Palaeogeography, Palaeoclimatology, Palaeoecology 206:303-310.

Feranec, R. S. 2005. Growth rate and duration of growth in the adult canine of *Smilodon gracilis*, and inferences on diet through stable isotope analysis. Bulletin of the Florida Museum of Natural History 45:369-377.

Feranec, R. S., and L. R. G. DeSantis. 2014. Understanding specifics in generalist diets of carnivorans by analyzing stable carbon isotope values in Pleistocene mammals of Florida. Paleobiology 40:477-493.

Fox-Dobbs, K., J. K. Bump, R. O. Peterson, D. L. Fox, and P. L. Koch. 2007. Carnivore-specific stable isotope variables and variation in the foraging ecology of modern and ancient wolf populations: case studies from Isle Royale, Minnesota, and La Brea. Canadian Journal of Zoology 85:458-471.

Freeman, P. W., and C. A. Lemen. 2007. An experimental approach to modeling the strength of canine teeth. Journal of Zoology 271:162-169.

Friedli, H., H. Lötscher, H. Oeschger, U. Siegenthaler, and B. Stauffer. 1986. Ice core record of the $^{13}C/^{12}C$ ratio of atmospheric CO_2 in the past two centuries. Nature 324:237-238.

Fuller, B. T., S. M. Fahrni, J. M. Harris, A. B. Farrell, J. B. Coltrain, L. M. Gerhart, J. K. Ward, R. E. Taylor, and J. R. Southon. 2014. Ultrafiltration for asphalt removal from bone collagen for radiocarbon dating and isotopic

analysis of Pleistocene fauna at the tar pits of Rancho La Brea, Los Angeles, California. Quaternary Geochronology 22:85-98.

Hubbe, A., M. Hubbe, and W. A. Neves. 2009. New Late-Pleistocene dates for the extinct megafauna of Lagoa Santa, Brazil. Current Research in the Pleistocene 26:154-156.

Gonyea, W. J. 1976. Behavioral implications of saber-toothed felid morphology. Paleobiology 2:332-342.

Grine, F. E. 1986. Dental evidence for dietary differences in *Australopithecus* and *Paranthropus*: a quantitative analysis of permanent molar microwear. Journal of Human Evolution 15:783-822.

Janis, C. M., and P. B. Wilhelm. 1993. Were there mammalian pursuit predators in the Tertiary? dances with wolf avatars. Journal of Mammalian Evolution 1:103-125.

Jones, B. D., and L. R. G. DeSantis. 2017. Dietary ecology of ungulates from the La Brea tar pits in southern California: a multi-proxy approach. Palaeogeography, Palaeoclimatology, Palaeoecology 466:110-127.

Kelly, J. F. 2000. Stable isotopes of carbon and nitrogen in the study of avian and mammalian trophic ecology. Canadian Journal of Zoology 78:1-27.

Kohn, M. J. 2010. Carbon isotope compositions of terrestrial C3 plants as indicators of (paleo) ecology and (paleo) climate. Proceedings of the National Academy of Sciences 107: 19691-19695.

Kohn, M. J., M. P. McKay, and J. L. Knight. 2005. Dining in the Pleistocene—who's on the menu? Geology 33:649-652.

Kurtén, B., and E. Anderson. 1980. Pleistocene Mammals of North America. Columbia University Press, New York, 442 pp.

Lee-Thorp, J. A., J. C. Sealy, N. J. van der Merwe. 1989. Stable carbon isotope ratio differences between bone collagen and bone apatite, and their relationship to diet. Journal of Archaeological Science 16:585-599.

McHenry, C. R., S. Wroe, P. D. Clausen, K. Moreno, and E. Cunningham. 2007. Supermodeled sabercat, predatory behavior in *Smilodon fatalis* revealed by high-resolution 3D computer simulation. Proceedings of the National Academy of Sciences of the United States of America 104:16010-16015.

Meachen-Samuels, J. A. 2012. Morphological convergence of the prey-killing arsenal of sabertooth predators. Paleobiology 38:1-14.

Meachen-Samuels, J. A., and B. Van Valkenburgh. 2009. Craniodental indicators of prey size preference in the Felidae. Biological Journal of the Linnean Society 96:784-799.

Meachen-Samuels, J. A., and B. Van Valkenburgh. 2010. Radiographs reveal exceptional forelimb strength in the sabertooth Cat, Smilodon fatalis. PLoS One 5:e11412.

Merriam, J. C., and C. Stock. 1932. The Felidae of Rancho La Brea. Carnegie Institution of Washington Publication No. 422:1-231.

Nowak, R. M. 2005. Walker's Carnivores of the World. Johns Hopkins University Press, Baltimore, Maryland.

O'Keefe, F. R., E. V. Fet, and J. M. Harris. 2009. Compilation, calibration, and synthesis of faunal and floral radiocarbon dates, Rancho La Brea, California. Natural History Museum of Los Angeles County Contributions in Science 518:1-16.

Prevosti, F. J., and F. M. Martin. 2013. Paleoecology of the mammalian predator guild of southern Patagonia during the latest Pleistocene: ecomorphology, stable isotopes, and taphonomy. Quaternary International 305:74-84.

Reid, R. E. B. and P. L. Koch. 2013. Did interference competition between grizzly bears and coyotes prevent Holocene coastal coyotes from consuming marine foods? Journal of Vertebrate Paleontology, Program and Abstracts, 2013, pp. 196-197.

Schoeninger, M. J., M. J. DeNiro, and H. Tauber. 1983. Stable nitrogen isotope ratios of bone collagen reflect marine and terrestrial components of prehistoric human diet. Science 220:1381-1383.

Schubert, B. W., P. S. Ungar, and L. R. G. DeSantis. 2010. Carnassial microwear and dietary behavior in large carnivorans. Journal of Zoology 280:257-263.

Scott, R. S., P. S. Ungar, T. S. Bergstrom, C. A. Brown, B. E. Childs, M. F. Teaford, and A. Walker. 2006. Dental microwear texture analysis: technical considerations. Journal of Human Evolution 51:339-349.

Scott, R. S., P. S. Ungar, T. S. Bergstrom, C. A. Brown, F. E. Grine, M. F. Teaford, and A. Walker. 2005. Dental microwear texture analysis shows within-species diet variability in fossil hominins. Nature 436:693-695.

Therrien, F. 2005. Mandibular force profiles of extant carnivorans and implications for the feeding behaviour of extinct predators. Journal of Zoology 267:249-270.

Trayler, R. B., R. G. Dundas, K. Fox-Dobbs, and P. K Van De Water. 2015. Inland California during the Pleistocene—Megafaunal stable isotope records reveal new paleoecological and paleoenvironmental insights. Palaeogeography, Palaeoclimatology, Palaeoecology 437:132-140.

Ungar, P. S., C. A. Brown, T. S. Bergstrom, and A. Walker. 2003. Quantification of dental microwear by tandem scanning confocal microscopy and scale-sensitive fractal analyses. Scanning 25:185-193.

van der Merwe, N. J., and E. Medina. 1989. Photosynthesis and $^{13}C/^{12}C$ ratios in Amazonian rain forests. Geochimica et Cosmochimica Acta 53:1091-1094.

van der Merwe, N. J., and E. Medina. 1991. The canopy effect, carbon isotope ratios and foodwebs in Amazonia. Journal of Archaeological Science 18:249-259.

Van Valkenburgh, B. 1988. Incidence of tooth breakage among large, predatory mammals. American Naturalist 131:291-302.

Van Valkenburgh, B. 1996. Feeding behaviour in free-ranging, large African carnivores. Journal of Mammalogy 77:240-254.

Van Valkenburgh, B. 2009. Costs of carnivory: tooth fracture in Pleistocene and recent carnivorans. Biological Journal of the Linnean Society 96:68-81.

Van Valkenburgh, B., and F. Hertel. 1993. Tough times at La Brea: tooth breakage in large carnivores of the Late Pleistocene. Science 261:456-459.

Van Valkenburgh, B., and F. Hertel. 1998. The decline of North American predators during the Late Pleistocene; pp. 357-374 in J. J. Saunders, B. W. Styles, G. F. Baryshnikov (eds.), Quaternary Paleozoology in the Northern Hemisphere. Illinois State Museum Science Papers, Volume 27.

Van Valkenburgh, B., and T. Sacco. 2002. Sexual dimorphism, social behavior, and intrasexual competition in large Pleistocene carnivorans. Journal of Vertebrate Paleontology 22:164-169.

Van Valkenburgh, B., M. F. Teaford, and A. Walker. 1990. Molar microwear and diet in large carnivores: inferences concerning diet in the sabretooth cat, Smilodon fatalis. Journal of Zoology 222:319-340.

Walker, A., H. N. Hoeck, and L. Perez. 1978. Microwear of mammalian teeth as an indicator of diet. Science 201:908-910.

Wang, Y., and T. E. Cerling. 1994. A model of fossil tooth and bone diagenesis: implications for paleodiet reconstruction from stable isotopes. Palaeogeography, Palaeoclimatology, Palaeoecology 107:281-289.

Wroe, S., C. McHenry, and J. Thomason. 2005. Bite club: comparative bite force in big biting mammals and the prediction of predatory behaviour in fossil taxa. Proceedings of the Royal Society of London, Biological Sciences 272:619-625.

Yamaguchi, N., A. Cooper, L. Werdelin, and D. W. Macdonald. 2004. Evolution of the mane and group-living in the lion (Panthera leo): a review. Journal of Zoology 263:329-342.

11 The Postcranial Morphology of *Smilodon*

MARGARET E. LEWIS

Introduction

The sabertooth cat *Smilodon* is one of the most recognizable figures in the pantheon of prehistoric animals and has been a figure of scientific interest since it was first discovered in Brazil (Lund, 1842). Among the machairodonts and other sabertooth mammals, *Smilodon* is, perhaps, the best known both because of the extensive craniodental and postcranial remains of this genus, and the fact that this genus is found at iconic sites, such as Rancho La Brea in California.

Currently, there are three species of *Smilodon* recognized: *S. gracilis*, *S. fatalis*, and *S. populator*. All three have been reported from South America, while only *S. gracilis* and *S. fatalis* are known from North America (Berta, 1985; Kurtén and Werdelin, 1990; Rincón et al., 2011). *Smilodon gracilis* is first found in the Pliocene of North America and may be ancestral to (or share a common ancestor with) the other species. *Smilodon gracilis* continued into the Middle Pleistocene in North America and appeared in the Early to Middle Pleistocene of South America (Rincón et al., 2011). *Smilodon fatalis* first appeared in North America in the Middle Pleistocene and in the latest Pleistocene made an incursion into northwestern South America (Kurtén and Werdelin, 1990; Prevosti et al., 2013), while *S. populator* is known from the Early through the Late Pleistocene of South America (Berta, 1985; Kurtén and Werdelin, 1990; Rincón et al., 2011; Prevosti and Soibelzon, 2012). All three species are represented by postcranial material at a variety of sites.

History of the Study of *Smilodon* Postcrania

Postcranial material of *Smilodon* has been known since the discovery of the genus, even if describing it in detail was not always a priority. The initial description of the genus *Smilodon* and its type species, *S. populator*, included isolated teeth and associated metapodials (Lund, 1842). Based on Argentinian material, Gervais and Ameghino (1880) distinguished *Smilodon* from other machairodonts through a variety of dental characters and the lack of an entepicondylar foramen, found in the humerus of most felids.

Smilodon gracilis was named by Cope (1880) based on a canine. However, Cope did a relatively extensive review of all known material, both craniodental and postcranial, causing him to refer *Trucifelis fatalis* Leidy, 1868 to *Smilodon* and to comment on perceived craniodental and postcranial changes through time in sabertooth species. Cope noted that *Smilodon* had retractile claws, reaffirmed the lack of an entepicondylar foramen in South American *Smilodon*, and included a figure of an entire skeleton and a humerus (his figures 12 and 15) belonging to what is now known as *S. populator*. Cope (1879, 1880) noted the powerful "feet" of *Smilodon populator* and later (1895) noted that there were postcranial specimens found alongside the type of *S. gracilis* that were likely to belong to this species.

Thus, among the generally recognized species today, only *Smilodon fatalis* (Leidy, 1868) was not based on material that included postcrania. Slaughter (1963) diagnosed *S. fatalis* using five dental characters and no postcranial characters, while revising the many species of *Smilodon* recognized at the time. Berta (1985, 1987) viewed *S. fatalis* as a junior synonym of *S. populator* due to the fragmentary nature of the material on which *S. fatalis* is based.

The definition of the genus *Smilodontopsis* Brown, 1908 from Conard Fissure, Arkansas, included postcranial features that distinguished this genus from *Smilodon*: an entepicondylar foramen on the humerus and relatively longer metatarsals with somewhat round shafts (Brown, 1908). The genus also shared some features of the manus with *Smilodon*. Although Brown used postcranial features to distinguish *Smilodontopsis* from other machairodontine genera (e.g., scapholunar morphology), he distinguished his two new species of *Smilodontopsis* primarily on differences in upper carnassial morphology. Because postcranial variation within extant and fossil species was little understood at the time, the postcranial differences that he noted played a role in his decision to name a new genus while recognizing great similarity to other material now placed in *Smilodon*. He did correctly identify three morphs among the Pleistocene sabertooths of North America based on the material known at the time: *Trucifelis*, *Dinobastis*, and *Smilodontopsis* (= his two species + *S. gracilis*). Today, we identify these three morphs as *Smilodon fatalis*, *Homotherium serum*, and *Smilodon gracilis,* and would also add the more recently discovered *Xenosmilus hodsonae* (Martin et al., 2000). Slaughter (1963) and Churcher (1984) later referred both of Brown's species of *Smilodontopsis* to *Smilodon fatalis*.

The first detailed description of postcranial material of *Smilodon* was made by Merriam and Stock (1932) as part of their monograph on the Felidae of Rancho La Brea. They provided measurements, photos, and descriptions of each element within the skeleton, as well as comparisons to *Panthera atrox*. Numbers of specimens from this site are so over-whelming that the authors could provide a substantial commentary on what was typical for a given element and provide size ranges for some elements. Their descriptions and conclusions provided the basis for much of how adaptations in North American *Smilodon* have been characterized since that time.

Another extensive study of the postcrania was undertaken by Méndez-Alzola (1941). This study covered Argentinian material now assigned to *S. populator*. Like Merriam and Stock, Méndez-Alzola provided extensive measurements and descriptions for all elements. His study was very influential in the understanding of this genus within South America.

Berta (1985) was the first to do a detailed comparison of North and South American material of *Smilodon*. She included lists of postcranial and craniodental specimens found in museums in Argentina, Bolivia, Brazil, Ecuador, and Peru, along with measurements of many long bones. Within her emended diagnosis of the genus, she noted that *Smilodon* differs from all other machairodonts but shares with *Megantereon* a suite of derived craniodental characters and short, stocky limbs and feet.

Over time, the use of relative postcranial proportions, overall size, and robusticity has played an increasingly important role in taxonomic discussions, with less emphasis on nonmetric characters. For example, Kurtén and Werdelin (1990) utilized postcrania in their differentiation of the three species of *Smilodon* and examined proportions within the limbs and between the thoracic and pelvic limbs. Significant unpublished postcranial material exists from the Talara tar seeps in Peru that includes almost all skeletal elements (Seymour, 2015; chapter 3, this volume). Using data supplied by C. S. Churcher, Kurtén and Werdelin (1990) determined that this material belongs in *Smilodon fatalis*, rather than *S. populator,* based on proportions of the skull and the elongated proportions of pedal bones.

A second example of the use of proportions and overall size can be found in the debate over the genus *Smilodontidion* (Kraglievich, 1948). Churcher (1967) suggested that the level of epiphyseal fusion in the pelvic limb (MACN Pv 6802) indicated that the individual was not fully mature and synonymized this material with *Smilodon populator*. Although Berta (1985) supported this conclusion, Prevosti and Pomi

(2007) suggested that this individual could not have grown to the size of *S. populator* given the high level of epiphyseal fusion. In their comparison of the postcrania of all three recognized species of *Smilodon* to MACN Pv 6802, Prevosti and Pomi noted that some elements of the specimen (e.g., tibia and length of the calcaneum) are already larger than *S. gracilis*, while others (e.g., patella and length of the astragalus) fall within the size range of this species. Most, but not all, measurements of *S. populator* are larger than those of MACN Pv 6802. In fact, MACN Pv 6802 overlaps all three species in various measurements. Because this material has derived craniodental features found in *Smilodon*, Prevosti and Pomi (2007) supported the referral of *Smilodontidion* to *Smilodon* but note the lack of diagnostic characters in the specimen that support the referral to a specific species.

We could hypothesize that species of *Smilodon* overlap enough in variation that their postcrania are distinguished primarily by differences in proportion and robusticity. However, it is more likely that additional features, like the lack of an entepicondylar foramen in *S. populator* (versus its presence in *S. gracilis*, *S. fatalis*, and felids in general) are waiting to be discovered. Given the more extreme proportions and overall size of *S. populator*, this species may be the most likely to have identifiable synapomorphies reflecting differences in behavior and ecology, as well as size-related shape changes.

ABBREVIATIONS

AMNH, American Museum of Natural History, New York; **ANSP,** Academy of Natural Sciences, Philadelphia; **FMNH,** Field Museum of Natural History, Chicago; **SAM,** South African Museum, Cape Town; **NMNH,** National Museum of Natural History, Washington, DC; **NHMUK,** Natural History Museum, London; **TMM,** Texas Memorial Museum, Austin.

Materials and Methods

Although this chapter, for the most part, is a review of what is known about the postcrania of *Smilodon*, some analyses were performed. Standard linear measurements were taken of various extant and fossil felid postcrania. Data on extant taxa were collected at the AMNH, ANSP, NHMUK, FMNH,

SAM, and NMNH, while data on fossil taxa were collected at the TMM and the AMNH, and were taken from the literature (Merriam and Stock, 1932; Méndez-Alzola, 1941; Berta, 1987, 1995; Prevosti and Pomi, 2007; Castro and Langer, 2008; Martin et al., 2011). Data on the femur of an unpublished specimen of *Smilodon* from South Carolina were kindly provided by John Babiarz (see chapter 5, this volume). Maximum lengths rather than the more appropriate functional lengths (articular surface to articular surface) of long bones were used so that published measurements of specimens could be used. (Functional length is preferable because it measures the actual distance that force is traveling and the portion of the bone that is contributing to the overall length of the limb.) All data were log-transformed. Linear regressions were performed on data from the following extant felids for comparison to extinct forms: lions (*Panthera leo*), jaguars (*P. onca*), tigers (*P. tigris*), leopards (*P. pardus*), snow leopards (*P. uncia*), and pumas (*Puma concolor*).

The Axial Skeleton

Numerous studies have examined the killing bite and prey acquisition in *Smilodon*. For this reason, there has been a greater focus on the cervical region than on other regions of the axial skeleton. Initially, sabertooth felids were thought to stab their prey, requiring strength in neck depression (e.g., Matthew, 1910; Merriam and Stock, 1932; Simpson, 1941). In contrast, Akersten (1985) hypothesized that *Smilodon* utilized a canine shear-bite, which also required great strength in the neck. Heald (1986:7) reported evidence of "severe compressive, hyperextension, hyperflexion and twisting injuries" to the cervical region of *S. fatalis*, indicating heavy usage of this region of the body.

Merriam and Stock (1932) suggested that the large mastoid process in *Smilodon* indicates well-developed mastoid portions of the *M. brachiocephalicus* and *M. sternocephalicus* muscles. However, Akersten (1985) noted that the wing of the atlas is elongated posteriorly and ventrally deflected. This morphology, along with the large muscle scars on the posterolateral portions of the mastoid processes suggested to Akersten a situation more similar to that described by Davis (1964) for the panda, where the robust atlantomastoid muscles are responsible for this morphology. This atlantomastoid musculature would have increased leverage to depress the head with primary flexion occurring at the atlantooccipital joint. The large neural spine on the axis also suggested to Akersten (1985) an enlarged *M. obliquus capitis caudalis,* while the enlarged cervical transverse processes suggested powerful *M. scalenus* muscles. Further anatomical research by Antón et al. (2004) confirmed the importance of the atlantomastoid muscles in head flexion in machairodontines.

Kurtén and Werdelin (1990) noted that *Smilodon populator* differs from other sabertooths in the configuration of the occipital plane, mastoid plane, and basicranial axis, leading them to suggest that *S. populator* may have held its head lower than other species of *Smilodon*. The morphology of cervical vertebrae in *S. populator* appear to support this (pers. observ.).

Antón and Galobart (1999) wondered why the neck of machairodonts was so elongated, which would seemingly make the cervical region weaker and more difficult to control. They reevaluated previous hypotheses and provided a summary of morphological predictions for the stabbing hypothesis versus the canine shear-bite hypothesis. Machairodontine cervical morphology indicates complex usage of the neck (Antón and Galobart, 1999). As the prey was stabilized and bitten, the neck could twist to a great degree to respond to struggling prey and prevent contact between the upper canines and bone. The identified ability to control the neck in a variety of positions would be critical for the canine shear-bite hypothesis, in contrast to just the ventral movements required for the stabbing hypothesis (Antón and Galobart, 1999). Current hypotheses regarding the implications of specific cervical features can be found in Table 11.1.

The robusticity of the cervical region continues in other areas of the axial skeleton. Centra tend to be ventrally keeled (Merriam and Stock, 1932). In *Smilodon*, as in *Homotherium*, the thoracic and lumbar neural spines are strongly inclined caudally (Dalquest, 1969). The thoracic and lumbar vertebrae are robust, with the lumbar vertebrae featuring well-developed anapophyses and metapophyses (Dalquest, 1969). The vertebral column of *S. gracilis* is

distinguished from that of Rancholabrean *S. fatalis* by size only (Berta, 1987, 1995). Brown (1908) noted that the lumbar vertebrae of *S. fatalis* appear to have more gracile anapophyses than those of *S. populator* and that the lumbar region of *S. populator* appears less feline. The shape of the transverse process varies in *S. fatalis* and its orientation differs from *Panthera* (see Merriam and Stock, 1932, for examples). *Smilodon* had a shortened lumbar region in comparison to extant felids (Martin, 1980; Wroe, 2008). The sacrum was also less tapered than in modern felids or *Homotherium* and had small anterior zygopophyses (Dalquest, 1969; Berta, 1995), but a relatively large area of articulation with the ilium (Merriam and Stock, 1932).

While the number of cervical, thoracic, lumbar, and sacral vertebrae in *Smilodon fatalis* are similar to that of other felids, there is variation. *Smilodon fatalis* normally (but not always) has C7, T13, L7, S3, Ca13± (Merriam and Stock, 1932), while Méndez-Alzola (1941) reported C7, T14, L6, S3, Ca? for his one specimen of *S. populator*. The caudal region of *Smilodon* is rather short; there are generally fewer than 15 caudal vertebrae, each of which is reduced in size (Merriam and Stock, 1932). In contrast, extant large felids may have up to 28 caudal vertebrae (Ewer, 1973).

All skeletal elements of *Smilodon* are fairly evenly represented at Rancho La Brea, with the exception of sternebrae and small vertebrae (Spencer et al., 2003). However, there are enough of these elements recovered to provide a general picture of the bones. According to Merriam and Stock (1932), the manubrium is similar to that of *Panthera atrox* in transverse width, but is deeper dorsoventrally. The sternebrae are also deep and the xiphoid process is robust. Méndez-Alzola (1941) noted that there are nine elements within the sternum for his Argentinian specimen. The ribs are short and robust in comparison to *P. atrox* (Merriam and Stock, 1932). Several hyoids of *Smilodon* are known from Rancho La Brea (Tejada-Flores, unpub. ms.) and from Talara, Peru (Seymour, 2015; chapter 3, this volume).

The Thoracic Limb

Researchers have long noted the enlarged thoracic limb relative to the pelvic limb in most machairodontines

Table 11.1. Features of the cervical region (and related cranial areas) in *Smilodon* in comparison to extant felids (Dalquest, 1969; Akersten, 1985; Berta, 1995; Antón and Galobart, 1999; Antón et al., 2004) with current hypotheses of their function (Antón and Galobart, 1999; Antón et al., 2004, and present chapter). Many of these features may be found in other machairodonts.

Element	Bony Feature	Implied Soft Tissue Feature	Implied Function
Cranium	Enlarged mastoid process that projects antero-ventrally, almost touching the postglenoid process	Enlarged obliquus capitis cranialis muscles (contra Matthew, 1910)	Increased strength in head flexion (bilateral contraction) and lateral flexion (unilateral contraction) at the atlanto-occipital joint
Cranium	Reduction of size and ventral projection of paroccipital process	Increased distance between origin and insertion of digastric muscle	Increased gape
C1–C7	Elongated neck	Requires increased musculature to maintain or increase strength of muscles	Increased flexibility
Atlas	Transverse process (wing) elongated posteriorly and ventrally deflected	Enlarged obliquus capitis cranialis muscles	Increased strength in head flexion (bilateral contraction) and lateral flexion (unilateral contraction) at the atlanto-occipital joint
Axis	Large neural spine with elongated caudal projection	Enlarged obliquus capitis caudalis	Increased strength in rotation at the atlanto-axial joint; Prevent atlas wings from being drawn anteriorly during atlanto-mastoid muscle contraction
C3	Enlarged hyperapophysis (positioned closer to central axis than in *Homotherium*); often asymmetrical	Enlarged multifidus?	Strength in cervical extension (bilateral contraction); possible asymmetrical preference for lateral flexion on one side (unilateral contraction)
C3–C7	Enlarged transverse processes	Enlarged scalenes, longissi-mus cervicis, intertrans-versalis cervicis, longus colli, and longus capiti muscles	Strength in ventral and lateral flexion, extension, and overall stabilization of the neck and depression of the head
C3–C7	Slit-like transverse foramen; variable presence of transverse foramen in C7	Reduced vertebral artery	Arterial blood to brain supplied primarily through internal carotid and/or presence of unknown artery

(e.g., Schaub, 1925), with *Smilodon* being no exception. It was recognized early on that the distal elements of *Smilodon* were shortened relative to the proximal elements (e.g., Merriam and Stock, 1932). While *Smilodon* is not the only felid—nor even the only machairodontine—to have distally shortened limbs, the shortening and concomitant robusticity within the limb is fairly extreme within the family and subfamily (Lewis and Lague, 2010).

Four robust felid clavicles from Rancho La Brea were thought to belong to *S. fatalis* simply because there are many more specimens of *Smilodon* than *Panthera atrox* at this locality (Merriam and Stock, 1932). However, these clavicles have been demonstrated to most likely belong to *P. atrox* (Hartstone-Rose et al., 2012). In contrast, Hartstone-Rose et al. discovered that felid clavicles at La Brea could be broken into two morphs, with the larger morph being most similar to extant *Panthera leo* clavicles. The small, more vestigial morph was presumed to belong to *Smilodon*.

The scapula differs greatly from that of *Panthera* and other extant felids. Merriam and Stock (1932) noted that the scapula of *S. fatalis* is narrower anteroposteriorly and higher dorsoventrally than in felines. The overall shape of the borders differs, with the inferior angle where the vertebral and axillary borders meet being greater in *Smilodon*. The free margin of the spine does not curve over the infraspinous fossa, as in extant felids. There is a prominent notch in the lateral border of the glenoid fossa below the base of the spine in *S. gracilis* and *S. fatalis* that is small or missing in *Panthera atrox* and *P. onca* (Merriam and Stock, 1932; Berta, 1995). Méndez-Alzola (1941) does not mention this feature in *S. populator*.

The humerus also differs from that of felines. The articular surface of the head is somewhat flatter and the greater tuberosity extends further posteriorly than in *P. atrox*, while the lesser tuberosity is not as well developed (Merriam and Stock, 1932). The overall shape of the greater tuberosity and the deltoid tuberosity differ as well. Berta (1987, 1995) noted that the lateral profile of the humerus is diagnostic of Smilodontini; the junction of the greater tuberosity and the anterior border of the shaft is nearly orthogonal. This connection is arcuate in the Felinae.

As mentioned previously, both *S. fatalis* and *S. gracilis* are said to have an entepicondylar foramen of the humerus while *S. populator* does not (e.g., Gervais and Ameghino, 1880; Cope, 1880; Brown, 1908; Merriam and Stock, 1932). However, Méndez-Alzola (1941) clearly describes one in his Argentinian specimen (see his Figure 2), and Merriam and Stock (1932) noted one individual at Rancho La Brea that does not have this foramen and documented variation in the foramen in *S. fatalis*. Burmeister (1881) thought that its absence in South American individuals was due to individual variation.

The presence of the entepicondylar foramen is known to vary within some taxonomic groups, and even within a species (e.g., Landry, 1958). Felids and a variety of other mammals typically have an entepicondylar foramen, while others have lost it

(e.g., canids). When this foramen is present, the median nerve and sometimes the brachial artery pass through it. Landry (1958) suggested that this structure serves as a retinaculum for the median nerve. As a consequence of this function, Landry hypothesized that certain species could afford to lose the foramen either because they had reduced the amount of humeral abduction and no longer exposed their median nerve (e.g., more cursorial species), or they had shorter axillae with tightly fitting skin that served as its own retinaculum for the nerve (e.g., extant catarrhines, including humans). However, Landry's hypothesis does not explain why this feature varies among closely related species known to have similar behavioral adaptations (e.g., two species of extant hamsters in Vymazalová et al., 2015) or within a single species (e.g., the extinct primate *Catopithecus browni* in Seiffert et al., 2000). This feature needs to be documented in greater detail in all specimens of *Smilodon* (e.g., presence/absence and level of expression of the foramen, when present) to determine whether it is indeed more common in *S. fatalis* and *S. gracilis* than in *S. populator*. If that is the case, then it might indicate increased cursoriality in *S. populator* (or the use of some other means of keeping the median nerve in place) and be useful for distinguishing this species from the others. However, given the variation in presence/absence already documented and the variation within the foramen itself when present, the ability to use this character to infer function and/or for taxonomic assignments may be limited in *Smilodon*. While it is possible that the variation in *Smilodon* indicates that this genus is a taxon on the verge of losing the need to keep the median nerve in place, this possibility is not something that can be tested at present as our understanding of the function of this foramen and its distribution is woefully inadequate.

Within *Smilodon*, it is not just the distal elements that are shortened relative to the proximal elements; the humerus is actually relatively short for a bone of its overall size (Lewis and Lague, 2010). Or, conversely, the bone is robust overall relative to its length. Part of this enlarged overall size means that the diaphysis is more robust, relative to overall length of the bone (Merriam and Stock, 1932; Sorkin,

2008; Lewis and Lague, 2010; Meachen-Samuels and Van Valkenburgh, 2010).

Inside the humerus, the cortical bone is also thicker (Anyonge, 1996; Meachen-Samuels and Van Valkenburgh, 2010), reflecting thoracic limbs that were loaded more heavily in bending and axial compression than in extant felids of similar size. This thick cortical bone, along with the overall robusticity of the shaft, means that *S. fatalis* humeri were more resistant to nonaxial bending and to bending in both the mediolateral and craniocaudal planes relative to bone length than those of extant felids (Meachen-Samuels and Van Valkenburgh, 2010). These studies support the idea that the forelimb was used to a greater degree in prey grappling and immobilization than in extant felids. However, humeral cross-sectional geometry has been demonstrated to be better at predicting body mass than prey size or locomotion within extant felids (Meachen-Samuels, 2010).

The ulna is robust and relatively straight and short with a broad distal end (Merriam and Stock, 1932; Berta, 1995). Merriam and Stock (1932) noted the enlarged and often longer attachments for the *M. triceps brachii* than in *Panthera atrox*. Bohlin (1940; see his figure 4) suggested that this massive extensor of the forearm would have been used to brace the thoracic limb in extension against the body as the head pulled back in a slicing motion. (Granted, he believed this motion was used during feeding on carcasses, as he had reconstructed *Smilodon* as having been more of a scavenger than a hunter.) The coronoid process is proximolaterally directed in contrast to the distally directed process in felines (Berta, 1995). There is a large scar for the insertion of *M. flexor digitorum profundus*, although this can be seen in other machairodonts as well. In general, all of the muscle attachments are large.

Like the ulna, the radius is short and robust with a robust styloid process. The radius of *S. fatalis* is even more heavily built than that of *Panthera atrox*, a larger animal (Meachen-Samuels, 2012). The radial shaft is slightly recurved (Berta, 1995). In *S. fatalis* and *S. populator,* the distal half of the bone is particularly robust (Merriam and Stock, 1932; Castro and Langer, 2008).

Merriam and Stock (1932) noted that the head of the radius of *S. fatalis* is not as wide in its maximum length as it is in extant large felids, but it is longer along the axis perpendicular to the long axis. They found the same relatively narrow maximum width and relatively wide minimum width in the distal radial articular surface. Lewis and Lague (2010) found a similar morphology in *S. gracilis* and noted that this shape is quite different from other machairodonts and extant species studied. In contrast to other machairodonts, *Smilodon* had a relatively thicker anteroposterior diameter of both the proximal and distal articular surfaces relative to the overall size of the bone than would be predicted by extant felid morphology. The increased anteroposterior width of both articular surfaces in *Smilodon* is most likely related to aiding in distributing their relatively heavier body mass (see discussion on body mass elsewhere in this chapter). This increased size is apparent even in regressions of maximum radial length versus maximum anteroposterior diameter of the radial head (Fig. 11.1).

The shape of the scapholunar reflects the robusticity of the distal radius, particularly in the dorsopalmar direction. The surface for the radial styloid process is broad, reflecting the robusticity of the process. In contrast, the cuneiform is relatively smaller than in *Panthera atrox* (Merriam and Stock, 1932). The pisiform is robust, a feature seen in other machairodontines. The unciform is large relative to other machairodonts. This increased size serves to lengthen the fourth metacarpal such that the third and fourth metacarpals end at roughly the same point, which Brown (1908) noted differs from the condition in *Megantereon cultridens*. Merriam and Stock provide detailed comparisons of carpals of *S. fatalis* to those of *P. atrox*.

Smilodon metacarpals are relatively short, and extremely so in *S. populator* (Méndez-Alzola, 1941). The metacarpals of *S. populator* and *S. fatalis* are similar in length, but those of *S. populator* have much wider midshafts (Kurtén and Werdelin, 1990). The first metacarpal of *S. fatalis* is relatively shorter and more robust than in *P. atrox*. The distal articular surface has a distinct median keel and the surface extends farther than in *P. atrox* (Merriam and Stock,

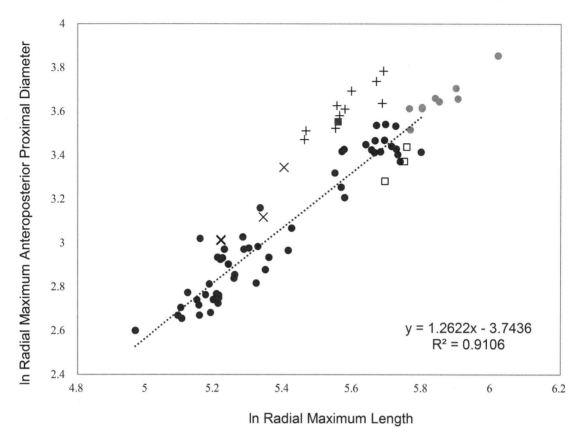

Figure 11.1. Linear regression of maximum length of the radius versus maximum anteroposterior diameter of the radial head (measured perpendicular to the maximum transverse width). Both measurements were natural logged. Linear regression based on extant species of *Panthera* and *Puma*. Note that while *Smilodon fatalis* is most like extant taxa, as all species of *Smilodon* increase in radial length, the head becomes disproportionately thicker anteroposteriorly. Or, conversely, the maximum length is short relative to anteroposterior head diameter. *Xenosmilus* is similar in this respect. This condition is in contrast to *Panthera atrox*, which follows the trend based on extant species. *Homotherium serum* has a somewhat elongated radial length for its head thickness. ● = *Panthera* and *Puma* (with lighter circles being the extinct *P. atrox*), X = *Smilodon gracilis*, + = *Smilodon fatalis*, □ = *Homotherium serum*, ■ = *Xenosmilus hodsonae*.

1932). The second metacarpal of *S. gracilis* varies in the amount of projection of the facet for the trapezium and has a sharp depression in the proximal articular surface; *Panthera onca* has a more gently concave articular surface (Berta, 1995).

Brown (1908) noted that the proximal facets on the first proximal phalanx are nearly equal in size and thus differ from extant felids (but are somewhat similar to those of *Megantereon cultridens*). *Smilodon* had an enlarged first terminal phalanx coupled with a shortened, robust proximal phalanx of the pollex (i.e., an enlarged dew claw). Other machairodonts have an enlarged first terminal phalanx as well, although the overall size and proportions vary from

taxon to taxon (Lewis and Lague, 2010 and references therein). Both the manual and pedal claws are retractile (Gonyea, 1976).

The Pelvic Limb

Early studies of machairodontines asserted that the pelvic limbs were weak relative to the thoracic limb (e.g., Schaub, 1925; Bohlin, 1940). However, this idea has long been refuted; the pelvic limb only appears weak in comparison to the forelimb and is similar to what would be predicted for a non-machairodontine felid of that size (e.g., Merriam and Stock, 1932; Simpson, 1941; Anyonge, 1993; Christiansen and Harris, 2005; Lewis and Lague, 2010). In fact, Prevosti and Pomi (2007) noted that the diaphyses of the femur, tibia, and metatarsals of MACN Pv 6802 (type of *Smilodontidion* that has been sunk into *Smilodon*) are more robust relative to overall size than in *Puma concolor*, *Homotherium*, *Machairodus*, *Nimravides*, *Panthera*, and *Megantereon*.

Few studies of *Smilodon* have investigated the morphology of the ossa coxae. Merriam and Stock (1932) and Berta (1995) noted a variety of differences between *S. fatalis* and *P. atrox*. The anterior and posterior inferior iliac spines are more prominent in *S. fatalis* and the anterior one is placed more dorsally on the iliac crest creating a sharper ridge. The proportions differ as well, with a narrower pubis and wider ischium. The ossa coxae of *S. gracilis* are smaller and have more gracile proportions than *S. fatalis* (Berta, 1995). All three species have a broader articular surface within the acetabulum than in extant felids (Méndez-Alzola, 1941; Berta, 1995).

As in the thoracic limb, the more distal elements of the pelvic limb are shortened relative to the proximal elements (Merriam and Stock, 1932). The femora of *S. gracilis* and *S. fatalis* are slightly shorter for their overall femoral size, but not significantly so (Lewis and Lague, 2010). The femur of *Smilodon*, like the humerus, has thick cortical bone (Anyonge, 1996; Meachen-Samuels and Van Valkenburgh, 2010). However, the femur of *S. fatalis* is similar in estimates of compressive and bending strength to extant felids (Meachen-Samuels and Van Valkenburgh, 2010). The femur is bowed anteroposteriorly to a great degree (Merriam and Stock, 1932; Baskin, 2005).

According to Berta (1987, 1995), the femur of *S. gracilis* is relatively shorter and more gracile at both the proximal and distal ends than in *S. fatalis*. Berta (1987, 1995) noted that *S. gracilis* has a tuberosity on the anterior margin of the intertrochanteric fossa that is characteristic of Smilodontini and lacking in *P. onca*. However, this tuberosity, which marks the insertion of *M. obturator externus*, is also present to a lesser degree in other large machairodontines like *Homotherium* (pers. observ.). The lesser trochanter is placed more medially and the intercondylar notch is deeper than in *P. onca* (Berta, 1995).

Like other machairodontines, *Smilodon* has a large femoral head relative to overall femoral size in comparison to lions and tigers (Lewis and Lague,

2010). While the relative femoral head size of *S. gracilis* was within the upper range of proportions predicted for an extant felid of similar size, *S. fatalis* had a significantly larger femoral head than predicted (Lewis and Lague, 2010). Even a simple linear regression of the anteroposterior diameter of the femoral head versus maximum length in *Panthera*, *Puma*, and machairodontines shows this pattern (Fig. 11.2): *S. gracilis* is more like extant *Panthera* and *Puma*. *Panthera atrox* is larger than expected, with *S. fatalis* and *S. populator* being slightly larger still, relative to femoral length. Not surprisingly, *Xenosmilus* had the relatively largest femoral head.

In contrast, all machairodontines tended to have relatively small femoral distal articular surfaces for the overall size of the femur in comparison to extant felids. In general, the larger the machairodont, the relatively smaller the distal articular surface. Once again, *S. gracilis* was more similar to extant felids in morphology. While still small relative to extant taxa, *S. gracilis* was not as small relatively as *S. fatalis* and other large machairodonts. This trend in machairodonts is in contrast to what is seen in extant felids, where tigers (*Panthera tigris*) tend to have proportionally large distal articular surfaces (Lewis and Lague, 2010).

The patella is more robust and is particularly thick anteroposteriorly in *S. fatalis*. The shape of the patella corresponds with the patellar surface of the femur, which is slightly deeper in *S. fatalis* than in *P. atrox* (Merriam and Stock, 1932). However, Méndez-Alzola (1941) reported that the patellar surface in *S. populator* is more quadrangular and less concave than in extants and notes that the proportions of the patella are the same as in extants. The proximodistal depth of the patella varies in *S. gracilis*, with material from McLeod being much deeper than that found at the Leisey Shell Pits (Berta, 1987).

Smilodon had a short tibia relative to overall tibial size that differed significantly from extant felids (Lewis and Lague, 2010). Extant jaguars also have significantly short tibiae, although not to the extreme found in *Smilodon* (Lewis and Lague, 2010). *Smilodon fatalis* also had a significantly larger overall size of the proximal end of the tibia due primarily to a lateral condyle that has a large anteroposterior diameter (Lewis and Lague, 2010). Berta (1987, 1995)

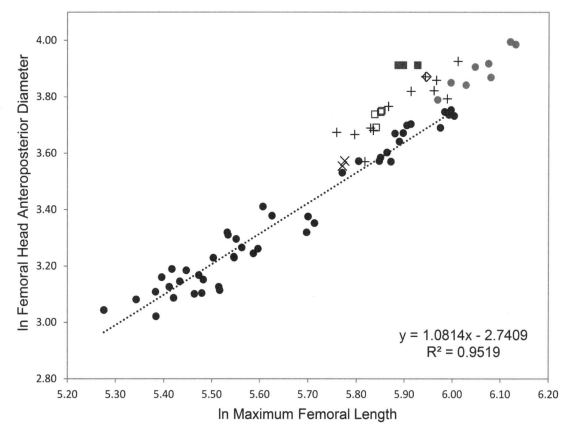

Figure 11.2. Linear regression of the maximum length of the femur versus the maximum anteroposterior diameter of the femoral head. Both measurements were natural logged. Linear regression based on extant species of *Panthera* and *Puma*. Note that while *Smilodon fatalis* is most like extant taxa in relative head thickness, *S. fatalis*, *S. populator*, and *Homotherium* occupy a trajectory above and perpendicular to that of extant felids. *Xenosmilus* has the relatively thickest heads. *Panthera atrox* has the longest femur, but appears to occupy a trajectory between that of *S. fatalis-S. populator-Homotherium* and extant felids. ● = *Panthera* and *Puma* (with lighter circles being the extinct *P. atrox*), X = *Smilodon gracilis*, + = *Smilodon fatalis*, ◊ = *Smilodon populator*, □ = *Homotherium serum*, ■ = *Xenosmilus hodsonae*.

noted that the proximal end of the tibia of *S. gracilis* differs from that of *S. fatalis* in that the posterior border of the lateral condyle projects more sharply. The distal tibia has a stepped posterior margin and grooves for the *M. flexor hallucis longus*, *M. flexor digitorum longus*, and *M. tibialis caudalis* are prominent (Berta, 1987, 1995). Méndez-Alzola (1941) noted that the medial condyle of *S. populator* is more inclined medially than in extant species.

The fibula is overall more robust than in extant felids (Merriam and Stock, 1932; Méndez-Alzola, 1941). The lateral malleolus of the fibula projects less prominently than in *P. onca* (Berta, 1995).

The astragalus in all species of *Smilodon* is what one would predict in a relatively heavy species: the

trochlea is broad and less deeply grooved than in non-machairodontine felids. Merriam and Stock (1932) noted that most astragali of *S. fatalis* have an astragalar foramen, while those belonging to *P. atrox* do not. Specimens of *S. populator* in both Argentina and Brazil also have this foramen (e.g., Méndez-Alzola, 1941; Rodrigues et al., 2004; Castro and Langer, 2008), although the Brazilian material from Abismo Iguatemi, Brazil, had additional foramina (Castro and Langer, 2008). The astragalar foramen is found in a wide variety of mammals past and present (Ameghino, 1904).

Merriam and Stock (1932) noted that there are long-necked and short-necked astragali (their figure 104 A and B). The long-necked specimen looks much like *S. gracilis* (compare to figure 14 in Berta, 1987), while their short-necked specimen looks a little like *S. populator* (compare to figure 1 in Méndez-Alzola, 1941). The astragalus of *S. populator* is similar in morphology to that of *Homotherium*, which is characterized by a short neck, wide head, and relatively more squared-off trochlea (pers. observ.). If indeed both forms belong to *S. fatalis*, it would be interesting to see whether there was change through time from the long-necked to the short-necked form, particularly as *S. fatalis* increases in size through time (see body size discussion).

The calcaneum of *Smilodon* is relatively short and has a navicular facet. This facet has been noted in *Homotherium crenatidens*, *H. serum*, *Megantereon*, *Xenosmilus*, and *Smilodon* (Ballesio, 1963; Rawn-Schatzinger, 1992; Sardella, 1998; Werdelin and Lewis, 2001; Martin et al., 2011) but is only variably present in African species of *Dinofelis* (Werdelin and Lewis, 2001). Rawn-Schatzinger (1992) suggested that this facet might be a machairodontine feature as she could not find one on *Panthera* calcanei. The orientation of the sustentaculum tali differs from African species of *Dinofelis*, *Homotherium*, *Panthera*, and *Megantereon*, with *Smilodon* having the most horizontal facet relative to the long axis of the bone (Werdelin and Lewis, 2001).

Two forms of ectocuneiforms were found (Merriam and Stock, 1932): one with a fused, hook-like process, the other with the hook being reduced or absent. This dual morphology has also been noted in specimens from Talara, Peru (Churcher, pers. comm. in Shaw and Tejada-Flores, 1985). Further analysis of the Rancho La Brea material revealed further variation in form, including the apparent truncation of the plantar process in more recent specimens (Shaw and Tejada-Flores, 1985). Shaw and Tejada-Flores noted that some extant felids had a plantar process that was unfused to the body of the ectocuneiform and, therefore, hypothesized that there were two sites of ossification within the ectocuneiform. They suggest that *Smilodon* individuals with truncated plantar processes might simply have had an unfused plantar process that did not preserve in the fossil record.

The metatarsals of *S. gracilis* are relatively short but longer and more gracile than the metacarpals (Berta,

1987, 1995). *Smilodon fatalis* had relatively longer metatarsals than did the more robust *S. populator* (e.g., Brown, 1908; Merriam and Stock, 1932). A rudimentary first metatarsal is often fused in abduction to the entocuneiform, although it may fuse to the second metatarsal (Merriam and Stock, 1932). On the fourth metatarsal, the facet for the fifth metatarsal is relatively larger than in *P. atrox*, *P. leo*, or *P. tigris* (Merriam and Stock, 1932). Phalanges are robust.

Sexual Dimorphism

Several studies have looked for evidence of sexual dimorphism in *Smilodon*. Berta (1985) was unable to sort her measurement data by sex. Kurtén and Werdelin (1990) provided some statistical suggestion of sexual dimorphism, particularly with respect to third metatarsal lengths of *S. fatalis* from Talara, Peru. Although their results were preliminary, Kurtén and Werdelin suggested that *Smilodon* was sexually dimorphic but not to the same degree as *Puma* or *Panthera*. Most studies based solely on craniodental material, however, have shown little to no sexual dimorphism (Van Valkenburgh and Sacco, 2002; Meachen-Samuels and Binder, 2010; but see Christiansen and Harris, 2012).

Postcranial Scaling and Limb Proportions

Within extant and extinct felids, there are three parallel scaling trends in the postcrania: (1) long-limbed extant taxa and *Homotherium*, (2) short-limbed extant taxa, *Smilodon*, and *Dinofelis*, and (3) all other extant taxa, including lions and tigers (Lewis and Lague, 2010). In general, felid long bone lengths scale isometrically with body mass (Anyonge, 1993; Christiansen and Harris, 2005). Felid humeral circumference increases disproportionately with body mass, while felid femoral circumference scales isometrically (Anyonge, 1993; Christiansen and Harris, 2005). Thus, one would predict an increase in humeral robusticity in larger-bodied felids (Lewis and Lague, 2010). Despite isometric scaling in length and circumference in the femur within felids as a whole, the overall shape of the femur of *S. gracilis* is slightly more similar to that predicted for felids of the same overall femoral size than is the larger

S. fatalis (Lewis and Lague, 2010). This reflects the different scaling pattern within short-limbed species.

Species of *Smilodon* have changed their limb proportions and robusticity through time. Based on an extensive study of material in North and South America, Berta (1985) suggested that *Smilodon* increased significantly in limb lengths from the Ensenadan to the Lujanian in South America, a point also made by Kurtén and Werdelin (1990). Ensenadan *S. populator* individuals are similar in hind limb proportions to Rancholabrean individuals, despite the smaller size of the Rancholabrean material (Berta, 1985). Kurtén and Werdelin (1990) also suggested that the thoracic limb of *S. populator* was somewhat longer, relative to pelvic limb length, than in *S. fatalis*. The forelimb of *S. gracilis* was more gracile than that of *S. fatalis* (Berta, 1987, 1995). Kurtén and Werdelin (1990) demonstrated that distal segments were shorter in *S. populator* than in other species of *Smilodon*. The manus of this species was extremely short.

The metacarpals and metatarsals of different species of *Smilodon* differ in robusticity. Material of *S. fatalis* from Rancho La Brea was compared with the same species from Talara, Peru, *S. populator*, and *S. gracilis* (Kurtén and Werdelin, 1990). *Smilodon populator* was unique in the great mediolateral width of the shaft, despite similarities in overall length to *S. fatalis*. *Smilodon gracilis*, in contrast, were relatively narrower in shaft width. Other studies of different material have found *S. fatalis* to have slightly longer metatarsals than *S. populator* (e.g., Brown, 1908; Merriam and Stock, 1932).

Size-related shape trends in extant felids have been demonstrated not to be the best predictors of morphology in extinct larger felids (Lewis and Lague, 2010). This has implications for studies of body mass reconstructions in felids that have relied on scaling trends in extant species due to the known body mass of extant individuals (Lewis and Lague, 2010).

Reconstructing Body Mass from the Postcrania and Its Implications

When variation among species is examined, even a cursory examination of the material shows that *S. populator* is much larger than *S. gracilis*, while *S. fatalis*

falls somewhere in between. However, even within a species, size can vary. *Smilodon gracilis* appears to increase in size through time, with the largest individuals located at Port Kennedy, Pennsylvania and McLeod, Florida (Berta, 1987, 1995). *Smilodon fatalis* may also have increased in size over time. A study of mandibular morphology of *S. fatalis* at Rancho La Brea indicated changes in gape and bite force, and an increase in body size over time, with the least amount of variation occurring in the youngest material (Meachen et al., 2014). See Table 11.2 for various body mass reconstructions.

The two largest studies were done by Anyonge (1993) and Christiansen and Harris (2005). Anyonge (1993) utilized postcranial measurements and cross-sectional geometry to reconstruct the body mass of various extinct carnivorans, including *S. fatalis*. His reconstruction (347–442 kg) was within the larger range of modern ursids. Christiansen and Harris (2005) reconstructed the body mass of the three species of *Smilodon* based on postcranial linear measurements and articular surface areas. In their study, *Smilodon gracilis* was comparable in mass to extant jaguars (55–100 kg), while *S. populator* ranged from 220 kg to 360 kg. and *S. fatalis* ranged from 160 kg to 280 kg. Christiansen and Harris suggested that some individuals of *S. populator* may have exceeded 400 kg, placing this species far above the range of the largest extant felids. A later study (Torregrosa et al., 2010) that adjusted body mass calculations for nasal aperture area resulted in body mass predictions falling between those of Anyonge (1993) and Christiansen and Harris (2005) for this species. *Smilodon populator* may be one of several extinct species that represent the upper limits of the body size of carnivorous mammals, assuming typical mammalian metabolic rates and energetic costs of transport (Carbone et al., 2007).

As a result of study after study documenting the robusticity and overall large body mass of all species of *Smilodon*, researchers have suggested a series of potential prey species. Based on regression of prey to predator body mass using extant felids, *Smilodon* may have preyed routinely on herbivores larger than 300 kg or even those over 600 kg at a minimum (Van Valkenburgh and Hertel, 1998). Van Valkenburgh and Hertel (1998) suggested that *Smilodon* had a greater

Published body mass reconstructions of *Rhizosmilodon* and *Smilodon* based on postcranial material with a few dental
nates for reference

cies	Mean (Range) kg	Locality	Age	Basis	Reference
fiteae	(56–85)	Whidden Creek, Florida (Palmetto Fauna; Upper Bone Valley, FL)	latest Hemphillian (early Pliocene)	Humerus, tibia; following Christiansen and Harris, 2005	Wallace and Hurlbert, 2013
racilis	85 (80–90)	Santa Fe River Beds, Florida	Blancan (early Pleistocene)	Dental	Kurtén, 1965
racilis	80 (70–90)	Port Kennedy Cave, Pennsylvania	Irvingtonian (middle Pleistocene)	Dental	Kurtén, 1965
racilis	(67–111)	Inglis IA, Leisey Shell Pit, Florida	latest Blancan and late early Irvingtonian (early Pleistocene)	Three femoral specimens; following Shaw and Tejada- Flores, 1985	Berta, 1995
racilis	(55–100)	Florida	early Pleistocene	Postcrania: linear measurements and articular surface areas	Christiansen and Harris, 2005
atalis	92 (85–103)	Conard Fissure, Arkansas		Dental	Kurtén, 1965
atalis	146.3 (135–154)	Rancho La Brea, California	11–36 kyBP, Rancho-labrean (late Pleistocene)	Five femoral specimens; following Alexander et al., 1979	Shaw and Tejada-Flores, 1985
atalis	155	Rancho La Brea, California	Rancholabrean (late Pleistocene)	One femur; following Shaw and Tejada-Flores, 1985	Berta, 1995
atalis	(160–280)	Rancho La Brea, California	Rancholabrean (late Pleistocene)	Postcrania: linear measurements and articular surface areas	Christiansen and Harris, 2005
atalis	338 (242–469)	Rancho La Brea, California	Rancholabrean (late Pleistocene)	Previously published equations adjusted for size of nasal aperture	Torregrosa et al., 2010
atalis	(347–442)	Rancho La Brea, California	Rancholabrean (late Pleistocene)	Postcrania: linear measurements and cross-sectional geometry	Anyonge, 1993
populator	295.8	Argentina	late Pleistocene	Previously published equations adjusted for size of nasal aperture	Torregrosa et al., 2010
populator	285	Abismo Iguatemi, São Paulo State, Brazil	14.58 kyBP Lujanian (late Pleistocene)	Femur (MZSP-PV 01), following Anderson et al., 1985, and Anyonge, 1993	Castro and Langer, 2008
populator	300	Luján, Buenos Aires Province, Argentina	Lujanian (late Pleistocene to early Holocene)		Fariña, 1996
populator	352 (127–745)	Luján, Buenos Aires Province, Argentina	Lujanian (late Pleistocene to early Holocene)	Numerous previously published craniodental and postcranial equations	Fariña et al., 1998
populator	(220–360) largest possibly >400	Argentina, Brazil	Late Pleistocene	Postcrania: linear measurements and articular surface areas	Christiansen and Harris, 2005
populator	(288–517)	Southern Patagonia, Chile	latest Pleistocene	Dental, humerus, tibia: new equations Van Valken-burgh (1990) and Christiansen and Harris (2005) data. Femur after Anyonge (1993)	Prevosti and Martin, 2013

Note: All estimates rounded to the nearest whole number.

Rhizosmilodon fiteae is a Pliocene member of the tribe Smilodontini in North America.

likelihood of encountering herbivores within their Size Class II and III (100–599 kg) than larger herbivores due to their gregarious nature and the negative relationship between body size and population density in herbivores. Thus, they suggested, North American *Smilodon* may have preyed on horses, llamas, and bovids, and occasionally on juveniles, if not the adults, of larger herbivores such as steppe and giant bison, western camel, proboscideans, and giant sloths. However, later isotope studies at Rancho La Brea suggest that while equids and mastodons were not important components of the *S. fatalis* diet, ruminants were (Coltrain et al., 2004; Feranec, 2004). Within South America, *S. populator* has been suggested to eat taxa such as notoungulates, glyptodons, ground sloths, litopterns, camelids, and equids (Berman, 1994; Prevosti and Vizcaíno, 2006; Prevosti and Martin, 2013; Bocherens et al., 2016).

Foot Posture

Ginsburg (1961) is often quoted for having said that later sabertooth species were digitigrade due to their possession of a "lion-like" humerus ("humérus léonin"; 1961:18), long metapodials, reduced first digit, and gracile tibialis posterior muscle. While he did not name all of the species that he studied, from this description and his text, it is clear that he was focused on European taxa and did not include *Smilodon* (or any species of *Megantereon*, for that matter) in his study. This led to his assertion that digitigrady first appeared in the Pontian epoch (latest Miocene) and that machairodonts thereafter were digitigrade. Portions of his description of *Sansanosmilus* (Barbourofelidae) from the European Middle to early Late Miocene sound much more similar to the anatomy of *Smilodon*: limb proportions similar to jaguars (i.e., shortened distal elements, including metapodials), large deltoid tuberosity on the humerus, a short, wide astragalus with a relatively flattened head (particularly in *S. populator*), and a short calcaneum where the sustentaculum is located relatively anteriorly on the bone. Ginsburg reconstructed *Sansanosmilus* as plantigrade. Ginsburg (1961) recognized that there were semidigitigrade and semiplantigrade species and noted that jaguars do not take full advantage of their digitigrady. However, he did not provide a means of recognizing these

intermediate categories, nor did he distinguish between differences in manual and pedal postures. Thus, given the characteristics that he associated with plantigrady, he would most likely have reconstructed *Smilodon*, or at least *S. populator*, as plantigrade.

Gonyea (1976) questioned whether the plantigrade species used by Ginsburg (1961) were actually digitigrade instead. He cited previous studies and his own work indicating that some of the species included could not hyperextend enough to be truly plantigrade and had digitigrade footprints. He found that *Smilodon* would have been able to extend the carpus to a greater degree than that of extant felids, but not enough for a fully plantigrade stance. Not having intermediate categories, Gonyea characterized the manus of *Smilodon* and similar taxa as digitigrade.

Baskin (2005), however, felt that Gonyea's work masked important differences in posture within these families. Instead, he suggested that Oligocene nimravids, *Barbourofelis*, and *Smilodon* should be considered to be subdigitigrade. This designation reflected his suggestion that their metapodials were not as perpendicular to the substrate as in fully digitigrade forms, but were still not in full contact with the substrate. This orientation would be in line with a less cursorial, more ambulatory form of locomotion and mirrored conclusions drawn about dirk-toothed species, in general, by Martin (1980).

Locomotion and Prey Procurement in *Smilodon*

Smilodon is known for its enlarged thoracic limb relative to the pelvic limb. In general, *Smilodon* has been hypothesized by many researchers to have hunted large prey relative to their body size (e.g., Merriam and Stock, 1932; Bohlin, 1940; Gonyea, 1976; Martin, 1980; Anyonge, 1996). However, just as the functions of the sabers have been debated through time, so have the functions of this enlarged limb and other components of the postcrania.

Bohlin (1940) examined the published figures of *Smilodon* and concluded that the thoracic limb flexors, or at least the *M. biceps brachii*, appeared to be relatively weak. This was problematic for him as he could not imagine how such an enlarged limb would function without a robust biceps. Like Marinelli (1938), Bohlin (1940) noted the similarity in limb proportions to hyenas, which were thought at the time to be primarily scavengers. This proportional similarity led Bohlin (1940) to suggest that *Smilodon* and other machairodonts were primarily scavengers that used their massive forelimbs in extension to brace against the carcass as the skull with enlarged canines was pulled backward to rip the carcass apart.

Most researchers, however, envisioned *Smilodon* as an active predator. Simpson (1941) hypothesized that the thoracic limb could be used to deliver the killing blow after the canines were used to stab the prey. Bohlin (1940) had pointed out, however, that the canines might not have been able to withstand the forces generated in a stabbing bite. Thus, changes in our understanding of canine function necessitated changes in hypotheses about thoracic limb usage. Large extant cats use their thoracic limbs to grapple with and restrain prey. Researchers began to realize that sabertooth taxa must have immobilized their prey with their robust thoracic limbs prior to using the teeth (e.g., Ginsburg, 1961; Gonyea, 1976; Martin, 1980). Immobilization would have been particularly important to avoid damaging the canines (Akersten, 1985). Slashing bites, rather than a suffocating bite that endangered the canines, would mean that the forelimbs of machairodontines took on a greater role in immobilization because the bite would not provide an additional means of stabilization of prey as it does in extant large felids (Meachen-Samuels and Van Valkenburgh, 2010). Not all species of *Smilodon* would have engaged prey in the same way. Based on differences in muscle attachment positions in the humerus, the forelimb of *S. gracilis* was probably capable of greater mobility than in *S. populator* (Berta, 1987). Support for the importance of the forelimb for immobilizing prey includes the finding that sabertooth species (not just sabertooth felids) with longer, thinner canines have more robust forelimbs (Meachen-Samuels, 2012). Presumably, differences in functional demands related to prey procurement not only drove the evolution of cranial shape diversity in machairodontines (Slater and Van Valkenburgh, 2008) but also differences in postcranial morphology (Lewis and Lague, 2010).

Based on analyses of *Smilodon fatalis* postcrania at Rancho La Brea, Anyonge (1996) suggested that *Smilodon* was possibly ambulatory, as well as an ambush predator. Anyonge examined brachial and crural indices and cross-sectional geometry of the humerus and femur. His research suggests that *Smilodon* was somewhat more bear-like in cross-sectional geometry than cat-like. (Other researchers have suggested other similarities to ursids, as well [e.g., Lund, 1842; Wroe, 2008].) Given its short metapodials and short tail, Anyonge (1996) suggested, *Smilodon* was probably slower in locomotion than other ambush predators.

Additional evidence of locomotor abilities comes from the ectocuneiform. Less cursorial felids have a robust plantar process with a pronounced hook for the insertion of tibialis posterior, while more cursorial species like cheetahs (*Acinonyx jubatus*) and extant canids have a reduced tibialis posterior and a reduced plantar process (Shaw and Tejada-Flores, 1985). Shaw and Tejada-Flores noted that the plantar process in *Smilodon* is most similar to that of extant tigers and jaguars (*Panthera tigris* and *P. onca*) and probably reflects ambush predatory behavior, not habitat (contra Gonyea, 1976).

As noted previously, the manual first terminal phalanx was enlarged on a short, but robust, proximal phalanx. While this is not unique for a machairodontine felid (Lewis and Lague, 2010 and references therein), these proportions indicate the importance of manual dexterity, and particularly pollical strength in *Smilodon*. This is further supported by the large sagittal keel in the pollical proximal phalanx, as well as the more extensive articular surface than in *P. atrox* as noted by Merriam and Stock (1932).

A series of morphological adaptations support the hypothesis that *Smilodon* shifted the center of gravity caudally to allow the head and thoracic limb to engage with prey. As reviewed earlier in this paper, *Smilodon* has a shortened lumbar region. If the forelimb is playing a greater role in prey grappling and restraint, one would predict a shortened lumbar region to aid in stabilizing the hindquarters while immobilizing prey. Weight would be distributed to a greater degree on the pelvic limb, while the thoracic limb was occupied with immobilizing prey (not unlike distribution of mass in ursids during semi-bipedal grappling behavior). Wroe (2008) suggested that *Smilodon* lumbar vertebrae are ursid-like in being shorter craniocaudally than most felids, but still have transverse processes oriented in the manner of felids, giving them better acceleration and less stability than ursids. Stability may have been accomplished via the various robust intervertebral articulations. In fact, Martin (1980) noted that the lumbar vertebrae are more "firmly interlocked" in dirk-toothed forms (i.e., sabertooth forms with relatively long, narrow, and unserrated or finely serrated canines), including *Smilodon*, which he attributed to the great stresses placed on the lumbar vertebrae during prey capture. Rothschild and Martin (2011) noted a series of pathologies in *Smilodon* and other dirk-toothed species that would follow from grappling with prey while placing the weight on the pelvic limb. Behavioral implications of research on pathologies are discussed later in this chapter.

A second line of evidence revolves around aspects of the morphology of the femur, astragalus, calcaneum, and metatarsals. As noted by Lewis and Lague (2010), *Smilodon* spp. had relatively large femoral heads in comparison to extant felids, a feature that is consistent with greater load-bearing and/or greater flexibility. The large tuberosity for the *M. obturator externus* that was originally noted by Berta (1987, 1995) indicates that this strong lateral rotator (and weaker extensor and adductor) may have helped stabilize the coxal joint and resist antagonistic movements during encounters with prey. The shape of the astragalus and calcaneum and the shortened, robust metatarsals are all consistent with more ambulatory behavior and possibly the ability to withstand heavy loading; these bones are not designed for fast running. However, the femur of *S. fatalis* is similar in estimates of compressive and bending strength to extant felids despite the thicker cortical bone (Meachen-Samuels and Van Valkenburgh, 2010), suggesting that loading might not exceed that seen in lions. Lions may place weight on their hindquarters during prey grappling, interspecific competition, and mating. Could *Smilodon* have been able to place more weight on the pelvic limb or place a similar amount of weight for a more

prolonged time than in extants given its short, robust lumbar region, large femoral head, and pedal morphology with subdigitigrade posture? The reconstruction of compressive and bending strength in the femur does not fit this hypothesis. Further analyses of the pelvic limb as a whole are needed.

Implications of Postcranial Pathology in *Smilodon*

Reported pathologies within *Smilodon* postcranial material include both trauma- and nontrauma-induced pathologies. Several studies have noted pathologies in the cervical, thoracic, and lumbar regions and the appendicular skeleton of material at Rancho La Brea (e.g., Moodie, 1923; Merriam and Stock, 1932; Heald, 1986; Rothschild and Martin, 2011; Brown et al., 2017). Predators risk injury not only from prey defense mechanisms and dangers within the habitat (Mukherjee and Heithaus, 2013), but also from inter- and intraspecific competition for prey/carcasses and the physiological demands of prey capture itself. Competition for mates, habitat/territory, and other resources can also cause injury. Studies indicate that among extant carnivores, a significant number of injuries are caused during intraspecific conflicts (e.g., West et al., 2006).

Some specimens of *Smilodon* from Rancho La Brea exhibit carnivore tooth marks (Spencer et al., 2003); it was not reported whether the marks came from conspecifics or other species. Rothschild and Martin (2011) reviewed a series of published injuries from Rancho La Brea that appear to have been caused by conspecifics. An axis (photo in Heald, 1986) with a large tooth puncture to the spinous process of the axis shows signs of healing and evidence of infection in the surrounding bone. Not surprisingly, adult male lions were found to be significantly more likely to die within a year if they were wounded on the forehead or the neck (West et al., 2006). Given the hypothesized critical role of the cervical region within *Smilodon* during prey acquisition, a dorsal neck wound could be severely debilitating.

Smilodon also appears to have been injured during prey acquisition. Heald (1986) suggested that the impact of attack created compressive forces that caused damage to vertebral centra and discs and the development of osteoarthritis seen in the skeleton.

Shermis (1983) described a severely injured subadult at Rancho La Brea that may have been kicked or crushed by an herbivore. The pelvic limb was essentially nonfunctional, yet healing had occurred. Rothschild and Martin (2011) reviewed a series of injuries that may have occurred while weight was shifted to the pelvic limb and the thoracic limb was engaged in grappling.

Brown and colleagues (2017) examined the distribution of traumatic injuries in *Smilodon*. The lumbar vertebrae and thoracic vertebrae adjacent to the scapula were the most commonly injured elements most likely due to torsion in these areas during grappling with prey. Injuries also clustered in the area around the acetabulum. Origins and insertions of rotator cuff muscles were also sites of trauma, consistent with grappling with large prey. Healed fractures and similar traumas were clustered at muscle attachment sites or structurally weak areas in *Smilodon*, while bite marks were more randomly distributed. The pattern of injuries in *Smilodon* was quite different than those in the presumed pursuit predator *Canis dirus*, which supported the different methods of prey acquisition in these species (Brown et al., 2017). The pattern of trauma distribution in *Smilodon* in the lumbar vertebrae and around the acetabulum is also consistent with shifting the center of gravity posteriorly during prey grappling.

Radiological study (Bjorkengren et al., 1987) of numerous vertebral specimens from Rancho La Brea indicated that some individuals from *S. fatalis* were suffering from ankylosing spondylitis, while others were suffering from diffuse idiopathic skeletal hyperostosis (DISH). A third group was suffering from the aftereffects of trauma to the vertebral column with possible subsequent infection. Additional causes of paravertebral ossification, hyperfluorosis, and hypervitaminosis A were ruled out as the cause of the pathologies (but see Rothschild and Martin, 2011).

Ankylosing spondylitis is a chronic disease involving the sacroiliac joints and axial skeleton that results in progressive stiffness of the vertebral column. This inflammatory disease has been found in a variety of extant and extinct vertebrates, including humans (Spencer et al., 1980). In humans, possession of the allele HLA-B27 is associated with increased risk of this disease, although it appears to be a polygenic disorder with unknown environmental triggers.

DISH is a systemic, noninflammatory disease that involves the ossification of soft tissues, including ligaments and enthuses. Longitudinal spinal ligaments are commonly affected. DISH has been reported in humans, neanderthals, and domestic dogs (Mader et al., 2013), but was not known to occur in domestic cats until recently (Bossens, et al., 2016). DISH may also have an underlying genetic cause as it is more common in some populations of humans and some breeds of dogs than in other populations or breeds, respectively (Kranenburg et al., 2010). Like ankylosing spondylitis, this condition is more common in males of all species studied.

Ankylosing spondylitis and DISH do not usually become symptomatic in humans until either early adulthood (ankylosing spondylitis, 20s to 30s) or later (DISH, 50s). They are both progressive, with the severity of symptoms varying among individuals. Thus, the discovery of these diseases in individuals at Rancho La Brea may indicate relatedness among these individuals, as sufferers would probably have passed on the associated alleles prior to feeling the effects of the disease.

Once advanced, ankylosing spondylitis or DISH would have impacted the ability of an individual of *Smilodon* to capture prey and/or mate. Both conditions can result in postural abnormalities and biomechanical changes in the musculoskeletal system. Prey species immobilized in the tar seeps at La Brea might have seemed an easier meal for afflicted predators than their usual fare. However, given rates of entrapment at one every 50 to 70 years (Marcus and Berger, 1984; Friscia et al., 2008), the scenario of trapped prey attracting afflicted individuals of *Smilodon* is probably unlikely.

Finally, a study of Harris lines in Pleistocene mammals from Rancho La Brea indicated that none of the extinct mammals, including *S. fatalis*, differed significantly from modern healthy mammals (Duckler and Van Valkenburgh, 1998). Harris lines form when growth is interrupted for any reason. Although Harris lines were found in *Smilodon*, they did not occur more frequently than in healthy carnivores.

Habitat Preferences

Several studies of the postcrania have placed *Smilodon* in closed habitats, or at least habitats that had enough cover for ambush predation (Gonyea, 1976; Anyonge, 1996; Martin et al., 2000; Meloro et al., 2013; Schellhorn and Sanmugaraja, 2015). While most of these studies focused on *S. fatalis*, at least one includes *S. populator* (Meloro et al., 2013). Comparison of African and North American machairodontines and extant felids suggests that *Smilodon* and *Megantereon* preferred mixed or closed habitats but does not rule out the possibility that they used more open habitats (Werdelin and Lewis, 2001). Kurtén and Werdelin (1990) suggested that the possession by *S. populator* of a thoracic limb that was longer than the pelvic limb might mean that this species was more adapted to the open plains than *S. fatalis*.

High carbon isotope values for *S. populator* from 24 to 10 kyBP suggest consumption of prey from an open, dry C_3 environment such as the Late Pleistocene open Pampas ecosystem and a lack of prey such as *Hippidion* or *Toxodon* that occupied more densely vegetated areas (Bocherens et al., 2016). Data from *S. fatalis* at Rancho La Brea also shows a preference for species that fed on C_3 plants (Feranec, 2004), while high nitrogen isotope values suggest that they targeted ruminants preferentially (Coltrain et al., 2004).

It must be remembered that felids tolerate a wide variety of habitats. Lions (*Panthera leo*) have limb lengths that classify them as open or possibly mixed habitat (e.g., Lewis, 1997). Within Asia, lions are restricted in range and inhabit dry, deciduous forest, while in Africa they are found in many habitats and are only missing from tropical rainforests and the interior of the Sahara Desert (Bauer et al., 2015). Jaguars (*P. onca*), with their closed habitat limb lengths, may inhabit areas with a variety of coverage but are most abundant where there is dense forest cover with permanent water sources (Caso et al., 2008).

More importantly, a single species may prefer different habitats for different activities and find the most appropriate microhabitats within the larger habitat. A study of prides inhabiting primarily either woodland or plains in the Serengeti (Hopcraft et al., 2005) showed that all lions tend to kill prey in areas where they have good cover from which to ambush prey and not in the open, short grass areas where prey abundance is greatest. Plains and woodland lions use different types of cover specific to their habitat. Scavenging takes place in the most open area, the short grass plains, as carcasses can be located easily from the tops of kopjes. Both groups of lions spend more time resting in areas with good cover within their respective habitats.

Thus, limb proportions are useful for primary preferences but do not cover the range of habitats that a felid species may inhabit in general or that a single individual or social group may utilize on a finer-grained level. In short, more mixed/open habitat prey avoid areas with dense cover as predators are easily hidden there (e.g., Sinclair, 1985), while ambush predators select areas where prey are easy to catch (e.g., Hopcraft et al., 2005). In the case of *Smilodon*, where morphology suggests ambush predation and isotope analyses and studies of the sites suggest that prey came from more open areas, we may be seeing behavior similar to the lions described above. *Smilodon* may be finding enough cover to ambush more open-adapted prey successfully and its limb proportions may relate more to prey procurement than habitat preference.

Carnivore Guilds and Extinction

Species of *Smilodon*, regardless of their temporal or geographic location, were likely to have been apex predators given their overall body size and dental morphology. All three species are commonly found with other large predators. For example, *S. gracilis* is often found with jaguars and *Miracinonyx* at various sites (Van Valkenburgh et al., 1990). In Florida, *S. gracilis* is found in association with *Homotherium* and *Miracinonyx*, as well as wolves and tremarctine bears (Berta, 1995). *Smilodon fatalis* overlapped in various locations with *Panthera atrox*, *Homotherium*, *Arctodus*, black bears, jaguars, pumas, dire wolves, timber wolves, bobcats, and coyotes (e.g., Merriam and Stock, 1932; Schubert et al., 2010). *Smilodon populator* is found alongside jaguars, pumas, a variety of smaller felids and canids, and *Arctotherium* (Prevosti and Vizcaíno, 2006; Prevosti and

Martin, 2013; Rodrigues et al., 2014; Bocherens et al., 2016). When just hypercarnivores are examined, North America had six felids (including *Smilodon*) and a wolf-like canid in the Pleistocene. In contrast, 7 out of 16 canids were hypercarnivores, with *Smilodon* arriving by the Middle to Late Pleistocene (Van Valkenburgh, 2007).

Although guild structure differed across time and geography, most guilds included multiple potential competitors for *Smilodon*. For example, probable prey sizes of the various North American carnivorans in the Late Pleistocene have been estimated and may have overlapped (Van Valkenburgh and Hertel, 1998; Van Valkenburgh et al., 2016). *Panthera atrox* and *Smilodon fatalis* may have preferred similar habitats and species of similar size based on their size and locomotor behavior (Anyonge, 1996; Van Valkenburgh and Hertel, 1998). Anyonge (1996) speculated that competition was minimized by temporal rather than spatial separation. In contrast, some carnivore guilds in South America had more obvious ecological separation among hypercarnivores. *Smilodon* hunted much larger prey than the smaller *Puma concolor*, while both *Smilodon* and *Panthera onca* could capture most large mammals in the region studied (Prevosti and Martin, 2013). In other areas of southern South America, an additional hypercarnivorous wolf-like species, *Canis nehringi*, could have also taken large prey (Prevosti and Vizcaíno, 2006). However, *Smilodon* likely had the ability to prey on the largest megafauna with greater regularity than the other species (Prevosti and Vizcaíno, 2006).

As discussed previously, various researchers note that species of *Smilodon* increase in body size through time and that each subsequent species was more massive on average than the last. Clearly, some stressor was driving the evolution of larger body size in this genus. Little to no sexual dimorphism suggests that male-male competition was not driving the increased body size. Possibilities therefore include interspecific and/or intraspecific competition for other resources (e.g., prey, carcasses, territory). Increased body size allows the capturing of larger prey, but is also a means of ascending in rank within the predator guild. Carnivorans can ensure their dominance in competitive interactions through

increasing the number of individuals within their group and/or by increasing individual body size. When surrounded by large groups of dire wolves and either solitary or grouped *Panthera atrox*, large body size (whether as part of a group or not) might convey an advantage in fending off kleptoparasites, competing for carcasses with conspecifics, and holding on to prime habitat with cover for ambush predation and resting in the shade. Unfortunately, as predator body mass increases, the minimum prey body mass that it is physiologically cost effective to capture also increases. No matter whether the predator guild was destabilized (e.g., Ripple and Van Valkenburgh, 2010) or whether climate change or human overhunting occurred or some combination of all three, once the medium to large herbivores were disappearing, a massive, hypercarnivorous predator living among other large predators would have had few options.

While machairodonts disappeared from Africa and Asia somewhere between 1 to 0.5 ma (Werdelin and Lewis, 2005; Louys, 2014), *Smilodon* persisted until roughly 11-12 kyBP in North America and 11-9.8 kyBP in South America (e.g., Marcus and Berger, 1984; Hubbe et al., 2009, 2013; Barnosky and Lindsey, 2010). On both continents, *Smilodon* overlapped at least temporally, if not geographically, with humans, although the impact of humans may have differed from region to region (e.g., Ripple and Van Valkenburgh, 2010; Hubbe et al., 2013). The extinction of *Smilodon* in North and South America is generally considered to be a part of the larger megafaunal extinction occurring on each continent at that time.

Megafaunal extinctions happened on different continents at different times, meaning that the modern predator guilds are more recent in some regions than in others. Within eastern Africa, the guild was reduced to its modern components sometime after 0.9 Ma, but long before the end of the Pleistocene (e.g., Werdelin and Lewis, 2005). The predator guild of Eurasia was reduced to modern components sometime after the Last Glacial Maximum, although a variety of species may have survived in refugia and recolonized Europe (e.g., Sommer and Benecke, 2006; Lewis, 2017). Large-bodied carnivorans persisted the longest in the Americas. Possible refugia for *Smilodon* have been identified in Brazil (Hubbe et al., 2013), although there is no evidence that they recolonized the region.

Thus, by 10 ka, the North and South American predator guilds had declined sharply in species diversity and presumably in overall number of individuals. Specialists were more likely to become extinct than generalists (Van Valkenburgh and Hertel, 1998), a situation also noted in African large-bodied carnivoran extinctions (e.g., Werdelin and Lewis, 2005; Lewis and Werdelin, 2007). A variety of large-bodied North American carnivorans became extinct: *Arctodus simus*, *Canis dirus*, *Panthera atrox*, *Smilodon fatalis*, and *Homotherium serum* (Van Valkenburgh and Hertel, 1998; Van Valkenburgh 2009). This decrease in carnivore richness mirrored an even more dramatic drop in herbivore diversity with 78% of herbivores larger than 30 kg in mass becoming extinct between 12 and 10 ka. Afterward there were 11 herbivores, only 3 of which had greater than 300 kg body mass (in contrast to the 27 species over 300 kg that had existed previously). Van Valkenburgh and Hertel (1998) suggest that extinction of 100-599 kg herbivores may have had a greater impact on *Smilodon* than the extinction of the larger herbivores (see previous discussion of body size and prey preference). Herbivores in this range went from 25 to 4 species during the same time interval. As noted by Van Valkenburgh and Hertel (1998), large predators simply cannot survive on smaller prey because they must expend more energy to capture the prey than is provided by the prey.

Conclusions

Of the three generally recognized species of *Smilodon*, *S. gracilis* and *S. fatalis* are more similar in postcranial morphology to each other than either is to *S. populator*. While studies have increasingly relied on proportions to identify postcranial material to taxon, it is possible that a true systematic first hand study of material at numerous sites for each species could uncover more postcranial characters. Such an undertaking is beyond the scope of this chapter.

Postcranial material provides evidence of a wide variety of behaviors in *Smilodon* spp. that distinguishes them from living species. The cervical region was strong and mobile, capable of aiding the precise placement of a bite and dealing with a variety of stresses. The thoracic limb had great strength and would have been more than capable of immobilizing prey species. The lumbar and pelvic limb may have supported the weight so that the thoracic limb was free, although the amount of weight and length of time it could be sustained is unclear.

The extinction of *Smilodon* coincided with the extinction of other large predators and herbivores in the Americas. Regardless of the initial cause, even with the removal of other large predators it is unlikely that *Smilodon* could have survived on the reduced number of herbivores of the requisite size.

ACKNOWLEDGEMENTS

The author would like to thank the editors, G. McDonald, L. Werdelin, and C. Shaw, for the invitation to be a part of this volume and for their wonderful editorial comments and suggestions. Thank you to John Babiarz for offering unpublished femoral data of *Smilodon*, to Mike Lague and John Babiarz for discussions related to the manuscript, and to two anonymous reviewers for their comments and suggestions. In addition, the author thanks M. Avery, J. Babiarz, J. Flynn, L. Gordon, E. Lundelius, R. MacPhee, N. Mudida, N. Simmons, R. Thorington, and E. Westwig for providing access to specimens. Funding for research on North American carnivorans came from a Stockton Sabbatical Subvention Grant. Extant data collection was supported by NSF and Stockton Distinguished Faculty Fellowships.

REFERENCES

Akersten, W. A. 1985. Canine function in *Smilodon* (Mammalia, Felidae, Machairodontinae). Natural History Museum of Los Angeles County, Contributions in Science 356:1-22.

Alexander, R. M., A. S. Jayes, G. M. O. Maloiy, and E. M. Wathuta. 1979. Allometry of the limb bones of mammals from shrews (*Sorex*) to elephants (*Loxodonta*). Journal of Zoology 189:305-314.

Ameghino, F. 1904. La perforación astragaliana en los mamíferos: no es un carácter originariamente primitivo. Anales del Museo Argentino de Buenos Aires. Serie III 4:349-460.

Anderson, J. F., A. Hall-Martin, and D. A. Russell. 1985. Long-bone circumference and weight in mammals, birds and dinosaurs. Journal of Zoology 207:53-61.

Antón, M., and A. Galobart. 1999. Neck function and predatory behavior in the scimitar toothed cat *Homotherium latidens* (Owen). Journal of Vertebrate Paleontology 19:771-784.

Antón, M., M. J. Salesa, J. F. Pastor, I. M. Sánchez, S. Fraile, and J. Morales. 2004. Implications of the mastoid anatomy of larger extant felids for the evolution and predatory behaviour of sabretoothed cats (Mammalia, Carnivora, Felidae). Zoological Journal of the Linnean Society 140:207-221.

Anyonge, W. 1993. Body mass in large extant and extinct carnivores. Journal of Zoology 231:339-350.

Anyonge, W. 1996. Microwear on canines and killing behavior in large carnivores: saber function in *Smilodon fatalis*. Journal of Mammalogy 77:1059-1067.

Ballesio, R. 1963. Monographie d'un *Machairodus* du gisement villafranchien de Senèze: *Homotherium crenatidens* Fabrini. Travaux du Laboratoire de Géologie de la Faculté des Sciences de Lyon 9:1-129.

Barnosky, A. D., and E. L. Lindsey. 2010. Timing of Quaternary megafaunal extinction in South America in relation to human arrival and climate change. Quaternary International 217:10-29.

Baskin, J. A. 2005. Carnivora from the Late Miocene Love Bone bed of Florida. Bulletin of the Florida Museum of Natural History 45:413-434.

Bauer, H., C. Packer, P. F. Funston, P. Henschel, and K. Nowell. 2015. *Panthera leo*. The IUCN Red List of Threatened Species 2015:e.T15951A79929984. http://dx.doi.org/10.2305/IUCN.UK.2015-4.RLTS.T15951A79929984.en. Accessed May 27, 2016.

Berman, W. D. 1994. Los carnívoros continentales (Mammalia, Carnivora) del Cenozoico en la provincia de Buenos Aires. Ph.D. dissertation, Universidad Nacional de La Plata, La Plata, 432 pp.

Berta, A. 1985. The status of *Smilodon* in North and South America. Natural History Museum of Los Angeles County, Contributions in Science 370:1-15.

Berta, A. 1987. The sabercat *Smilodon gracilis* from Florida and a discussion of its relationships (Mammalia, Felidae, Smilodontini). Bulletin of the Florida State Museum, Biological Sciences 31:1-63.

Berta, A. 1995. Fossil carnivores from the Leisey Shell Pits, Hillsborough County, Florida. Bulletin of the Florida Museum of Natural History 37:463-500.

Bjorkengren, A. G., D. J. Sartoris, S. Shermis, and D. Resnick. 1987. Patterns of paravertebral ossification in the prehistoric saber-toothed cat. American Journal of Roentgenology 148:779-782.

Bocherens, H., M. Cotte, R. Bonini, D. Scian, P. Straccia, L. Soibelzon, and F. J. Prevosti. 2016. Paleobiology of sabretooth cat *Smilodon populator* in the Pampean Region (Buenos Aires Province, Argentina) around the Last Glacial Maximum: Insights from carbon and nitrogen stable isotopes in bone collagen. Palaeogeography, Palaeoclimatology, Palaeoecology 449:463-474.

Bohlin, B. 1940. Food habit of the machaerodonts, with special regard to *Smilodon*. Bulletin of the Geological Institute, Uppsala 28:156-174.

Bossens, K., S. Bhatti, I. Van Soens, I. Gielen, and L. Van Ham. 2016. Diffuse idiopathic skeletal hyperostosis of the spine in a nine-year-old cat. Journal of Small Animal Practice 57:33-35.

Brown, B. 1908. The Conard Fissure, a Pleistocene bone deposit in northern Arkansas: with descriptions of two new genera and twenty new species of mammals. Memoirs of the American Museum of Natural History 9:155-208.

Brown, C., M. Balisi, C. A. Shaw, and B. Van Valkenburgh. 2017. Skeletal trauma reflects hunting behaviour in extinct sabre-tooth cats and dire wolves. Nature Ecology and Evolution 1:1-7. doi 10.1038/s41559-017-0131.

Burmeister, H. 1881. Atlas de la "description physique de la République Argentine" contenant des vues pittoresques et des figures d'histoire naturelle composées par le Dr. H. Burmeister, Le texte traduit en français avec le concours de E. Daireaux. Deuxième section. Mammifères.-Erläuterungen zur Fauna Argentina, enthaltend ausführliche Darstellungen neuer oder ungenügend bekannter Säugethiere . . . Ie Lief. Die Bartenwale der argentinischen Küsten. PE Coni.

Carbone, C., A. Teacher, and J. M. Rowcliffe. 2007. The costs of carnivory. PLoS Biology 5:e22.

Caso, A., Lopez-Gonzalez, C., Payan, E., Eizirik, E., de Oliveira, T., Leite-Pitman, R., Kelly, M. & Valderrama, C. 2008. *Panthera onca*. The IUCN Red List of Threatened Species 2008: e.T15953A5327466. http://dx.doi.org/10.2305/IUCN.UK.2008.RLTS.T15953A5327466.en. Accessed May 27, 2016.

Castro, M. C., and M. C. Langer. 2008. New postcranial remains of *Smilodon populator* Lund, 1842 from south-central Brazil. Revista Brasileira de Paleontologia 11:199-206.

Christiansen, P., and J. M. Harris. 2005. Body size of *Smilodon* (Mammalia, Felidae). Journal of Morphology 266:369-384.

Christiansen, P., and J. M. Harris. 2012. Variation in craniomandibular morphology and sexual dimorphism in pantherines and the sabercat *Smilodon fatalis*. PLoS One 7(10):e48352.

Churcher, C. S. 1967. *Smilodon neogaeus* en las barrancas costeras de Mar del Plata, provincia de Buenos Aires.

Publicaciones del Museo Municipal de Ciencas Naturales de Mar del Plata 1:245-262.

Churcher, C. S. 1984. The status of *Smilodontopsis* (Brown, 1908) and *Ischyrosmilus* (Merriam, 1918): a taxonomic review of two genera of sabretooth cats (Felidae, Machairodontinae). Royal Ontario Museum, Life Sciences Contributions, 140:1-59.

Coltrain, J. B., J. M. Harris, T. E. Cerling, J. R. Ehleringer, M.-D. Dearing, J. Ward, and J. Allen. 2004. Rancho La Brea stable isotope biogeochemistry and its implications for the palaeoecology of Late Pleistocene, coastal southern California. Palaeogeography, Palaeoclimatology, Palaeoecology 205:199-219.

Cope, E. D. 1879. On the genera of Felidae and Canidae. Proceedings of the Academy of Natural Sciences of Philadelphia 31:168-194.

Cope, E. D. 1880. On the extinct cats of America. American Naturalist 14:833-858.

Cope, E. D. 1895. The fossil Vertebrata from the fissure at Port Kennedy, Pa. Proceedings of the Academy of Natural Sciences, Philadelphia 47:446-450.

Dalquest, W. W. 1969. Pliocene carnivores of Coffee Ranch. Bulletin of the Texas Memorial Museum 15:1-8.

Davis, D. D. 1964. The Giant Panda: A Morphological Study of Evolutionary Mechanisms. Fieldiana: Zoology Memoirs 3:1-339.

Duckler, G. L., and B. Van Valkenburgh. 1998. Exploring the health of Late Pleistocene mammals: the use of Harris lines. Journal of Vertebrate Paleontology 18:180-188.

Ewer, R. F. 1973. The Carnivores. Cornell University Press. Ithaca, New York. 500 pp.

Fariña, R. A. 1996. Trophic relationships among Lujanian mammals. Evolutionary Theory 11:125-134.

Fariña, R. A., S. F. Vizcaíno, and M. S. Bargo. 1998. Body mass estimations in Lujanian (Late Pleistocene-Early Holocene of South America) mammal megafauna. Mastozoología Neotropical 5:87-108.

Feranec, R. S. 2004. Isotopic evidence of saber-tooth development, growth rate, and diet from the adult canine of *Smilodon fatalis* from Rancho La Brea. Palaeogeography Palaeoclimatology Palaeoecology 206:303-310.

Friscia, A. R., B. Van Valkenburgh, L. Spencer, and J. Harris. 2008. Chronology and spatial distribution of large mammal bones in Pit 91, Rancho La Brea. Palaios 23:35-42.

Gervais, H., and F. Ameghino. 1880. Les mammifères fossiles de l'Amérique du Sud. pp. 512-645 in A. J. Torcelli (ed.), Obras Completas y Correspondéncia

Científica, Volume 2. Taller de Impresiones Oficiales, Buenos Aires.

Ginsburg, L. 1961. Plantigradie et digitigradie chez les carnivores fissipedes. Mammalia 25:1-21.

Gonyea, W. J. 1976. Behavioral implications of saber-toothed felid morphology. Paleobiology 2:332-342.

Hartstone-Rose, A., R. C. Long, A. B. Farrell, and C. A. Shaw. 2012. The clavicles of *Smilodon fatalis* and *Panthera atrox* (Mammalia, Felidae) from Rancho La Brea, Los Angeles, California. Journal of Morphology 273:981-991.

Heald, F. P. 1986. Paleopathology at Rancho La Brea. Anthroquest 36:6-7.

Hopcraft, J. G. C., A. Sinclair, and C. Packer. 2005. Planning for success: Serengeti lions seek prey accessibility rather than abundance. Journal of Animal Ecology 74:559-566.

Hubbe, A., M. Hubbe, and W. A. Neves. 2009. New Late-Pleistocene dates for the extinct megafauna of Lagoa Santa, Brazil. Current Research in the Pleistocene 26:154-156.

Hubbe, A., M. Hubbe, and W. A. Neves. 2013. The Brazilian megamastofauna of the Pleistocene/Holocene transition and its relationship with the early human settlement of the continent. Earth-Science Reviews 118:1-10.

Kraglievich, L. J. 1948. *Smilodontidion riggii*, n. gen. n. sp. Un nuevo y pequeño esmilodonte en la fauna pliocena de Chapadmalal. Revista del Museo Argentino de Ciencias Naturales, "Bernardino Rivadavia," Ciencias Zoologica 1:1-44.

Kranenburg, H. C., L. A. Westerveld, J. J. Verlaan, F. C. Oner, W. J. A. Dhert, G. Voorhout, H. A. W. Hazewinkel, and B. P. Meij. 2010. The dog as an animal model for DISH? European Spine Journal 19:1325-1329.

Kurtén, B. 1965. The Pleistocene Felidae of Florida. Bulletin of the Florida State Museum, Biological Sciences 9:215-273.

Kurtén, B., and L. Werdelin. 1990. Relationships between North and South American *Smilodon*. Journal of Vertebrate Paleontology 10:158-169.

Landry, S. O. 1958. The function of the entepicondylar foramen in mammals. American Midland Naturalist 60:100-112.

Leidy, J. 1868. Notice of some vertebrate remains from Harden Co., Texas. Proceedings of the Academy of Natural Sciences, Philadelphia 20:174-176.

Lewis, M. E. 1997. Carnivoran paleoguilds of Africa: implications for hominid food procurement strategies. Journal of Human Evolution 32:257-288.

Lewis, M. 2017 An uneasy history: carnivoran and hominin relationships over the last four million years. pp. 29-61 in M. Petraglia, N. Boivin, & R. Crassard (eds.),

From Colonisation to Globalisation: Species Movements in Human History. Papers from the Fyssen Foundation Conference, Paris, October, 2013. Cambridge University Press, Cambridge, UK.

Lewis, M. E., and L. Werdelin. 2007. Patterns of change in the Plio-Pleistocene carnivorans of eastern Africa: implications for hominin evolution; pp. 77-105 in R. Bobe, Z. Alemseged, and A. K. Behrensmeyer (eds.), Hominin Environments in the East African Pliocene: An Assessment of the Faunal Evidence. Springer, Dordrecht, The Netherlands.

Lewis, M. E., and M. R. Lague. 2010. Interpreting sabre-tooth cat (Carnivora, Felidae, Machairodontinae) postcranial morphology in light of scaling patterns in felids; pp. 411-465 in A. Goswami and A. Friscia (eds.), Carnivoran Evolution: New Views on Phylogeny, Form and Function. Cambridge University Press, Cambridge, UK.

Louys, J. 2014. The large terrestrial carnivore guild in Quaternary Southeast Asia. Quaternary Science Reviews 96:86-97.

Lund, P. W. 1842. Blik paa Brasiliens Dyreverden för sidste jordomvæltning. Fjerde afhandling: Fortsættelse af pattedyrene. Det Kongelige Danske videnskabernes selskabs naturvidenskabelige og mathematiske afhandlinger 9:137-208.

Mader, R., J.-J. Verlaan, and D. Buskila. 2013. Diffuse idiopathic skeletal hyperostosis: clinical features and pathogenic mechanisms. Nature Reviews Rheumatology 9:741-750.

Marcus, L. F., and R. Berger. 1984. The significance of radiocarbon dates for Rancho La Brea; pp. 159-183, in P. S. Martin and R. G. Klein (eds.), Quaternary Extinctions: A Prehistoric Revolution. University of Arizona Press, Tucson, Arizona.

Marinelli, W. 1938. Der Schädel von Smilodon, nach der Funktion des Kieferapparates analysiert. Palaeobiologica 6:246-272.

Martin, L. D. 1980. Functional morphology and the evolution of cats. Transactions of the Nebraska Academy of Sciences 8:141-154.

Martin, L. D., J. P. Babiarz, and V. L. Naples. 2011. The osteology of a cookie-cutter cat, *Xenosmilus hodsonae*; pp. 43-98 in V. L. Naples, L. D. Martin, and J. P. Babiarz (eds.), The Other Saber-tooths: Scimitar-tooth Cats of the Western Hemisphere. Johns Hopkins University, Baltimore, Maryland.

Martin, L. D., J. P. Babiarz, V. L. Naples, and J. Hearst. 2000. Three ways to be a saber-toothed cat. Naturwissenschaften 87:41-44.

Matthew, W. D. 1910. The phylogeny of the Felidae. Bulletin of the American Museum of Natural History 28:289-316.

Meachen-Samuels, J. A. 2010. Comparative scaling of humeral cross-sections of felids and canids using radiographic images. Journal of Mammalian Evolution 17:193-209.

Meachen-Samuels, J. A. 2012. Morphological convergence of the prey-killing arsenal of sabertooth predators. Paleobiology 38:1-14.

Meachen-Samuels, J., and W. J. Binder. 2010. Sexual dimorphism and ontogenetic growth in the American lion and sabertoothed cat from Rancho La Brea. Journal of Zoology 280:271-279.

Meachen-Samuels, J. A., and B. Van Valkenburgh. 2010. Radiographs reveal exceptional forelimb strength in the sabertooth cat, *Smilodon fatalis*. PLoS One 5:e11412.

Meachen, J. A., F. O'Keefe, and R. Sadleir. 2014. Evolution in the sabre-tooth cat, *Smilodon fatalis*, in response to Pleistocene climate change. Journal of Evolutionary Biology 27:714-723.

Meloro, C., S. Elton, J. Louys, L. C. Bishop, and P. Ditchfield. 2013. Cats in the forest: predicting habitat adaptations from humerus morphometry in extant and fossil Felidae (Carnivora). Paleobiology 39:323-344.

Méndez-Alzola, R. 1941. El *Smilodon bonaërensis* (Muñiz): estudio osteológico y osteométrico del gran tigre fósil de La Pampa comparado con otros félidos actuales y fósiles. Annales del Museo Argentino de Ciencias Naturales 40:135-252.

Merriam, J. C., and C. Stock. 1932. The Felidae of Rancho La Brea. Carnegie Institution of Washington Publication No. 422:1-231.

Moodie, R. L. 1923. Paleopathology: An Introduction to the Study of Ancient Evidences of Disease. University of Illinois Press, Urbana, Illinois, 567 pp.

Mukherjee, S., and M. R. Heithaus. 2013. Dangerous prey and daring predators: a review. Biological Reviews 88:550-563.

Peters, C. R., R. J. Blumenschine, R. L. Hay, D. A. Livingstone, C. W. Marean, T. Harrison, M. Armour-Chelu, P. Andrews, R. L. Bernor, R. Bonnefille, and L. Werdelin. 2008. Paleoecology of the Serengeti-Mara ecosystem; pp. 47-94 in A. R. E. Sinclair, C. Packer, S. A. R. Mduma, and J. M. Fryxell (eds.), Serengeti III. Human Impacts on Ecosystem Dynamics. University of Chicago Press, Chicago, Illinois.

Prevosti, F. J., and F. M. Martin. 2013. Paleoecology of the mammalian predator guild of Southern Patagonia during the latest Pleistocene: Ecomorphology, stable isotopes, and taphonomy. Quaternary International 305:74-84.

Prevosti, F. J., and L. Pomi. 2007. Revisión sistemática y antigüedad de *Smilodontidion riggii* (Carnivora, Felidae, Machairodontinae). Revista del Museo Argentino de Ciencias Naturales nueva serie 9:67-77.

Prevosti, F. J., and L. Soibelzon. 2012. Evolution of the South American carnivores (Mammalia, Carnivora): a paleontological perspective; pp. 102-123 in B. D. Patterson and L. P. Costa (eds.), Bones, Clones, and Biomes: The History and Geography of Recent Neotropical Mammals. University of Chicago Press, Chicago, Illinois.

Prevosti, F. J., and S. F. Vizcaíno. 2006. Paleoecology of the large carnivore guild from the Late Pleistocene of Argentina. Acta Palaeontologica Polonica 51:407-422.

Prevosti, F. J., F. M. Martin, and M. Massone. 2013. First Record of *Smilodon* Lund (Felidae, Machairodontinae) in Tierra Del Fuego Island (Chile). Ameghiniana 50:605-610.

Rawn-Schatzinger, V. 1992. The scimitar cat *Homotherium serum* Cope: osteology, functional morphology, and predatory behavior. Illinois State Museum Reports of Investigations 47:1-80.

Rincón, A. D., F. J. Prevosti, and G. E. Parra. 2011. New saber-toothed cat records (Felidae, Machairodontinae) for the Pleistocene of Venezuela, and the Great American Biotic Interchange. Journal of Vertebrate Paleontology 31:468-478.

Ripple, W. J., and B. Van Valkenburgh. 2010. Linking top-down forces to the Pleistocene megafaunal extinctions. BioScience 60:516-526.

Rodrigues, P. H., F. J. Prevosti, J. Ferigolo, and A. M. Ribeiro. 2004. Novos materiais de Carnivora para o Pleistoceno do estado do Rio Grande do Sul, Brasil. Revista Brasileira de Paleontologia 7:77-86.

Rodrigues, S., L. S. Avilla, L. H. Soibelzon, and C. Bernardes. 2014. Late Pleistocene carnivores (Carnivora, Mammalia) from a cave sedimentary deposit in northern Brazil. Anais da Academia Brasileira de Ciências 86:1641-1655.

Rothschild, B. M., and L. D. Martin. 2011. Pathology in saber-tooth cats; pp. 35-41 in V. L. Naples, L. D. Martin, and J. P. Babiarz (eds.), The Other Saber-Tooths: Scimitar-Tooth Cats of the Western Hemisphere. Johns Hopkins University, Baltimore, Maryland.

Sardella, R. 1998. The Plio-Pleistocene Old World dirk-toothed cat *Megantereon* ex gr. *cultridens* (Mammalia, Felidae, Machairodontinae), with comments on taxonomy, origin and evolution. Neues Jahrbuch für Geologie und Paläontologie, Abhandlungen 207:1-36.

Schaub, S. 1925. Ueber die osteologie von *Machaerodus cultridens* Cuvier. Eclogae Geologicae Helvetiae 19:255-266.

Schellhorn, R., and M. Sanmugaraja. 2015. Habitat adaptations in the felid forearm. Paläontologische Zeitschrift 89:261-269.

Schubert, B. W., R. C. Hulbert, B. J. MacFadden, M. Searle, and S. Searle. 2010. Giant short-faced bears (*Arctodus simus*) in Pleistocene Florida USA, a substantial range extension. Journal of Paleontology 84:79-87.

Seiffert, E. R., E. L. Simons, and J. G. Fleagle. 2000. Anthropoid humeri from the Late Eocene of Egypt. Proceedings of the National Academy of Sciences of the United States of America 97:10062-10067.

Seymour, K. L. 2015. Perusing Talara: Overview of the Late Pleistocene fossils from the tar seeps of Peru; 97-109 pp. in J. M. Harris (ed.), La Brea and Beyond: The Paleontology of Asphalt-Preserved Biotas, Natural History Museum of Los Angeles County, Los Angeles, California.

Shaw, C. A., and A. E. Tejada-Flores. 1985. Biomechanical implications of the variation in *Smilodon* ectocuneiforms from Rancho La Brea. Natural History Museum of Los Angeles County, Contributions in Science 359:1-8.

Shermis, S. 1983. Healed massive pelvic fracture in a *Smilodon* from Rancho La Brea, California. PaleoBios 1:121-126.

Simpson, G. G. 1941. The function of saber-like canines in carnivorous mammals. American Museum Novitates 1130:1-12.

Sinclair, A. R. E. 1985. Does interspecific competition or predation shape the African ungulate community? Journal of Animal Ecology 54:899-918.

Slater, G. J., and B. Van Valkenburgh. 2008. Long in the tooth: evolution of sabertooth cat cranial shape. Paleobiology 34:403-419.

Slaughter, B. H. 1963. Some observations concerning the genus *Smilodon*, with special reference to *Smilodon fatalis*. Texas Journal of Science 15:68-81.

Sommer, R. S., and N. Benecke. 2006. Late Pleistocene and Holocene development of the felid fauna (Felidae) of Europe: a review. Journal of Zoology 269:7-19.

Sorkin, B. 2008. A biomechanical constraint on body mass in terrestrial mammalian predators. Lethaia 41:333-347.

Spencer, D. G., R. D. Sturrock, and W. W. Buchanan. 1980. Ankylosing spondylitis: Yesterday and today. Medical History 24:60-69.

Spencer, L. M., B. Van Valkenburgh, and J. M. Harris. 2003. Taphonomic analysis of large mammals recovered from the Pleistocene Rancho La Brea tar seeps. Paleobiology 29:561-575.

Torregrosa, V., M. Petrucci, J. A. Pérez-Claros, and P. Palmqvist. 2010. Nasal aperture area and body mass in felids: Ecophysiological implications and paleobiological inferences. Géobios 43:653-661.

Van Valkenburgh, B. 2007. Deja vu: the evolution of feeding morphologies in the Carnivora. Integrative and Comparative Biology 47:147-163.

Van Valkenburgh, B. 2009. Costs of carnivory: tooth fracture in Pleistocene and Recent carnivorans. Biological Journal of the Linnean Society 96:68-81.

Van Valkenburgh, B., F. Grady, and B. Kurtén. 1990. The Plio-Pleistocene cheetah-like *Miracinonyx inexpectatus* of North America. Journal of Vertebrate Paleontology 10:434-454.

Van Valkenburgh, B., M. W. Hayward, W. J. Ripple, C. Meloro, and V. L. Roth. 2016. The impact of large terrestrial carnivores on Pleistocene ecosystems. Proceedings of the National Academy of Sciences of the United States of America 113:862-867.

Van Valkenburgh, B., and F. Hertel. 1998. The decline of North American predators during the Late Pleistocene; pp. 357-374 in J. J. Saunders, B. W. Styles, and G. F. Baryshnikov (eds.), Quaternary Paleozoology in the Northern Hemisphere. Illinois State Museum Scientific Papers, Springfield, Illinois.

Van Valkenburgh, B., and T. Sacco. 2002. Sexual dimorphism, social behavior, and intrasexual competition in large Pleistocene carnivorans. Journal of Vertebrate Paleontology 22:164-169.

Vymazalová, K., L. Vargova, and M. Joukal. 2015. Variability of the pronator teres muscle and its clinical significance. Romanian Journal of Morphology and Embryology 56:1127-1135.

Wallace, S. C., and R. C. Hulbert, Jr. 2013. A new machairodont from the Palmetto Fauna (Early Pliocene) of Floridae, with comments on the origin of Smilodontini (Mammalia, Carnivora, Felidae). PLoS One 8:e56173.

Werdelin, L., and M. E. Lewis. 2001. A revision of the genus *Dinofelis* (Mammalia, Felidae). Zoological Journal of the Linnean Society 132:147-258.

Werdelin, L., and M. E. Lewis. 2005. Plio-Pleistocene Carnivora of eastern Africa: species richness and turnover patterns. Zoological Journal of the Linnean Society 144:121-144.

West, P. M., H. MacCormick, G. Hopcraft, K. Whitman, M. Ericson, M. Hordinsky, and C. Packer. 2006. Wounding, mortality and mane morphology in African lions, *Panthera leo*. Animal Behaviour 71:609-619.

Wroe, S. 2008. How to build a mammalian super-predator. Zoology 111:196-203.

12 *Smilodon* Paleopathology: A Summary of Research at Rancho La Brea

CHRISTOPHER A. SHAW AND C. S. WARE

Rancho La Brea and Collections

That Rancho La Brea is unique in a variety of ways is well documented (Akersten et al., 1983; Shaw and Quinn, 1986; Stock, 1992; Shaw, 2007). It is *the* classic carnivore trap, where 90% of the remains of large mammals (> 18 kg) represent carnivorans that were initially attracted to the site by large herbivores which had became entrapped and perished in the asphalt seeps. Alive or recently deceased, mired animals baited the petroleum trap, which ensnared bears, wolves, coyotes, felines, and sabertooth cats looking for an easy feeding opportunity. Equally susceptible to entrapment were large meat-eating birds: storks, teratorns, vultures, owls, hawks, and eagles. Over a period of at least 35,000 years, the asphalt seeps cyclically spun their deadly webs, perhaps not continually but certainly with a period-icity that resulted in tens of thousands of animals being entrapped and preserved. In fact, only one episode of entrapment involving ten large mammals and birds per decade can readily account for all the fossils, including plants and animals (conservatively estimated at over 5 million specimens), recovered from Rancho La Brea in the past 125 years of excavations. The asphalt that ensnared them also served to beautifully preserve the plant and animal parts by saturating the dead tissues, leaving them impervious to alteration by ground water and resulting in most fossils not only being physically but also chemically unchanged. Sample sizes of some species, rarely found at other fossil localities, are enormous (e.g., the dire wolf, *Canis dirus*, the most common large mammal, represented by over 3,500 individual animals, and the golden eagle, *Aquila chrysaetos*, the most common large bird, with over 1,000 individuals recovered). Furthermore, the number of identified species (over 650) at Rancho La Brea makes it the largest and most diverse late Pleistocene terrestrial biota and most intensely studied terrestrial paleoecosystem worldwide known to date.

At Rancho La Brea, the sabertooth cat, *Smilodon fatalis* (Leidy, 1868), is the second most common large mammal recovered from these asphalt deposits. Excavations conducted by the Southern California Academy of Sciences between 1907 and 1910 and those by the Natural History Museum of Los Angeles County between 1913 and 1930 exhumed over one million vertebrate fossils (called the Hancock Collection). These comprise the bulk of the collec-tions housed at the George C. Page Museum of La Brea Discoveries (recently renamed The La Brea Tar Pits and Museum) in Los Angeles (Shaw et al., 2008). In the 15 years following termination of the excava-tions in 1915, the Hancock Collection was cleaned and catalogued, and sorted into wooden boxes for long-term storage while various research projects were initiated. Although over 100 distinct asphalt deposits were excavated, Rancho La Brea is considered one locality, and the collections have been organized by (1) taxon, (2) skeletal element, and (3) site (termed "pit"). Minimum numbers of individuals (MNI) of *Smilodon* from Rancho La Brea have been published: Stock (1929) estimated 1,500 individuals, while Marcus (1960) recorded 1,029 and Miller (1968) counted over 2,100 and estimated an additional 200 in the collections at the University of California Museum of Paleontology.

In the intervening years, many more specimens of *Smilodon* have been collected during excavation and salvage programs, which has led to speculation that perhaps more than 3,000 individuals are now represented, but this can only be resolved by taking a new census. This requires examining not only the large and small collections acquired through excavation and salvage programs but also counting materials that have been on loan or exchanged to other institutions. Skeletal elements of *S. fatalis* in the Hancock Collection number approximately 166,000; over 5,100 of these (which is less than 1% of the total sample) exhibit some sort of paleopathologic condition (often with multiple abnormalities observed on a single specimen) and are maintained as a separate collection at the museum (Shaw, 2001; Shaw et al., 2008). The MNI of injured animals may never be fully realized, given the circumstance of entrapment, burial, disarticulation, trampling, and mixing of skeletal elements at Rancho La Brea. In addition, an injury to one bone can affect the proper functioning of other bones, causing pathological conditions that can only be studied if the skeleton is found articulated or associated, a rare phenomenon at Rancho La Brea. Additional excavations and salvages since 1969 have added to the sample size of *Smilodon* skeletal material along with its attendant pathological elements, which remain to be evaluated, characterized, and studied thoroughly.

Paleopathological Research

Roy Lee Moodie (1880-1934) is considered a pioneer of American paleopathology (McNassor, 2001b; Cook, 2012; Waldron, 2015) and paleohistology (Schultz, 2012). He was the first to describe pathological skeletal elements of *Smilodon fatalis* (in addition to

several other extinct taxa) recovered from Rancho La Brea. Between 1918 and 1925, he published several review articles (Moodie, 1918a, 1918b, 1918c) and two books (Moodie, 1923a, 1923b) regarding the antiquity of pathologies found in vertebrates through geologic time. His prime pathologic examples from the Pleistocene Epoch involved a small sample of dire wolf (*Canis dirus*) and sabertooth cat (*Smilodon fatalis*) remains from Rancho La Brea that had been presented to him by Elmer S. Riggs at the Field Museum in Chicago. During this period, Moodie published descriptions of half a dozen Rancho La Brea skeletal elements that illustrated infections, necrosis, spondylitis, hyperostosis, and trauma. Another Rancho La Brea specimen, also provided by Riggs, led Moodie (1922) to be the first to describe the brain morphology in *S. fatalis*.

In 1925-1926 Moodie began working with the much larger Hancock Collection at the Los Angeles Museum of History, Science, and Art (the present-day Natural History Museum of Los Angeles County), sorting through thousands of skeletal elements and segregating pathological specimens for further study (Young and Cooper, 1926). By 1928 he was living in Los Angeles and accessing the collections on a regular basis. Henry Wilde, an employee of the Los Angeles Museum, organized and set aside a modest number of pathological skeletal elements and thus began a separate collection of like-affected specimens (Heald, 1986). From 1926 until his death in 1934, Moodie published a series of short, technical papers that document specific pathologies exhibited by several taxa found in the Rancho La Brea collections. For *Smilodon*, he published on fractures and fusions in vertebrae (Moodie, 1926, 1927, 1930b); ossification of rib costal cartilages (Moodie, 1926); dental caries, pyorrhea, osteomyelitis, impactions, and abscesses in upper and lower jaws (Moodie, 1929a, 1929b); facial asymmetry by loss of upper canines (Moodie, 1929a, 1929c); the first report of apical closure of tooth roots, not only in *Smilodon* but in all other carnivoran taxa as well (Moodie, 1930a); hypertrophy and infection caused by violent trauma (Moodie, 1930b); luxations and subluxations (Moodie, 1930c) (Fig. 12.1); and, finally, a study of sacralization of the last lumbar vertebra with consequential fusion to the sacrum (Moodie, 1930d). Just days prior to his death, Moodie

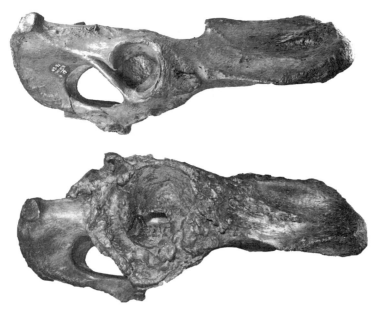

Figure 12.1. Right innominates of *Smilodon fatalis* from Rancho La Brea in lateral view. Top, normal specimen, LACMHC K2572; bottom, specimen exhibiting massive chronic infection and destruction of the joint capsule resulting from subluxation, LACMHC 131. Moodie (1930c) stated that this pelvis is "the most strikingly pathological specimen in the collection" that he had seen from Rancho La Brea. Photographs by C. A. Shaw.

completed a monograph on all of his observations of *Smilodon* paleopathology, a monograph that unfortunately, remains unpublished (McNassor, 2001b; Cook, 2012).

In the late 1920s, physicians Frank B. Young and Albert L. Cooper acquired pathologic specimens of *Smilodon fatalis* from Rancho La Brea that were incorporated into an exhibit presented in Dallas, Texas, at the 1926 meeting of the American Medical Association, and in 1927 at a meeting of the Radiological Society of North America in Milwaukee, Wisconsin. The two publications that resulted from these exhibitions (Young and Cooper, 1926, 1927) outline the history of pathologies through geologic time, similar to the review-style format implemented by Moodie in his publications between 1918 and 1923. The *Smilodon* skeletal elements used in the exhibitions included three specimens of fused cervical, thoracic, and lumbar vertebrae showing osteoarthritic changes consistent with spondylitis, which Young and Cooper (1927) believed was suggestive of "tuberculous spondylitis" or Pott's disease in humans. They also exhibited several skeletal elements of *Smilodon* that had fractures with infection, osteoarthritis, and osteomyelitis. Conceding that climate change with attendant changes in forage and overspecialization in some species could

have led to extinction in some groups, Young and Cooper (1926, 1927) contended that "fatal epidemic diseases" were responsible for the demise of the weakened and remnant populations. They concluded, therefore, that disease was not only equally responsible for extinctions of plants and animals during the Pleistocene, but perhaps throughout the history of life on Earth.

From 1913 to 1950, Chester Stock was the central figure in research on fossils recovered from Rancho La Brea (McNassor, 2001a). Discussing *Smilodon* in his seminal publication on this locality, Stock (1930) acknowledged the (then recent) studies by Moodie stating that, according to Moodie, most pathological conditions in the Rancho La Brea collections could be attributed to injury with subsequent infection. Stock (1930) mentioned the prevalence of healed fractures, luxations, arthritis, and "disturbances in normal bone development" in *Smilodon*. He includes a figure illustrating a series of four lumbar vertebrae fused together with an ossification along the dorsolateral neural arches (where the origin and insertion of the *M. multifidus lumborum* muscle series would attach) and compares and contrasts this with an unaffected lumbar series. This publication went through six subsequent editions and numerous printings virtually unchanged for over 60 years.

The seventh edition (Stock, 1992, which is a revision of the sixth edition by John M. Harris) updated and added new information throughout, including in the discussion of *Smilodon* pathologies and using the same illustrations. In their classic morphological study of felid skeletal remains from Rancho La Brea, Merriam and Stock (1932) mentioned, in passing, diseases resulting from injuries, but except for fusions between skeletal elements in the manus and pes, they devote little space or effort to describing pathologic *Smilodon* skeletal elements.

After Moodie's death, there was a long hiatus of research on the paleopathology collections from Rancho La Brea. From 1962 to 1975, these collections went through a period of cleaning, expansion, and reorganization (Heald, 1986) at the Natural History Museum of Los Angeles County. The work was done by Gretchen Sibley, who was employed in the museum's department of education. Her interest focused on osteoarthritic changes in bone, fractures with attendant infection, and bone tumors. She coauthored only one publication on paleopathology with a curator of vertebrate paleontology during this time (Macdonald and Sibley, 1969). This popular publication is illustrated with two impressive *Smilodon* specimens from the Rancho La Brea collection: (1) the fusion of three lumbar vertebrae exhibiting a massive hyperostosis stimulated by infection and (2) a fractured and healed pelvis, also showing significant hypertrophic bone as a result of major trauma and infection.

Once construction of the George C. Page Museum of La Brea Discoveries was completed in 1977, the entire Rancho La Brea fossil collection was transferred from the basement of the Natural History Museum of Los Angeles County and reorganized from the old wooden-box system into new, custom-designed steel-and-plastic-drawer storage facilities at the new museum. During the following two decades of curatorial activity, the Hancock Collection was intensely scrutinized for specimens exhibiting any possible pathologic condition; those skeletal elements that varied from "normal" were separated into the Pathology Collection for further analysis. This tripled the size of the original sample of pathologic specimens (Heald, 1986). Mary L. Romig, Mary J. Odano, and Christopher A. Shaw facilitated

the processing of this subset of the Hancock Collection and retired surgeon Fred P. Heald assessed the pathologies. As a team, they were able to assemble the present Pathology Collection, which is composed of over 5,100 bones of sabertooth cat (*Smilodon fatalis*), about 3,200 skeletal elements of dire wolf (*Canis dirus*), and around 1,000 pathologic items that represent several mammalian taxa, including ground sloths, coyotes, lion-like cats, rabbits, rodents, horses, camels, and bison. As the materials became more easily accessible, there was a resurgence of interest in the Rancho La Brea paleopathological collection at the museum.

Stewart Shermis, a paleopathologist in the Department of Anthropology at California State University, Long Beach, published three papers as senior author (and one other as coauthor) on some of the more spectacular pathologic conditions found in *Smilodon*. Shermis (1983) described a pelvic trauma (originally figured in Macdonald and Sibley, 1969) that involved many chronic diseases, including "dysraphism (an incomplete closure of a neural tube), muscle and bone asymmetry, dysplastic hip joint, osteoarthritis, and possible scoliosis and muscle atrophy." He makes an arbitrary association of a right femur that, along with the affected right side of the (nonassociated) pelvis, indicates the presence of chronic septic degenerative osteoarthritis of the joint capsule, which affected the surrounding musculature. Sometime after this publication by Shermis (1983), it was determined that the femur was from a different individual, and the femur was confidently matched to the pelvis studied by Moodie (1930c). At the time of Moodie's study, this particular femur was in the collection but went unrecognized as a match to any pelvis. Shermis (1983) stated that the severity of the conditions found in this one pelvis were singular evidence that *Smilodon fatalis* was a "pride-dwelling animal" or otherwise living in social groups.

Next, Shermis (1985a) examined a variety of oral diseases in *Smilodon*. He found that marginal periosteitis indicated chronic gingivitis from the incisors to the carnassials. In addition, Shermis (1985a) identified necrosis, osteomyelitis, and septic responses in upper and lower jaws, a pseudoarthrosis in one lower jaw, and healed fractures in bone and teeth, all of which he believed resulted from traumatic destruction of hard and soft tissues through aggressive use in procuring food. The same year, Shermis (1985b) published his analysis of seven *Smilodon* skulls that exhibited loss of maxillary canines due to avulsion fractures of the saber with attendant infection and resorption of alveoli resulting in facial asymmetry. Two of the skulls used in this study had previously been examined and published by Moodie (1929a, 1929c), who felt that the specific cause of the canine fracture was indeterminate. Shermis (1985b) asserted that the upper canines were used on both hard and soft tissues, and that occasional fractures could be expected as a hazard of predation or a result of intraspecific competition.

Finally, Shermis collaborated with several arthropathy specialists in analyzing patterns of paravertebral ossifications in the thoracic and lumber vertebrae of *Smilodon* (Bjorkengren et al., 1987). They were able to place each of 37 spinal fusions into one of three categories: (1) trauma with ensuing infection (17 specimens), (2) diffuse idiopathic skeletal hyperostosis (=DISH, 9 specimens), and (3) ankylosing spondylitis (11 specimens). In addition, a combination of these three patterns was found in another nine specimens. Although spondylitis (*sensu lato*) had been recognized by earlier investigators (Young and Cooper, 1927; Moodie, 1927, 1930b), this was the first publication to report DISH in *Smilodon*. A manuscript written by Shermis on the subject of "intraspecific combat" (that unfortunately was rejected for publication just prior to his untimely death) is cited multiple times in his other publications on *Smilodon* (Shermis, 1983, 1985a, 1985b). In it, he indicates that the oval perforation located in the area of the nasofrontal suture of a skull was produced by the puncture of a *Smilodon* upper canine, a feature noted and previously published by Miller (1983).

By 1993, the *Smilodon* skulls in the Hancock Collection had been inventoried twice in order to make sure that no theft of skulls had occurred from one decade to the next. During the second inventory, Richard F. Wheeler, an employee of the Museum at Rancho La Brea, began recording morphological data from the skulls into an electronic database. He noticed that peculiar depressions adjacent to the sagittal and occipital crests occurred on the parietals

(which he termed "parietal pits"). He observed that these depressions were variable in size and depth but common on over 50% of the more than 1,000 skulls examined (Wheeler, 1993a, 1993b), but he did not explain these phenomena. Fred P. Heald (pers. comm., 1994) suggested that the parietal depressions represent areas of chronic reinjury (multiple small periosteal tears or perhaps a subchondral lesion) in a small area of origin of the *M. temporalis*. Duckler (1997) postulated that these depressions in the parietals were most likely due to heightened local mechanical stresses in the fibers of the temporalis muscle. Rothschild and Martin (2006) claimed the cause of this condition is due to mechanical stress resulting in local microfractures and occasional distraction of the tendons at the muscle origins. In another published study on stress in animal populations using Harris lines in limb bones (Duckler and Van Valkenburgh, 1998), no evidence was found that indictes a decline in the health of *Smilodon fatalis* (or other megafaunal species recovered from Rancho La Brea) prior to the Pleistocene extinctions.

There is a plethora of publications authored by rheumatologist Bruce M. Rothschild and the late Larry D. Martin on bone pathology and paleopathology, but only a few references to *Smilodon*. Several review articles and two review-style books include mentions of DISH (Rothschild, 1987a; Rothschild and Martin, 1993, 2006, 2011), osteophyte formation in the patellar region (Rothschild, 1987b; Rothschild and Martin, 2011), and spondyloarthropathy (Rothschild and Martin, 1993, 2006, 2011). Rothschild and Martin (1993) felt that the oval perforation in the skull described by Miller (1983) is evidence for the "stabbing hypothesis" predation model, and even though they state there is evidence of healing, there is no suggestion of bone healing around the perforation (Miller, 1983). The origin of the parietal depressions published by Duckler (1997) as well as his study of Harris lines (Duckler and Van Valkenburgh, 1998) is also noted by Rothschild and Martin (2006, 2011).

Although Rothschild and Martin (2006) stated that short tails in the domestic Manx cat are due to a potentially "lethal mutation" that results in spina bifida and postulate that the presence of spina bifida in *Smilodon* might indicate a similar genetic mechanism for their abbreviated tail, the authors disagree with this analogy. Modern domestic animal breeds such as the Manx (*Felis silvestris domesticus*) are not viable analogs for discussing the short tail and presence of spina bifida in *S. fatalis* because (1) domestic cats and dogs are under the influence of human bioengineering, (2) not all Manx cats have short tails, (3) the Manx is not the only domestic cat with a short tail, (4) the incidence of spina bifida is low in both *S. fatalis* and in the Manx cat when a large sample is examined, (5) individual variation is a key factor in spina bifida and (6) other wild felids with short tails such as the Canada lynx (*Lynx canadensis*) and the bobcat (*L. rufus*) have a low incidence of spina bifida.

It is difficult to extrapolate a cause for a disease or pathology based on a small sample size, and *Smilodon fatalis* is not represented in large numbers at sites other than Rancho La Brea. Therefore, when a large sample is available from Rancho La Brea, and the incidence of pathologies is less than 1%, as stated above, the roles of individual variation and familial lineages become a more prominent causality than the 'short tail' hypothesis, which can lead to inaccurate citations and misleading information for future researchers. It is also crucial that all of the skeletal elements (such as the sacra of *S. fatalis*) be examined as an entire unit in a comparative osteological examination procedure. This is critical to establishing the relative incidence of a pathology within the context of the entire sample available for that bone. The examination of one or two elements at a time or the lack of examination of the entire elemental collection does not show the individual variation, the degree of pathology, nor the comparative statistical data necessary for a conclusive diagnosis.

Smilodon is a very popular fossil carnivore and has been discussed in multiple books, in magazines, on television, and in other media. The scientific facts are often lost, changed, or embellished, and this is damaging to the accuracy of scientific research. A good example is the publication of Mestel (1993), in which she presents some of the more spectacular items in the Rancho La Brea collection as illustrative of all the hallmarks that are *Smilodon* (stout and robust build in a big cat, with enormous teeth for dispatching large prey, a variety of traumatic injuries to neck, pelvis, and limbs, and a skull with professed

evidence of "intraspecific combat"), all with a narrative to inform and entertain the public. It is unfortunate that Rothschild and Martin (2011) used this article extensively as a platform to discuss *Smilodon* paleopathology and behavior.

Therefore, the existing literature on pathology may also contain inaccurate data, or data that is outdated or has not been amended in publication. Interviews with scientists in popular magazines (such as *Discover*) may not reflect correct scientific analysis or the accurate statements of the person being interviewed. These are not peer-reviewed articles but rather are edited by magazine employees, which may result in comments that are taken out of context and/or misinformation being allowed to be printed.

Occasionally, controversy arises even among professionals who attempt to differentiate between normal variables and actual pathology. Rothschild and Martin (2011) addressed a valid concern about how individual variation might be confused with paleopathology by referencing a study of biomechanics and variation in tarsals of *Smilodon* by Shaw and Tejada-Flores (1985). They erroneously diagnose these elements as pathologic when, in actuality, they exemplify normal individual variation.

Skeletal imperfections are frequently open to varying interpretations, such as the observation by Rothschild and Martin (2011:fig. 3.1B and caption) of an "osteoarthritic osteophyte" on the proximal tibia at the knee joint of a *Smilodon* skeleton from Rancho La Brea mounted by L. C. Bessom and C. A. Shaw in 1976, which is presently on exhibit at the museum. This specimen does not exhibit an osteophytic anomaly but is an example of a normal skeletal variant in this fossil felid.

From 1980 until his death in 2000, Fred P. Heald spent countless hours analyzing every specimen in the Pathology Collection at the museum, attempting to diagnose and understand the etiology of each injury and/or disease that was presented to him. In collaboration with Christopher A. Shaw and Mary L. Romig, Heald developed a scheme in which he divided the pathologies into six categories: (1) developmental anomalies, (2) chronic reinjuries (that is, evidence of "pulled or stressed muscles"), (3) dental disease, (4) traumatic injuries, (5) arthritis, and (6) punctures and/or infections that indicate possible or probable bite wounds (Heald, 1989; Heald and Shaw, 1991). What he found most interesting were (1) the severe traumatic injuries in *Smilodon fatalis* and *Canis dirus* that he felt demonstrated sociality in these extinct carnivorans, (2) the chronic reinjuries to muscles of the limbs (along with trauma) that may be used to reconstruct the posture or stance of a predator during acts of prey capture, and (3) any evidence, including punctures and/or infection, that could indicate bite wounds. In his sole descriptive publication, Heald (1986) used healed traumatic injuries and degenerative osteoarthritis in *Canis dirus* and a healed bite wound in *Smilodon* to promote research on the Pathology Collection at the Page Museum and to demonstrate survivability in seriously wounded animals prior to being entrapped and preserved at Rancho La Brea. Fred Heald's legacy is in the analysis and diagnoses he made on the thousands of pathological specimens he examined in the Pathology Collection at the Museum. Shaw worked closely with Heald throughout those two decades, beginning as his facilitator in the organization of pathological specimens and diagnostic information, and later as his protégé. This is reflected in the earlier publication of their organizational work on the Pathology Collection (Shaw, 1989) and later "musings" regarding social behavior in *Smilodon* (Heald and Shaw, 1991; Shaw et al., 1991; Shaw, 2001a, 2001b). The first recognized case of congenital scoliosis in an extinct mammal (*Smilodon fatalis*) was published by Shaw et al. (2008), which culminated in a study based on foundation work done before Heald died and Romig retired. Shaw and Carrie Howard have analyzed causes for facial asymmetry in *Smilodon fatalis* compared to that in *Canis dirus* in order to interpret and contrast hunting methods in these extinct species (Shaw and Howard, 2015).

Most recently, Balisi et al. (2013) and Brown et al. (2017) have used Heald's diagnostic information about injuries to infer differences in hunting mode and behavior in *Smilodon fatalis* (an ambush predator) and *Canis dirus* (a pursuit predator). In the most recent extensive (and beautifully illustrated) treatment on sabertooth animals, Mauricio Antón (2013) includes a short section on paleopathology in which he discusses and illustrates muscle injuries to the

M. deltoideus on the humerus (first briefly addressed by Turner and Antón, 1997) and *M. multifidus* series along the spine of *Smilodon*, as well as mentioning traumatic injuries to the chest, limbs, and feet. The skull with the oval perforation (Miller, 1983) is also illustrated, and Antón mentions evidence (Shaw, 1989) of a bite wound in a scapula caused by another individual of *Smilodon*. Following his discussion of paleopathology, Antón (2013) incorporated paleopathological evidence to support theoretical reconstructions of hunting techniques in *Smilodon* and other sabertooth carnivores, as many of our paleontologist and paleoartist colleagues have done in the past century.

In 2006, paleopathologist C. S. Ware began her scientific research on *Smilodon fatalis* pathologies in the museum collection at Rancho La Brea. Ware's examination of this extensive pathological collection of skulls and postcranial elements identified the presence of the following conditions: (1) periodontal disease, (2) tooth breakage and loss, (3) bone fractures, (4) ankylosing spondylosis, (5) blunt force trauma, (6) bony exostoses, (7) degenerative joint disease, (8) infection, (9) fused skeletal elements, (10) osteomyelitis, (11) pseudoarthrosis, (12) hyperostosis, (13) eburnation, (14) congenital defects, and (15) reinjury. In addition, she compared these pathologies in remains of the extinct dire wolf (*Canis dirus*) to that of the extant gray wolf (*Canis lupus*). Most recently, Ware focused on an anomaly shared by both species: lumbosacral transitional vertebrae (LSTV). This is a congenital defect in mammals where two distinct types of fusion (termed sacralization and lumbarization) are present and which can be observed during gross examination. Both types of fusion are the result of "border shifting" of the lumbar or sacral vertebrae (Barnes, 1994) in which the affected vertebra takes on the morphological profile of the adjacent vertebra. With sacralization, the shift occurs in a forward direction (or cranial shift) in which the seventh lumbar (L-VII) vertebra takes on the appearance of the first sacral (S-I) vertebra. In lumbarization, the shift occurs in a backward direction (or caudal shift) in which the S-I vertebra takes on the characteristics of the L-VII vertebra. In this profile there is often a dorsoventral twisting anomaly at the junction of the L-VII and the

Figure 12.2. Sacrum of *Smilodon fatalis* from Rancho La Brea; LACMHC 6946, showing the LSTV characteristics of the fused L-VII vertebra (left, ventral view; right, dorsal view) as compared with the normal sacrum of *Smilodon*, LACMHC 142107. Photographs by C. S. Ware.

S-I vertebrae (Fig. 12.2). This was the first time that LSTV had been reported in *S. fatalis*. Another component of LSTV that Ware is investigating causes paralysis and eventual hind limb malfunction, a condition called cauda equina syndrome, which is often associated with LSTV in modern felids. Ware is preparing multiple manuscripts which describe this congenital anomaly in *S. fatalis* in great detail and will be used to compare and contrast this condition in other large carnivores.

Behavioral Interpretations Incorporating Paleopathology

Paleopathology is the study of disease in the fossil record with an emphasis on the evolution and effect of disease over long expanses of time and across species lines. This provides an overall view of the general health of a species, its adaptation (or lack thereof) to disease, and the impact of disease on populations. In the fossil record the primary evidence of these factors comes from the study and documentation of skeletal material. The overall profile of a large carnivore such as *Smilodon fatalis* in chronological context is achieved only by implementing a multidisciplinary scientific investigative approach. As research has shown, *S. fatalis* lived with many types of trauma, disease, and injury, and had ailments including periodontal disease, broken and missing teeth, scoliosis and other vertebral anomalies, and chronic reinjury traumas. These conditions

are a direct result of hunting behavior, feeding strategies, and predator/prey interactions, along with congenital defects, infection, and disease.

The study of paleopathology adds significant insight into the overall character of extinct species such as *Smilodon fatalis* and *Canis dirus*. Previous investigators have used Rancho La Brea and the exquisitely preserved skeletal remains of *Smilodon* recovered from this locality to infer many kinds of behavior. Paleopathological *Smilodon* fossils from Rancho La Brea have been used both directly and indirectly to bolster support for ideas on the following topics (and many more): the blood sucking hypothesis (Matthew, 1901; Merriam and Stock, 1932); the scavenging hypothesis (Bohlin, 1940, 1947); the stabbing/slashing/ripping hypothesis (Matthew, 1901, 1910; Simpson, 1941; Kurtén, 1952; Schultz et al., 1970; Martin, 1980; Miller, 1983); the shear bite hypothesis (Akersten, 1985; Turner and Antón, 1997; Antón and Galobart, 1999; McHenry et al., 2007; Rothschild and Martin, 2011; Antón, 2013; Brown, 2014); the solitary cat hypothesis (Van Valkenburgh and Sacco, 2002; McCall et al., 2003; Kiffner, 2009); and the social/gregarious cat hypothesis (Merriam and Stock, 1932; Gonyea, 1976; Shermis, 1983; Akersten, 1985; Heald and Shaw, 1991; Carbone et al., 2008; Van Valkenburgh et al., 2009). As research moves forward regarding morphological interpretation and the inferred habits of sabertooth carnivore species, the collection of *Smilodon fatalis* skeletal elements from

Rancho La Brea, whether pathologic or not, will play a central role in our understanding of these truly magnificent creatures.

It appears that, despite all of the stressors evident from the pathologic specimens of *Smilodon* from Rancho La Brea, this was a very efficient, robust, and generally healthy apex predator during the late Pleistocene. The pathological skeletal elements show the difficulties of being a large carnivore and making a dangerous living as an ambush predator feeding on the largest of prey species. These pathologies also speak to the adaptive ability of an injured animal and survivability as a population while being able to perform successfully in the large predator role for tens of thousands of years. It is the hope of the authors that there will be a time and the circumstances in which another sabertooth animal like the iconic *Smilodon fatalis* could evolve and would exist to thrill and fascinate, as well as terrify, biologists in the distant future.

REFERENCES

Akersten, W. A. 1985. Canine function in *Smilodon* (Mammalia, Felidae, Machairodontinae). Natural History Museum of Los Angeles County, Contributions in Science 356:1-22.

Akersten, W. A., C. A. Shaw, and G. T. Jefferson. 1983. Rancho La Brea: status and future. Paleobiology 9:211-217.

Antón, M. 2013. Sabertooth. Indiana University Press, Bloomington, Indiana, 243 pp.

Antón, M. and A. Galobart. 1999. Neck function and predatory behavior in the scimitar toothed cat *Homotherium latidens* (Owen). Journal of Vertebrate Paleontology 19:771-784.

Balisi, M., C. Brown, B. Van Valkenburgh, and C. Shaw. 2013. What can paleopathology tell us about hunting modes? Journal of Vertebrate Paleontology, Program and Abstracts 33:82.

Barnes, E. 1994. Developmental Defects of the Axial Skeleton in Paleopathology. University Press of Colorado, Niwot, Colorado, 360 pp.

Bjorkengren, A. G., D. J. Sartoris, S. Shermis, and D. Resnick. 1987. Patterns of paravertebral ossification in the prehistoric saber-toothed cat. American Journal of Radiology 148:779-782.

Bohlin, B. 1940. Food habits of the machaerodonts, with special regard to *Smilodon*. Bulletin of the Geological Institutions of the University of Uppsala 28:156-174.

Bohlin, B. 1947. The saber-toothed tigers once more. Bulletin of the Geological Institutions of the University of Uppsala 32:11-20.

Brown, C., M. Balisi, C. A. Shaw, and B. Van Valkenburgh. 2017. Skeletal trauma reflects hunting behaviour in extinct sabre-tooth cats and dire wolves. Nature Ecology and Evolution 1:1-7. doi 10.1038/s41559-017-0131.

Brown, J. G., 2014. Jaw function in *Smilodon fatalis*: a reevaluation of the canine shear-bite and a proposal for a new forelimb-powered Class 1 lever model. PLoS One 9:e107456. doi:10.1371/journal.pone.0107456.

Carbone, C., T. Maddox, P. J. Funston, M. G. L. Mills, G. F. Grether, and B. Van Valkenburgh. 2009. Parallels between playbacks and Pleistocene tar seeps suggest sociality in an extinct sabretooth cat, *Smilodon*. Biology Letters 5:81-85. doi:10.1098/rsbl.2008.0526.

Cook, D. C. 2012. Roy L. Moodie: pioneer paleopathologist of deep time; pp. 70-81 in J. E. Buikstra and C. A. Roberts (eds.), The Global History of Paleopathology, Pioneers and Prospects. Oxford University Press, Oxford, UK.

Duckler, G. L. 1997. Parietal depressions in skulls of the extinct saber-toothed felid *Smilodon fatalis*: evidence of mechanical strain. Journal of Vertebrate Paleontology 17:600-609.

Duckler, G. L., and B. Van Valkenburgh. 1998. Exploring the health of late Pleistocene mammals: the use of Harris lines. Journal of Vertebrate Paleontology 18:180-188.

Gonyea, W. J. 1976. Behavioral implications of saber-toothed felid morphology. Paleobiology 2:332-342.

Heald, F. P. 1986. Paleopathology at Rancho La Brea. AnthroQuest 36:6-7.

Heald, F. P. 1989. Injuries and diseases in *Smilodon californicus* Bovard, 1904 (Mammalia, Felidae) from Rancho La Brea, California. Journal of Vertebrate Paleontology 9:25A.

Heald, F. and C. Shaw. 1991. Sabertooth cats; pp. 26-27, in J. Seidensticker and S. Lumpkin (eds.), Great Cats: Majestic Creatures of the Wild. Rodale Press, Emmaus, Pennsylvania.

Kiffner, C. 2009. Coincidence or evidence: was the sabretooth cat *Smilodon* social? Biology Letters 5:561-562.

Kurtén, B. 1952. The Chinese *Hipparion* fauna. Societas Scientiarum Fennica, Commentationes Biologicae 13:1-82.

Leidy, J. 1868. Notice of some vertebrate remains from Hardin County, Texas. Proceedings of the Academy of Natural Sciences of Philadelphia 20:174-176.

Macdonald, J. R., and G. Sibley. 1969. Paleopathological ponderings or how to tell a sick saber-tooth. Los Angeles County Museum of Natural History Quarterly 8:26-30.

Marcus, L. F. 1960. A census of the abundant large Pleistocene mammals from Rancho La Brea. Los Angeles County Museum of Natural History, Contributions in Science 38:1-11.

Martin, L. D. 1980. Functional morphology and the evolution of cats. Transactions of the Nebraska Academy of Sciences 8:141-154.

Matthew, W. D. 1901. Fossil mammals of the Tertiary of northeastern Colorado. Memoirs of the American Museum of Natural History 1:355-447.

Matthew, W. D. 1910. The phylogeny of the Felidae. Bulletin of the American Museum of Natural History 28:289-316.

McCall, S., V. Naples, and L. Martin. 2003. Assessing behavior in extinct animals: was *Smilodon* social? Brain Behavior and Evolution 61:159-164.

McHenry, C. R., S. Wroe, P. D. Clausen, K. Moreno, and E. Cunningham. 2007. Supermodeled sabercat, predatory behavior in *Smilodon fatalis* revealed by high-resolution 3D computer simulation. Proceedings of the National Academy of Sciences of the United States of America 104:16010-16015.

McNassor, C. 2001a. Chester Stock: A memorable fossil hunter 1892-1950. Terra 31:28-34.

McNassor, C. 2001b. Roy Lee Moodie 1880-1934. Terra 38:44.

Merriam, J. C., and C. Stock. 1932. The Felidae of Rancho La Brea. Carnegie Institution of Washington Publication No. 422:1-231.

Mestel, R. 1993. Saber-tooth tales. Discover 14:50-59.

Miller, G. J. 1968. On the age distribution of *Smilodon californicus* Bovard from Rancho La Brea. Los Angeles County Museum of Natural History, Contributions in Science 131:1-17.

Miller, G. J. 1983. Some new evidence in support of the stabbing hypothesis for *Smilodon californicus* Bovard. Carnivore 3:8-26.

Moodie, R. L. 1918a. Paleontological evidences of the antiquity of disease. Scientific Monthly 7:265-281.

Moodie, R. L. 1918b. Studies in paleopathology, I. General consideration of the evidences of pathological conditions found among fossil animals. Annals of Medical History 4:374-393.

Moodie, R. L. 1918c. Studies in paleopathology, II. Pathological evidences of disease among ancient races of man and extinct animals. Surgery, Gynecology and Obstetrics, 27:498-510.

Moodie, R. L. 1922. On the endocranial anatomy of some Oligocene and Pleistocene mammals. Journal of Comparative Neurology 34:343-379.

Moodie, R. L. 1923a. Paleopathology: An Introduction to the Study of Ancient Evidences of Disease. University of Illinois Press, Urbana, Illinois, 567 pp.

Moodie, R. L. 1923b. The Antiquity of Disease. University of Chicago Press, Chicago, Illinois, 148 pp.

Moodie, R. L. 1926. La paléopathologie des mammifères du Pléistocène. Biologie Médicale 16:431-440.

Moodie, R. L. 1927. Studies in paleopathology, XX.: Vertebral lesions in the sabre-tooth, Pleistocene of California, resembling the so-called myositis ossificans progressiva, compared with certain ossifications in the dinosaurs. Annals of Medical History 9:91-102.

Moodie, R. L. 1929a. Studies in paleodontology, XVI: the California sabre-tooth; the mandibular teeth and associated structures. Pacific Dental Gazette 37:317-321.

Moodie, R. L. 1929b. Studies in paleodontology, XVII: the California sabre-tooth; two impactions and an abscess. Pacific Dental Gazette 37:767-770.

Moodie, R. L. 1929c. Studies in paleodontology, XXV: the California sabre-tooth; facial asymmetry following loss of sabre. Pacific Dental Gazette 37:764-766.

Moodie, R. L. 1930a. Studies in paleodontology, XXII: apical closure of root canals in adult Pleistocene Carnivora. Pacific Dental Gazette 38:1-4.

Moodie, R. L. 1930b. Studies in paleopathology, XXV: hypertrophy in the sacrum of the sabre-tooth, Pleistocene of southern California. American Journal of Surgery 8:1313-1315.

Moodie, R. L. 1930c. Studies in paleopathology, XXVI: Pleistocene luxations. American Journal of Surgery 9:348-362.

Moodie, R. L. 1930d. Studies in paleopathology, XXVIII: the phenomenon of sacralization in the Pleistocene sabre-tooth. American Journal of Surgery 10:587-589.

Rothschild, B. M. 1987a. Diffuse idiopathic skeletal hyperostosis as reflected in the paleontologic record: dinosaurs and early mammals. Seminars in Arthritis and Rheumatism 17:119-215.

Rothschild, B. M. 1987b. Paleopathology of the spine of Cretaceous reptiles; pp. 97-99 in T. Appelboom (ed.), Art, History and Antiquity of Rheumatic Disease. Elsevier, Brussels.

Rothschild, B. M., and L. D. Martin. 1993. Paleopathology: Disease in the Fossil Record. CRS Press, Boca Raton, Florida, 386 pp.

Rothschild, B. M., and L. D. Martin. 2006. Skeletal impact of disease. New Mexico Museum of Natural History and Science, Bulletin 33:1-187.

Rothschild, B. M., and L. D. Martin. 2011. Pathology in saber-tooth cats; pp. 33-41 in V. L. Naples, L. D. Martin, J. P. Babiarz, and H. T. Wheeler (eds.), The Other Saber-tooths: Scimitar-toothed Cats of the Western Hemisphere. Johns Hopkins University Press, Baltimore, Maryland.

Schultz, C. B., M. R. Schultz, and L. D. Martin. 1970. A new tribe of saber-toothed cats (Barbourofelini) from the Pliocene of North America. Bulletin of the Nebraska State Museum 9:1-31.

Schultz, M. 2012. A short history of paleohistology; pp. 738-750 in J. E. Buikstra and C. A. Roberts (eds.), The Global History of Paleopathology, Pioneers and Prospects. Oxford University Press, Oxford, UK.

Shaw, C. A. 1989. The collection of pathologic bones at the George C. Page Museum, Rancho La Brea, California: a retrospective view. Journal of Vertebrate Paleontology 9:38A.

Shaw, C. A. 2001a. Old wounds: the paleopathology of Rancho La Brea. Terra 31:17.

Shaw, C.A. 2001b. The sabertoothed cats. Terra 38:26-27.

Shaw, C.A. 2007. The history, geology and paleontology of the La Brea tar pits; pp. 1-9 in Oil on Their Shoes: Famous and Little-Known Oil Seeps of Los Angeles and Ventura Counties, Pacific Section of the American Association of Petroleum Geologists Field Guide 2, 2007 National AAPG Convention, Long Beach, California.

Shaw, C. A., and A. E. Tejada-Flores. 1985. Biomechanical implications of the variation in Smilodon ectocuneiforms from Rancho La Brea. Natural History Museum of Los Angeles County, Contributions in Science 359:1-8.

Shaw, C. A., F. P. Heald, and M. L. Romig. 1991. Paleopathological evidence of social behavior in Smilodon fatalis from Rancho La Brea. Southern California Academy of Sciences, Annual Meeting, Abstract no. 19.

Shaw, C. A., and C. Howard. 2015. Facial asymmetry in the sabercat (Smilodon fatalis) and wolf (Canis dirus) from Rancho La Brea, Los Angeles, California. Paleobios suppl. 32 (Supplement to no. 1):15.

Shaw, C. A., and J. P. Quinn. 1986. Rancho La Brea: A look at coastal southern California's past. California Geology 39:123-133.

Shaw, C. A., M. L. Romig, and F. P. Heald. 2008. Congenital scoliosis in Smilodon fatalis (Mammalia, Felidae) from Rancho La Brea, California; pp. 383-388 in X. Wang and L. G. Barnes (eds.), Geology and Vertebrate Paleontology of Western and Southern North America: Contributions in Honor of David P. Whistler. Natural History Museum of Los Angeles County, Science Series 41.

Shermis, S. 1983. Healed massive pelvic fracture in a Smilodon from Rancho La Brea. Paleobios 1:121-126.

Shermis, S. 1985a. Alveolar osteitis and other oral diseases in Smilodon californicus. Ossa 12:187-196.

Shermis, S. 1985b. Canine fracture avulsion in Smilodon californicus. Bulletin of the Southern California Academy of Sciences 84:86-95.

Simpson, G. G. 1941. The function of saber-like canines in carnivorous mammals. American Museum Novitates 1130:1-12.

Stock, C. 1929. A census of the Pleistocene mammals of Rancho La Brea, based on the collections of the Los Angeles Museum. Journal of Mammalogy 10:281-289.

Stock, C. 1930. Rancho La Brea: A record of Pleistocene life in California. Los Angeles Museum Science Series 1:1-82.

Stock, C. 1992. Rancho La Brea: A record of Pleistocene life in California, revised by J. M. Harris, Natural History Museum of Los Angeles County Science Series 37:1-113.

Turner, A., and M. Antón. 1997. The Big Cats and Their Fossil Relatives: An Illustrated Guide to Their Evolution and Natural History. Columbia University Press, New York, 234 pp.

Van Valkenburgh, B., T. Maddox, P. J. Funston, M. G. L. Mills, G. F. Grether, and C. Carbone. 2009. Sociality in Rancho La Brea Smilodon: arguments favour evidence over coincidence. Biology Letters doi:10.1098/rsbl.2009.0261

Van Valkenburgh, B., and T. Sacco. 2002. Sexual dimorphism, social behavior, and intrasexual competition in large Pleistocene carnivorans. Journal of Vertebrate Paleontology 22:164-169.

Waldron, T. 2015. Roy Lee Moodie (1880-1934) and the beginnings of palaeopathology. Journal of Medical Biography 23:8-13.

Wheeler, R. F. 1993a. A systematic survey of Smilodon fatalis skulls from Rancho La Brea. Current Research in the Pleistocene 10:112-115.

Wheeler, R. F. 1993b. Morphological variation in Smilodon fatalis crania from Rancho La Brea. Abstract, Journal of Vertebrate Paleontology 13:62A.

Young, F. B., and A. L. Cooper. 1926. Evidence of diseases as shown in fossil and prehistoric remains: paleopathology. Transactions of Section on Pathology and Physiology of the American Medical Association, 1-11.

Young, F. B., and A. L. Cooper. 1927. A study in paleopathology. Radiology 8:230-240.

Index